NEW ADVANCES IN MATERIALS TECHNOLOGIES

Experimental Characterizations, Theoretical Modeling, and Field Practices

NEW ADVANCES IN MATERIALS TECHNOLOGIES

Experimental Characterizations, Theoretical Modeling, and Field Practices

Edited by

Hossein Hariri Asli

Ali Pourhashemi, PhD

Ann Rose Abraham, PhD

A. K. Haghi, PhD

AAP | APPLE ACADEMIC PRESS

First edition published 2024

Apple Academic Press Inc.
1265 Goldenrod Circle, NE,
Palm Bay, FL 32905 USA

760 Laurentian Drive, Unit 19,
Burlington, ON L7N 0A4, CANADA

CRC Press
2385 NW Executive Center Drive,
Suite 320, Boca Raton FL 33431

4 Park Square, Milton Park,
Abingdon, Oxon, OX14 4RN UK

Library and Archives Canada Cataloguing in Publication

Title: New advances in materials technologies : experimental characterizations, theoretical modeling, and field practices / edited by Hossein Hariri Asli, Ali Pourhashemi, PhD, Ann Rose Abraham, PhD, A.K. Haghi, PhD.
Names: Asli, Hossein Hariri, editor. | Pourhashemi, Ali, editor. | Abraham, Ann Rose, editor. | Haghi, A. K., editor.
Description: First edition. | Includes bibliographical references and index.
Identifiers: Canadiana (print) 20230597424 | Canadiana (ebook) 20230597491 | ISBN 9781774914847 (hardcover) | ISBN 9781774914854 (softcover) | ISBN 9781003459262 (ebook)
Subjects: LCSH: Materials—Mathematical models. | LCSH: Materials—Research—Methodology.
Classification: LCC TA404.8 .N49 2024 | DDC 620.1/10151—dc23

ISBN: 978-1-77491-484-7 (hbk)
ISBN: 978-1-77491-485-4 (pbk)
ISBN: 978-1-00345-926-2 (ebk)

About the Editors

Hossein Hariri Asli
Research Scholar, College of Engineering and PhD Candidate, Lamar University, Texas, USA

Hossein Hariri Asli, is currently a research scholar and PhD candidate at Lamar University, Beaumont, Texas, USA. He has over 50 publications in international journals and has authored several books. His research activities includes flood monitoring and mapping using the Internet of Things, networked sensors, and GIS in Southeast Texas. He is collaborating and sharing the elevation and water stage data with federal agencies, including the Texas Department of Transportation, Houston TranStar, and the National Weather Service.

Ali Pourhashemi, PhD
Professor, Chemical and Biochemical Engineering, CBU, Memphis, Tennessee, USA

Ali Pourhashemi, PhD, is a Professor in the Chemical and Biochemical Engineering Department at Christian Brothers University, Memphis, Tennessee, USA. He is involved with teaching, research, and supervising industrial internship programs. Dr. Pourhashemi's primary teaching areas include process design, heat transfer, fluid mechanics, ChE thermodynamics, and unit operations laboratory courses. He has authored several books and research publications in chemical engineering, materials sciences, and related areas and has been involved with various packaging projects.

Ann Rose Abraham, PhD
Assistant Professor, Department of Physics Sacred Heart College (Autonomous), Thevara, Kochi, Kerala, India

Ann Rose Abraham, PhD, is currently an Assistant Professor at the Department of Physics, Sacred Heart College (Autonomous), Thevara, Kochi, Kerala, India. Her PhD thesis was titled, "Development of Hybrid Mutliferroic Materials for Tailored Applications". She has expertise in the field of condensed matter physics, nanomagnetism, multiferroics, and polymeric nanocomposites, etc. She has research experience at various reputed national institutes including Bose Institute, Kolkata, India; SAHA Institute of Nuclear Physics, Kolkata, India; and UGC-DAE CSR Centre, Kolkata, India and has collaborated with various international laboratories. She is the recipient of a Young Researcher Award in the area of physics and Best Paper Awards–2020, 2021, a prestigious forum for showcasing intellectual capability. She served as assistant professor and examiner at the Department of Basic Sciences, Amal Jyothi College of Engineering, under APJ Abdul Kalam Technological University, Kerala, India. Dr. Abraham is a frequent speaker at national and international conferences. She has a good number of publications to her credit in many peer-reviewed high impact journals of international repute. She has authored many book chapters and edited more than 10 books with Taylor and Francis, Elsevier, etc. Dr. Abraham received her MSc, MPhil, and PhD degrees in Physics from the School of Pure and Applied Physics, Mahatma Gandhi University, Kerala, India.

A. K. Haghi, PhD
Research Associate,
Department of Chemistry, University of Coimbra, Portugal

A. K. Haghi, PhD, is a retired professor and has written, co-written, edited or co-edited more than 1000 publications, including books, book chapters, and papers in refereed journals with over 3800 citations and h-index of 32, according to the Google Scholar database. He is currently a research associate at the University of Coimbra, Portugal. Professor Haghi has received several grants, consulted for several major corporations, and is a frequent speaker to national and international audiences. He is Founder and former Editor-in-Chief of the *International Journal of Chemoinformatics and Chemical Engineering* and *Polymers Research Journal*. Professor Haghi has acted as an editorial board member of many international journals. He has served as a member of the Canadian Research and Development Center of Sciences & Cultures. He has supervised several PhD and MSc theses at the University of Guilan (UG) and co-supervised international doctoral projects. Professor Haghi holds a BSc in urban and environmental engineering from the University of North Carolina (USA) and holds two MSc degrees, one in mechanical engineering from North Carolina State University (USA) and another one in applied mechanics, acoustics, and materials from the Université de Technologie de Compiègne (France). He was awarded a PhD in engineering sciences at Université de Franche-Comté (France). He is a regular reviewer of leading international journals.

Contents

Contributors

Hossein Hariri Asli
Doctorate of Civil Engineering Candidate, Lamar University, Texas State University System, Beaumont, Texas, USA

Avinash V. Borgaonkar
Department of Mechanical Engineering, Pimpri Chinchwad College of Engineering (PCCOE), Pune, Maharashtra, India

Laura Crociani
Institute of Condensed Matter Chemistry and Technologies for Energy (ICMATE), National Council of Research (CNR), Padova, Italy

R. Geethu
Department of Basic Science, SCMS School of Engineering and Technology, Karukutty, Ernakulam, Kerala, India

Greeshma Sara John
Department of Physics, Nanotechnology and Advanced Materials Research Center, CMS College (Autonomous), Kottayam, Kerala, India

Athira Maria Johnson
Department of Physics, Nanotechnology and Advanced Materials Research Center, CMS College (Autonomous), Kottayam, Kerala, India

Sonali Kale
Department of Applied Sciences and Humanities, Pimpri Chinchwad College of Engineering (PCCOE), Pune, Maharashtra, India

V. I. Kodolov
Basic Research–High Educational Center of Chemical Physics and Mesoscopy, Udmurt Scientific Center, Ural Division, Russian Academy of Sciences, Russia; M.T. Kalashnikov Izhevsk State Technical University, Izhevsk, Russia

V. V. Kodolova-Chukhontzeva
Basic Research–High Educational Center of Chemical Physics and Mesoscopy, Udmurt Scientific Center, Ural Division, Russian Academy of Sciences, Russia; Peter Great St. Petersburg State Polytechnic University, St. Petersburg, Russia

K. V. Arun Kumar
Department of Physics, Nanotechnology and Advanced Materials Research Center, CMS College (Autonomous), Kottayam, Kerala, India

Parveen Lata
Associate Professor, Department of Mathematics, Punjabi University, Patiala, Punjab, India

Beena Mathew
School of Chemical Sciences, Mahatma Gandhi University, Kottayam, Kerala, India

Anju Nair
Department of Basic Science, SCMS School of Engineering and Technology, Karukutty, Ernakulam, Kerala, India

Najiya Nazar
St. Joseph's College for Women, Alappuzha, Kerala, India

Rony Rajan Paul
Department of Chemistry, CMS College, Kottayam, Kerala, India

Ashwini S. Pillai
St. Joseph's College for Women, Alappuzha, Kerala, India

Shital B. Potdar
Department of Chemical Engineering, National Institute of Technology, Warangal, Telangana, India

Mamatha Susan Punnoose
Bishop Chulaparambil Memorial College, Kottayam, Kerala, India

M. V. Santhosh
Department of Basic Science, SCMS School of Engineering and Technology, Karukutty, Ernakulam, Kerala, India

Jesna Sara Shaji
Department of Chemistry, Mar Thoma College, Tiruvalla, Pathanamthitta, Kerala, India

Satya Bir Singh
Department of Mathematics, Punjabi University, Patiala, Punjab, India

Sukhveer Singh
Assistant Professor, Punjabi University APS Neighborhood Campus, Dehla Seehan, Punjab, India

G. Sreekala
St. Joseph's College for Women, Alappuzha, Kerala, India

M. S. Sreeraj
Space Science Group, Department of Physics, Sacred Heart College, Thevara, Kochi, Ernakulam, Kerala, India

S. L. Sruthi
Department of Basic Sciences and Humanities, Rajagiri School of Engineering and Technology, Rajagiri Valley, Kakkanad, Kochi, Ernakulam, Kerala, India

S. G. Sumod
Space Science Group, Department of Physics, Sacred Heart College, Thevara, Kochi, Ernakulam, Kerala, India

P. Arjun Suresh
Department of Physics, Nanotechnology and Advanced Materials Research Center, CMS College (Autonomous), Kottayam, Kerala, India

N. V. Unnikrishnan
School of Pure and Applied Physics, Mahatma Gandhi University, Kottayam, Kerala, India

Yu. M. Vasil'chenko
Basic Research – High Educational Center of Chemical Physics and Mesoscopy, Udmurt Scientific Center, Ural Division, Russian Academy of Sciences, Russia; M.T. Kalashnikov Izhevsk State Technical University, Izhevsk, Russia

Abbreviations

AFM	atomic force microscopy
AgNPs	silver nanoparticles
AGS	silver gallium sulfide
AIL	anode interface layer
ANN	artificial neural network
ASE	amplified spontaneous emission
ASIC	application-specific IC
BAW	bulk acoustic wave
BBR	bending beam rheometer
BHJ	bulk heterojunction
Ce	cerium
CL	chemiluminescence
CNTs	carbon nanotubes
COF	friction coefficient
CP	conducting polymer
CP	conjugated polymer
CT	charge transfer
DHS	dynamic headspace
DMF	dimethylformamide
DPP	diketo pyrrolo-pyrrole
DSR	dynamic shear rheometer
Dy	dysprosium
EPBT	energy payback time
EQE	external quantum efficiency
ERFs	electrorheological fluids
ETL	electron transport layer
Eu	europium
FBG	fiber Bragg grating
FF	fill factor
FIA	flow injection analysis
FOPS	fiber-optic polarimetric sensor
FPW	flexural plate wave
FWHM	full width at half maximum

GIS	geospatial information system
HOMO	highest occupied molecular orbital
HREM	high-resolution electron microscopy
HTL	hole transport layer
IC	integrated circuit
INDEX	inside-needle dynamic extraction
IoT	internet of things
IQE	internal quantum efficiency
I_{sc}	short circuit current
ISE	ion-selective electrodes
ISFET	ion-sensitive field effective transistor
KdV	Korteweg de-Vries
La	lanthanum
LCST	lower critical solution temperature
LFE	linear dependencies of free energies
LSPR	localized surface plasmon resonance
LUMO	lowest unoccupied molecular orbital
MDA	multivariate data analysis
MEH-PPV	poly(2-methoxy-5(20-ethylhexyloxy) -1,4-phenylenevinylene
MEMS	microelectromechanical systems
MI	modulation instability
MIR	mid-infrared
MNPs	magnetic nanoparticles
MOS	metal oxide semiconductor
MOSFET	metal-oxide-semiconductor field effect transistor
MPD	mean profile depth
MREs	magneto-rheological elastomers
MRFs	magneto-rheological fluids
MRI	magnetic resonance imaging
MSCR	multiple stress creep recovery
MSMAs	magnetic shape memory alloys
MTM	Marshall testing machine
NATM	Nottingham Asphalt testing machine
NC	nanocube
NIR	near-infrared
NLSE	nonlinear Schrodinger equation
NPs	nanoparticles

NREL	National Renewable Energy Laboratory
ODCB	ortho-dichlorobenzene
OLED	organic light emitting diode
OPVs	organic photovoltaics
P&T	purge and trap
PAV	pressure aging vessel
PCBM	1-(3-methoxycarbonyl)propyl-1,1-phenyl-(6,6)C61
PCE	power conversion efficiencies
PDA	personal digital assistant
PEDOT:PSS	poly(3,4-ethylene-dioxythiophene):poly(styrene sulfonate)
PFN	pulse forming network
PHTBT	poly(2-dodecyl-4,7-bis(4-hexylthiophen-2-yl) -2H-benzo[d][1,2,3]triazole)
PLH	poly-*l*-histidine
PV	photovoltaics
PZT	lead zirconium titanate
QCM	quartz crystal microbalance
RAP	reclaimed asphalt pavement
RAS	reclaimed asphalt shingles
RBF	radial basis function
RREs	rare earth elements
RS	remote sensing
RV	rotational viscometer
SAW	surface acoustic wave
SBS	styrene-butadiene-styrene
SBSE	stir bar sorptive extraction
SEM	scanning electron microscopy
SERS	surface-enhanced Raman scattering
SH-APM	shear horizontal acoustic plate mode
Sm	samarium
SMA	shape memory alloys
SMA	stone matrix asphalt
SMMs	shape memory materials
SMPs	shape-memory polymers
SPME	solid-phase-microextraction
SPR	surface plasmon resonance
SWCNT	single-walled carbon nanotube
TMDs	transition metal dichalcogenides

TPD	temperature programmed desorption
TSM	thickness shear mode
UCST	upper critical solution temperature
UV	ultraviolet
VLSI	very large-scale integration
V_{oc}	voltage
Yb	ytterbium

Preface

The main objective of this new book is to examine the mathematical models and experimental methods for engineering materials and structures, and also to determine practical applications under a wide range of conditions, and to set up what is needed to produce a new generation of new materials. The diversity of such mathematical models and experimental methods applied to different types of advanced materials and structures and their behavior is discussed in this book.

The subjects covered in this book range from geospatial information system (GIS) and networked sensors modeling, mathematical modeling in fluid and solid mechanics, deformations in a nonlocal isotropic thermoelastic material, optical properties of solid materials, nanoscale, and modern energy materials and devices, to smart materials-based E-nose technology.

In each chapter, the explanations and the selection of references are presented in detail to provide the essential background for further studies and explore relevant literature for postgraduate research students.

The book is an essential reference that is useful for researchers, who specialize in advanced materials and structures, experimental mechanics of materials, mathematical modeling, and related fields of applied mathematics. The chapters presented are useful for the design and industrial applications of advanced materials and composites. This book is also suitable for postgraduate engineering students who want to gain an overview of the applied mechanics of materials.

PART I

Innovative Models and Practices for Engineering

CHAPTER 1

Reclaimed Asphalt Pavement (RAP) Based on Geospatial Information System (GIS) and Networked Sensors Modeling

HOSSEIN HARIRI ASLI

Doctorate of Civil Engineering Candidate, Lamar University, Texas State University System, Beaumont, Texas, USA

ABSTRACT

The generation of desirable asphalt with the aid of natural, non-toxic, available, and inexpensive materials improves the asphalt pavement quality. The use of oils as a natural and non-toxic agent can improve the properties of bitumen and asphalt. The geospatial information system (GIS) and networked sensors modeling is the best tool to observe the behavior of asphalt pavement against fatigue. In this work, first of all, the Fatigue life tests for samples of 2% and 4% of Rhamnolipid Biosurfactant were done by using historical data and guidance of the soil mechanics office laboratory. Based on the GIS, the critical zone with the most needs for the reclaimed asphalt pavement (RAP) was identified. The sample of 4% of Rhamnolipid Biosurfactant had the best fatigue performance. The RAP, including the sample of 4% of Rhamnolipid Biosurfactant was preferred for the critical zone. The regression method investigated the parameters of the fatigue index, mean profile depth (MPD), and the number of wheel passes for the RAP, including the sample of 4% of rhamnolipid

New Advances in Materials Technologies. Hossein Hariri Asli, Ali Pourhashemi, Ann Rose Abraham, & A. K. Haghi (Eds.)

biosurfactant. From the regression analysis, an acceptable correlation between the parameters was observed. Finally, the laboratory results were compared with the Regression analysis for the RAP, including the sample of 4% of Rhamnolipid Biosurfactant at the critical zone. The regression analysis showed that the power function had a suitable correlation on the scatter diagram. The fatigue index due to the sample of 4% Rhamnolipid Biosurfactant as laboratory results were very close to the Power function solutions.

1.1 INTRODUCTION

The roads are the national assets in all of the worlds, which provide communication possibilities among different points. Furthermore, the cost of road construction and pavement has increased over time in all of the countries. Hence, the maintenance of roads is a critical job that can be advantageous to our living environment. Therefore, the studies about the properties of the material used in road construction can lead to improving the mechanical properties of hot asphalt mixtures. Today, researchers of all over the world have conducted extensive tests to distinguish the causes of asphalt failure, which can be used to achieve further knowledge on pavement which benefits from the roads as a national treasure. It should be noted that the in-pavement design, traffic is in fact one of the most important parameters to be considered. As a matter of fact, the main parameters that we are obligated to consider are load repetition, moving loads criteria along with axle configuration, contact pressure and wheel load. The factors affecting pavement design include:

- Wheel load;
- Axle configuration;
- Contact pressure;
- Vehicle speed;
- Repetition of loads;
- Subgrade type;
- Temperature;
- Precipitation.

A more recent study [1] has shown the process for asphalt pavement installation includes:

- **1st:** Demolition and removal.
- **2nd:** Grading and Sloping.
- **3rd:** Making the sub-base prepared.
- **4th:** Binder and surface course.
- **5th:** Installing high quality asphalt surface.
- **6th:** Butt Joints and Transitions.

To complete the project:

- 1st: Porous asphalt. Porous asphalt has been around since the mid-1970s.
- 2nd: Perpetual pavement. Perpetual pavement is a combination of asphalt and the multilayer
- paving design process.
- 3rd: Quiet pavement.
- 4th: Warm-mix asphalt.
- 5th: Thin overlays.

The types of asphalt pavement includes:

- Porous asphalt. Porous asphalt has been around since the mid-1970s.
- Perpetual pavement. Perpetual pavement is a combination of asphalt and the multi-layer paving design process.
- Quiet pavement.
- Warm-mix asphalt.
- Thin overlays.

Most roads in the world are paved by hot mix asphalt, therefore studies about the properties of asphalt mixtures have significant importance. The use of different oils such as Rhamnolipid Biosurfactant in bitumen improves the mechanical behavior of asphalt, likewise, improving the properties of bitumen binder [2]. Along with previous research, in this work, by using different percentages of rhamnolipid biosurfactant (0, 2, and 4% of Rhamnolipid Biosurfactant ratios to the bitumen weight) as bitumen modifiers, the properties of bitumen, as well as hot mix asphalt, have been investigated. The flashpoint, dynamic shear rheometer (DSR), bending beam rheometer (BBR), and rotational viscosity tests were carried out on modified bitumen samples. The partial chromatography of bitumen is divided into four main groups with similar rheological properties and the same reactivity. These groups are discussed in subsections.

1.1.1 ASPHALTENE

The molecular mass of asphaltene as a polar compound is about 1,000 to 1,00,000. Whereas asphaltene is a viscous compound, which can stabilize the bitumen mixture. Asphaltene is a chemically similar compound to resins, except that asphaltene contains more awkward atoms such as nitrogen, oxygen, sulfur, and resins.

1.1.2 POLAR AROMATICS (RESIN COMPOUNDS)

The molecular mass of polar aromatics is about 500 to 50,000 and they are polar compounds. They have strong adhesion properties. They are used as a co-solvent for oil and asphaltene. When the bitumen oxidizes, it is the resin that takes in oxygen and turns into asphaltene. When bitumen turns into two separate phases of gel-cell, it is the ratio of asphaltene to the resin that provides the amount of gel to cell in the bitumen.

1.1.3 NON-POLAR AROMATICS (PETROLEUM COMPOUNDS)

The molecular mass of petroleum compounds is about 300 to 20,000. Aromatics in bitumen are polar and non-polar. Both types have aromatic rings and are dense in bitumen. Polar aromatics are blackish-brown while non-polar aromatics are yellow.

1.1.4 SATURATED COMPOUNDS

The molecular mass of saturated compounds is about 300 to 1,500. They are rather branched or unbranched aliphatic hydrocarbons, but may also have cyclic compounds. These compounds are non-polar or white viscous oils that contain most of the bitumen wax compounds. Research on bitumen shows that when bitumen are extracted with benzene, the following compounds are found in it:

- Aromatic hydrocarbons;
- Binaphthyls;
- Anthracene or phtharacene;

- Oxygenated compounds;
- Fatty acid methyl esters;
- Methyl aliphatic ketones.

The heating operations have also been performed on hydrogen-rich and low-hydrogen bitumen. This heat treatment has been performed between temperatures of 200 to 500 degrees Celsius. After this procedure, the test of products including the bitumen-soluble chloroform, vaporized bitumen, and insoluble residues was investigated. The non-aromatic components remained in the bitumen. There were also large amounts of unsaturated non-hydrocarbons in the evaporated bitumen. The study of bitumen structure components leads to a better understanding of bitumen, which is one of the main components of asphalt mixtures. The hydrocarbons obtained from the bitumen heating process were significantly different from the raw bitumen: higher gravity, higher asphaltene content, higher viscosity, and lower boiling point of distillation in accordance with the raw bitumen. Atomic bitumen had less hydrogen, and if we want to use this bitumen in refineries, the content of sulfur and nitrogen atoms must be reduced to increase the atomic H/C ratio. Bitumen fumes contain large amounts of aromatic polycyclic compounds. The other research shows that sulfur compounds are higher in bitumen and more concentrated than other molecules of the same weight. The most important factor in the structure and morphology of bitumen is the composition of bitumen. The structure of bitumen is a colloidal composition of asphaltene, which is stabilized by polar aromatics in large amounts of aromatic naphthenes and saturated compounds [3]. The research on the composition of different types of bitumen shows that asphaltenes make up 2% to 4%, polar aromatics 27% to 47%, aromatic petroleum products 19% to 51%, and saturated compounds 2% to 10% of bitumen hydrocarbons. The research on the ratio of hydrogen to carbon represents that all the investigated bitumen have six major compounds with 5 sub-branches [4]:

- Dicycloalkanes;
- Alkyl benzenes;
- Naphthenic benzenes;
- Diatromates; and
- Abundant.

Asphaltene obtained by repeated sedimentation method, including alkyl chains or alkylene bridges, often appears as a substitute in the coal molecular network. The heat of asphaltene showed that most of the gaseous compounds were formed by the cleavage of weak bonds such as alkyl-alkyl-ether and strong bonds such as bridges between semi-aromatic and methyl groups on aromatic rings in the asphaltene structure. The interpretations obtained by the analysis of elements in bitumen showed that the types of bitumen include: 82% to 88% carbon, 8% to 11% hydrogen, up to 1.5% oxygen, up to 1% nitrogen, up to 6% sulfur, and very little amount of metallic or non-metallic elements. In the bitumen structure, aromatics play a major role in their properties, though the role of saturated compounds cannot be ignored. The results of biosurfactant production in the presence of different carbon sources showed that gasoline and glycerol were well used by bacteria as a source of carbon and energy. The culture medium surface tension was reduced from 79 to 39 and 37. Biosurfactants also have surfactant properties, such as reduced surface tension and interfacial tension. The tendency to use Biosurfactants because of their advantages such as low toxicity, biodegradability, and their effectiveness in a wide range of pH and temperature has increased significantly. Notable emulsification of Rhamnolipid Biosurfactant from waste materials, as a green biomaterial, promises to reduce the problem of waste accumulation and hydrophobic oil contaminants. The results of the research showed that Biosurfactants increase the adhesion of sand to asphalt. Due to the non-toxic nature and the production from renewable sources, research has shown that the use of biosurfactants in various applications has increased. Rhamnolipids have two hydrophilic and hydrophobic agents. They have the property of cleaner for soil contaminants and hence their applications have increased.

In a study on PG 58-28 bitumen, the effect of natural oil extracted from wood waste on the behavior of high-temperature modified bitumen was investigated [5]. In this study, wood oil containing 15–30% moisture contained 5–8% humidity and modified by 4% Polyethylene polymer was used. The wood oil was applied to modify the bitumen and DSR and multiple stress creep recovery (MSCR) tests. These tests were used to obtain the groove mark which is a measure of high-temperature endurance of bitumen, bitumen recovery percentage, and irreversible creep percentage. The results of the DSR test shows an increase in the parameter and a decrease in the phase angle, and consequently, an increase in the groove sign with an increase in the percentage of modifier oil. In accordance with increasing criterion results in improved behavior at high bitumen temperatures, the use of waste wood bio-oil in bitumen modification can improve its behavior at high temperatures. By the MSCR test, the addition of wood oil reduced the irreversible creep and increased the bitumen recovery percentage, indicating an improvement in the strength and elasticity of the bitumen at high temperatures.

A more recent study [5] has shown statistical experiments were used to evaluate the percentage of improvement of properties. The results showed that wood oil had a significant effect on bitumen modification. Also, the results of MSCR and DSR tests were slightly different. In the DSR test, intact wood oil behaved worse compared to the other two cases due to the high moisture content of the intact wood oil. Meanwhile, in the MSCR test, the wood oil combined with the polymer, performed weaker, which could be due to polymer failure at high temperature.

In a study on the PG 15–25 hard bitumen, the effect of natural oils derived from oilseed and sunflower seeds was investigated [6]. In this study, samples containing 71.4% natural bitumen, 17.9% vegetable oils, and 10.7% bitumen were tested to investigate their temperature behavior. The Glass conductivity temperature obtained from the DSR results showed that these waste vegetable oils have a lower glass conductivity temperature than bitumen. The bitumen obtained by combining with these oils has a lower glass conductivity temperature. The evaluation of the DSR test shows that by reducing the glass conductivity temperature and adding vegetable oils used in the research to bitumen, the behavior at low bitumen temperature is improved. This event reduces the possibility of thermal failure at low temperatures which is the benefit of using these oils.

As discussed by several authors [7] evaluated the potential of using sunflower oil as a rejuvenator in the asphalt self-healing process in microcapsules in their research. In this study, two types of bitumen, PG 70–22 and PG 76–22 were combined with rejuvenating oil obtained from the sunflower plants with a weight percentage of 5% by weight of bitumen. The usable temperature interval (the interval between the maximum and the lowest temperature) at which bitumen is expected to behave appropriately is defined by the bitumen experiments. As the usable temperature range of bitumen, the use of sunflower oil has positively improved both the low and high-temperature behavior of bitumen.

A study [8] has shown investigated the heating process in order to produce bio-oil from pig manure and the adhesive. The manure oil refraction process was used for the modification of PG 64–22 bitumen.

The bitumen fracture temperature was calculated by BBR test. The behavior at low bitumen temperature and groove mark by DSR test to study the behavior at high bitumen temperature. The failure temperature decreased with increasing oil percentage, which indicates an improvement in cracking at low bitumen temperatures. The bitumen with a higher groove mark should have more resistance to grooving and therefore better performance at high temperatures. With the increasing oil percentage, the bitumen performance at all temperatures has decreased.

The other group of researchers [9] studied two asphalt adhesives under grades A and B with grades 60–80 and 40–60 and bitumen modified by styrene-butadiene-styrene (SBS) polymer called C, as control samples to investigate the effect of vegetable oil on aging resistance. They paid bitumen waste vegetable oil used as a bitumen additive includes frying oil wastes that are separated from the solid particles and water during the filtering process and in weight percentages of 3%, 4%, 5%, 6%, and 7% combined with bitumen. The aged bitumen samples were used by rotating thin enamel tests to simulate short-term aging and the pressure aging vessel (PAV) to simulate long-term aging of bitumen and penetration degree test, before and after aging in order to determine the strength of bitumen. The degree of penetration ratio was calculated as a measure of stability against bitumen aging. The penetration ratio has increased by the addition of vegetable oil, compared to the base bitumen, which indicates an improvement in the resistance of bitumen to aging. Recently site planners and public works officials have found that they offer the opportunity to manage storm water storage in an environmentally friendly way by removing the toxic agent of asphalt in order to improve water quality, recharge groundwater, and keep the flow of runoff in line with non-developed areas. They have been used successfully in a variety of climates for city streets.

RAP is in fact a practical substitute to virgin materials because it will lower the need to apply virgin aggregate. Recycling asphalt pavement however generates a cycle of reprocessing materials that effectively optimizes the application of natural resources. Meanwhile, it minimizes the amount of expensive new asphalt binder required in the production of asphalt paving mixtures. It should be noted that a high percentage RAP mixtures are obtained using processing and production practices, resulting in lower cost and significant energy savings.

Based on an analysis of pavements containing 30% RAP through the LTPP program (or the Long-Term Pavement Performance Program) it has been observed that the performance of pavements containing up to 30%

RAP is identical to that of pavements constructed from virgin materials with no RAP. Several computational modeling and studies indicate the value of using multiple variables in regression analyzes to predict performance of pavements containing up to 30% RAP. The results of this work raise the question of similar approach might be useful in predicting normal functional pavements materials and methods range on the basis of models that include multiple pavement and demographic predictor variables. There are many computational techniques to better control the pavement performance. By application of RAP as surface mixture, we face 2 major issues. The first issue is in fact the friction resistance in many states. We need to pay special attention every time the origin of the aggregates in the RAP are not clear. A second issue is the probability of using too much RAP which will definitely over-stiffen the surface course. It also make it more liable to cracking or to be loosed. Both issues can definitely be dependent to the reality that RAP is typically removed from old roadways and may contain different types of binders, chip seals, aggregates, patches, etc., all mixed in one stockpile [10–15, 17].

According to other researcher's efforts, this work investigated the method of asphalt pavements that have made possible to modify the design of new non-toxic systems by Rhamnolipid Biosurfactant. This method can be used more widely with a lot of benefits and lower cost. The non-toxic and environmentally friendly systems of RAP based on the GIS, networked sensors and regression method is one of the new methods for identifying the behavior of asphalt pavement due to fatigue. In order to predict the value of the fatigue phenomena as a dependent variable and mean profile depth (MPD) and the number of wheels passes as independent variables the regression analysis in this work was used.

1.2 MATERIALS AND METHODS

The Rhamnolipid Biosurfactant is a liquid contact Biofungicide used in agricultural, horticultural, and turf settings to prevent and control plant pathogens such as downy mildews, Pythium, and Phytophthora. Surfactants are in demand for a wide range of industrial applications as they increase solubility, foaming capacity and lower surface tensions. Hence, this work studied for using of Rhamnolipid Biosurfactant in bitumen and hot asphalt mixtures. Most of the main roads in the world are paved by hot mix asphalt. The aim of this work is to inform practitioners about the state of the use of oils as a bio-agent in order to improve the properties of bitumen and asphalt bitumen. This is a practical way for increasing the ability against fatigue for asphalt pavement mixtures. This work also tried to show that the Rhamnolipid Biosurfactant has a better behavior against fatigue compared to samples that have been made by the bitumen modified with other oils. The dateline for the data collection was from 09:00 a.m., 10/02/2017, until 20/07/2018. In order to improve the rheological and mechanical properties of asphalt, the effects of application of 2% and 4% of rhamnolipid biosurfactant as a new kind of additive have been investigated. The Rhamnolipid Biosurfactant in oil phase was prepared.

The procedure of work was divided into the following steps:

1. The run of fatigue life tests for the samples of 2% and 4% of rhamnolipid biosurfactant by the soil mechanic office laboratory.
2. Identification of the critical zone with the most needs to the RAP based on the GIS.
3. To select the sample of rhamnolipid biosurfactant with the best fatigue performance and to suggest it as the RAP for critical zone based on the results of steps 1 and 2.
4. To investigate the parameters due to curve fit modeling for fatigue index; MPD and the number of wheel passes for the RAP by the Regression method.
5. To compare the results of Step 1 with the regression analysis of Step 4 based on the identification of function with suitable correlation on scatter diagram.

1.2.1 RESEARCH FORMULATION

In this work the following considerations were made by using data for MPD and the number of wheels passes as independent variables with nomenclature "X" against fatigue index as a dependent variable with nomenclature "Y" (Eqn. (1)):

$$Y = f(X_1, X_2) \tag{1}$$

The curve estimation procedure was run based on the "Y" and "X" data which have been detected by the field tests. The Regression model was built based on the field tests data. The Scatter diagram drawing and correlations procedure was formed by estimating Regression statistics. The calculations based on the experiments and laboratory tests for three cases were as in Eqns. (2)–(4):

1. First Case:

$$F.I. = f(t) \tag{2}$$

   where; *F.I.* is the fatigue index; and *t* is the MPD (mm).
2. Second Case:

$$F.I. = f(n) \tag{3}$$

   where; *F.I.* is the fatigue index; and *n* is the number of wheel passes for Specimens.

   The most important effects based on regression and model summary concentrated on MPD.
3. Third Case:

$$F.I. = f(t, n) \tag{4}$$

   where; *F.I.* is the fatigue index; *t* is the MPD (mm); and *n* is the number of wheel passes for specimens.

1.2.2 RESEARCH STANDARDS

In this work, many standards were used. For example, the ASTM eLearning professional, comprehensive, and cost-effective training which focused on the construction material testing includes the tests of aggregates. The interactive eLearning courses covered essential content from the ASTM standard [18–22].

The granulation of aggregates has been selected by using a series of sieves for sieving surface layers, including aggregates that have passed through sieves of 19 and 12.5 mm. In order to be able to compare and examine the samples more accurately, all the aggregates were sieved in the laboratory so that there was no difference between the samples in terms of granulation. The ASTMC127 was used to determine the specific gravity of coarse grains. The ASTMC128 was used to determine the specific gravity of fine grains. The ASTMC188 was used to determine the filler specific gravity.

The aggregates in the stages of preparation, compaction, and operation of asphalt mixtures, due to stresses caused by the weight of rollers and vehicles are usually exposed to wear and breakage and may change the granulation due to crushing and breaking. To prevent this type of damage or reduce it to a minimum, the aggregates must be hard, strong, durable, and have sufficient strength. To measure these properties, different tests have been proposed by different standards. In this work, the Los Angeles abrasion test was used. The test procedure was described in AASHTO T 69.

The fracture rate of aggregates was determined by performing a fracture percentage test. Then the fracture rate of coarse-grained materials was determined. The ASTM D5821 standard was applied for explaining the fracture rate test procedure. The bitumen was pure bitumen with a performance grade of 64–16 equivalent to bitumen with a penetration grade of 60/70. The Superpave tests were performed on the desired binder. The selected binder was examined and evaluated by SHRP special tests.

Throughout asphalt production, due to high temperature and airflow, the adhesive material aged with both mechanisms. This aging was simulated in a short time by a rotating thin enamel test (AASHTO-T240 or ASTM-D2872).

After testing and examining the aggregates and bitumen used the Marshall method. Then the optimal bitumen percentage of the mixtures was determined. By using this optimal bitumen percentage, the necessary samples were prepared and evaluated according to the relevant standards. The actual specific gravity of Marshall samples was determined according to the ASTM-D2726 standard and used in the mixed design. By definition, the actual specific gravity of asphalt samples is equal to the actual weight of the sample in air. This parameter is determined to calculate the percentage of space of the compacted mixture and the percentage of space between the aggregates of dense mixtures, which are two important factors in the design of the mixture and the determination of the optimal bitumen.

The ASTM D-2041 standard, known as the theoretical maximum specific gravity of the Rice method, was used to calculate the maximum theoretical specific gravity of asphalt mixtures. According to this method, two kilograms of asphalt mixture were prepared according to the maximum nominal size of aggregates. The mixture was cooled to ambient temperature and placed in a special glass container under a vacuum pressure of 0.3 to 3.7 kPa for 2 to 15 minutes.

In order to determine the Marshall test endurance and flow, their height was measured. The Marshall endurance values correspond to the standard specimens had a height of 63.5 mm. The thickness of the constructed

Marshall specimen was less or more than the standard value, the Marshall endurance number for the standard specimen should be corrected. For this purpose, the correlation coefficients provided in ASTM-D1559 were used.

The dynamic stiffness modulus of asphalt samples was investigated by indirect traction method. The samples were first placed in the temperature control chamber for 24 hours to reach the test temperature homogeneously and then they placed in the loading frame of the Nottingham apparatus.

The fatigue of asphalt mixtures was obtained by indirect tensile testing. The rupture was determined by measuring the amount of vertical deformation of the specimen, and the criterion of fatigue, rupture, or vertical displacement was considered equal to 4 mm. The fatigue testing is usually performed in two ways: loading with constant stress and loading with constant strain. In the constant stress test, the strain increases with the number of load pulses. In the case of a certain tensile strain for any amount of stress, the relationship between tensile strain and the number of cycles leading to failure can be plotted. As a result of applying vertical compressive stress in the fatigue test, indirect tensile stress and strain occurred horizontally, and repeated loading eventually led to vertical cracks in the center of the specimen. In this work, the frequent axial load and uniaxial creep tests was used to determine different variables of asphalt mix in the amount of resistance to permanent deformation. The load was applied axially to the specimens, and two sensors continuously measured the permanent deformation according to the British standards BS-DD185 and BS-DD226.

1.2.3 RESEARCH APPARATUS

1.2.3.1 THE RAPID AGING CHAMBER TESTING APPARATUS

In this work, the effects of long-term wear were investigated by performing an aging chamber test, in which the asphalt mixture, in short-term wear conditions, was placed inside the greenhouse to simulate the conditions for long-term wear.

1.2.3.2 DYNAMIC SECTION RHEOMETER (DSR) TEST APPARATUS

Due to the dependence of bitumen behavior on temperature and loading time, the optimal test for bitumen adhesives should include both factors.

This capability has been developed in experimental machine tools called dynamic rheometers, dynamic section rheometer (DSR) or oscillating shear rheometers. The rheometers can evaluate both time and temperature effects. In this work, by performing (DSR) test, rheological properties (phase difference angle and shear modulus) were measured at medium and high temperatures. In this experiment, bitumen was placed between two parallel plates, one fixed and the other oscillating.

1.2.3.3 ROTATIONAL VISCOMETER (RV) TEST APPARATUS

The rotary viscometer determines the rotational viscosity by measuring the required torque while maintaining a constant rotational speed of the rotating plate immersed in the bitumen. In the case of bituminous adhesives, especially modified bitumen, could be easily pumped and moved in hot mixing equipment. The bitumen was easily mixed with stone materials at the mixing temperature.

1.2.3.4 BENDING BEAM RHEOMETER (BBR) TEST APPARATUS

It is not entirely reliable to perform a bitumen hardness test at a low temperature with a dynamic shear strength measurement rheometer. In this work, the bending beam rheometer (BBR) accurately evaluated the properties of adhesives at low pavement temperatures. The BBR test temperature depends on the lowest pavement operating temperature. The bituminous adhesive behaves more like an elastic solid. The two dynamic shear and flexural beam tests predicted the hardness behavior of the adhesives over a wide range of temperature changes. By performing BBR test, the amount of fluctuations and creep of the adhesive material under constant load and temperature were measured.

1.2.3.5 CREEP STIFFNESS MEASURING BY BENDING BEAM RHEOMETER (BBR) TEST APPARATUS

The creep stiffness measured by the BBR is not sufficient to fully describe the bitumen capacity against pre-fracture elongation. For example, some adhesives have high creep hardness, but can be stretched further before fracture. The researchers have developed a special specification system to adapt to these rigid but elastic adhesives. These materials have relatively high creep stiffness and have reasonable elastic behavior at low temperatures. The direct tensile test applies only to adhesives with a BBR creep hardness of between 300 and 600 MPa. If the creep hardness is less than 300 MPa, there is no need to perform the test (the case in the present work).

1.2.3.6 MODIFIED BITUMEN BI-SURFACTANT RHAMNOLIPID BY SILVERSON APPARATUS

The Silverson apparatus was used to modify bitumen with Rhamnolipid Biosurfactant. The bitumen and modifier were mixed at 140°C for 60 minutes at 2,500 rpm.

1.2.3.7 MARSHALL TESTING MACHINE (MTM)

The Marshall testing machine (MTM) was used to determine the strength and flow of asphalt mixtures. To make Marshall samples, mixtures of 1,200 g of asphalt with granular aggregates were prepared with bitumen percentages (4.5 to 7, by increase of 0.5%). To prepare each sample and to simulate heavy traffic, 75 cylinders were applied to each side of the samples.

1.2.3.8 NOTTINGHAM ASPHALT TESTING MACHINE (NATM)

The Nottingham asphalt testing machine (NATM) was used to perform non-destructive tests on asphalt samples in this work. This machine determines the mechanical properties of asphalt mixtures under dynamic loading conditions. The basis of the operation was to determine the mechanical behavior of the asphalt mixture. The operation was included stiffness tests by indirect tensile method due to repeated loading, uniaxial creep, indirect tensile method for fatigue and axial creep with repeated loading. The Poisson's ratio was typically considered to 0.35, which was the best value to show the behavior of asphalt materials [4, 23–33].

1.3 RESULTS

Today, the works of some research show that non-toxic agent like oils with organic base can improve the properties of bitumen and asphalt bitumen. The use of different oils in bitumen improves the mechanical function of asphalt, likewise, improving the properties of bitumen binder. The studies about the properties of asphalt mixtures due to the oils with organic base have significant importance for all of the world which roads are paved by hot mix asphalt. In this work, by using different percentages of rhamnolipid biosurfactant (samples of 2% and 4% of Rhamnolipid Biosurfactant), the properties of bitumen as well as hot mix asphalt have been investigated. The results of this work were derived by various experiments. The 10 samples were analyzed due to the tests that were performed by the same laboratory conditions.

The results of work can be divided into the following stages:

1. The fatigue life tests for samples of 2% and 4% of Rhamnolipid Biosurfactant run by using historical data and guidance of the soil mechanic office laboratory.
2. The sample of 4% of Rhamnolipid Biosurfactant had the best fatigue performance according to the results of Step 1.
3. Based on the GIS, the critical zone with the most needs to the RAP was identified.
4. The RAP include the sample of 4% of Rhamnolipid Biosurfactant was preferred for critical zone based on the laboratory results of Step 1.
5. The Regression method investigated the parameters of the Fatigue index, MPD and the number of wheel passes for the RAP include the sample of 4% of rhamnolipid biosurfactant during field test at critical zone.
6. The laboratory results of Step 1 were compared with the Regression analysis of Step 5 for the RAP include the sample of 4% of Rhamnolipid Biosurfactant at critical zone. The laboratory results of Step 1 were compared with the Power function which had suitable correlation on scatter diagram.

The description of results for the first stage are as follows:

- The fatigue life tests for samples of 2% and 4% of Rhamnolipid Biosurfactant run by using historical data and guidance of the soil mechanic office laboratory.

1.3.1 INVESTIGATING THE PROPERTIES OF STONE AND BITUMEN MATERIALS

Table 1.1 shows the results of the study on characteristics of stone materials used in this research.

TABLE 1.1 Specifications of Stone Products

	Test Method	Test Method
Los Angeles abrasion method, percent	ASTM C-131	17
Weight loss with sodium sulfate, percentage	ASTM C-88	0.7
Fracture percentage	ASTM D-5821	97
Percentage of broad and long aggregates	ASTM D-4791	4.2

In this work, in order to investigate the effect of different amounts of Biosurfactants on the rheological properties of bitumen, experiments for bitumen with a functional rating of 16–64 are considered as a control sample for bitumen modified with 2% and 4% Biosurfactant. According to the results, as predicted, the biosurfactant has softened the bitumen, and the main reason for this seems to be the increase in the amount of oil in the bitumen modified with this additive. This means that the bitumen at the operating temperature of 64°C cannot meet the minimum criterion of the groove. The preferred pavement, which is 1 kPa for the ratio of the complex modulus to the sine of the phase difference angle satisfies and reduces the performance of high-temperature bitumen by one degree. On the other hand, according to Figure 1.1, bitumen modified with a biosurfactant was able to reduce the viscosity (RV) of bitumen. This was very desirable because it reduced the mixing temperature in the asphalt plant. However, to ensure that the Biosurfactant-modified bitumen can be used in the asphalt plant. It was necessary to consider tests related to the control of asphalt samples.

According to the results of the dynamic section rheometer (DSR) test at medium temperature and the study of excellent pavement fatigue parameter ($G^*.\sin\delta$) and low temperature BBR test, it can be expected that biosurfactant can improve the performance of asphalt samples against fatigue and heat cracks. The results related to the advanced pavement fatigue cracks criterion were showed in Figure 1.2.

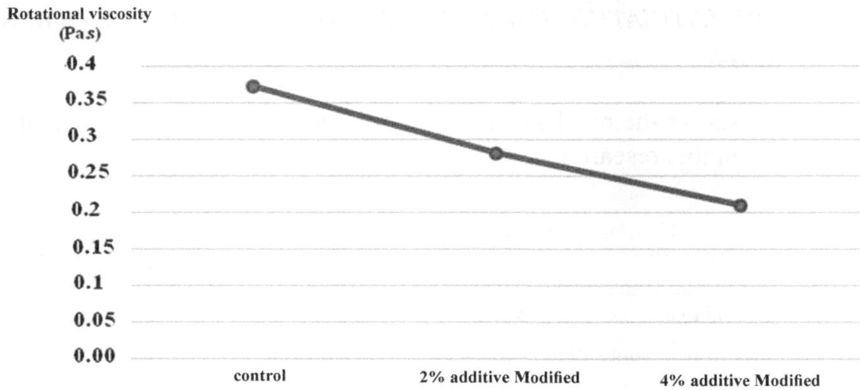

FIGURE 1.1 Viscosity changes of controlled and modified bitumen for different values of biosurfactant rhamnolipid.

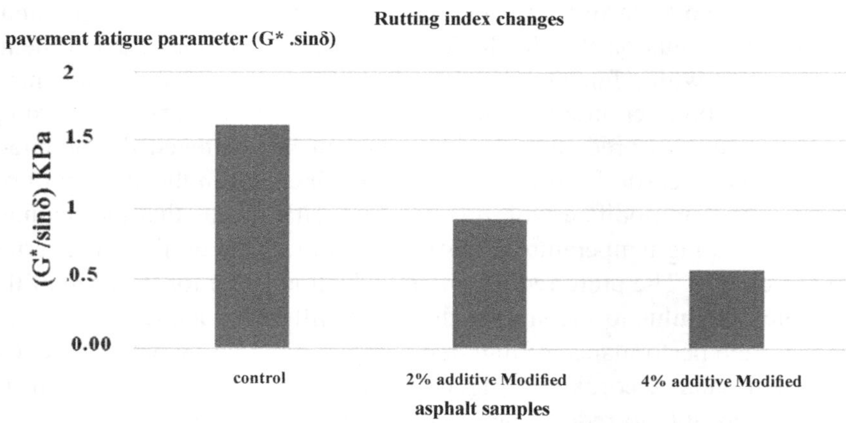

FIGURE 1.2 Rutting index changes of controlled and modified bitumen for different values of biosurfactant rhamnolipid.

1.3.2 RESULTS RELATED TO TESTS PERFORMED ON ASPHALT SAMPLES

At first, the parameters of the Marshall test to determine the optimal bitumen were obtained by using Marshall tests. Then, by using the optimal bitumen of asphalt samples to perform modulus tests of resilience, creep, repetitive axial load, and fatigue test by the indirect tensile method.

1.3.3 RESULTS OF MARSHALL TESTS FOR ASPHALT SAMPLES

The percentage of bitumen suitable for making asphalt mixtures is called the optimal bitumen percentage. The optimum bitumen percentage is determined by using Marshall Test curves. This is because the percentage of bitumen that causes the highest amount of asphalt strength may not be the same as the percentage of bitumen that causes the highest specific gravity or the amount of empty asphalt concrete space may not be adequate, so the bitumen used to make asphalt mixes The average value used is the maximum strength, the highest specific gravity and the most suitable amount of empty space in asphalt concrete. The optimal amount of bitumen obtained should be such that it is used with the technical specifications of asphalt concrete within the technical specifications of the case regulations. In this work, samples of asphalt mixtures made with different percentages of bitumen were tested and their percentage of empty space was determined. The related curves were plotted separately in different percentages. These curves showed the effect of bitumen on the technical characteristics of asphalt concrete. By the analysis of these curves, the optimal percentage for the asphalt mixture was obtained. As shown in Figure 1.3, the addition of Rhamnolipid Biosurfactant increased the Marshall strength of the modified samples compared to the control sample. The samples containing 2% Rhamnolipid Biosurfactant had the highest Marshall strength. By increasing Rhamnolipid Biosurfactant up to 4% (compared to the sample containing 2% Rhamnolipid Biosurfactant), the Marshall strength of the samples decreased. According to Figure 1.3, bitumen containing 2%, Rhamnolipid Biosurfactant has the lowest liquidation value.

The sample containing 4% of liquefaction Rhamnolipid Biosurfactant was similar to the control sample. Also, according to the results of Marshall's strength, the Marshall ratio index was calculated. This ratio is known as an indicator for measuring the resistance of the mixture to shear stress, permanent deformation, and groove. The higher Marshall ratio results in greater mixture stiffness, greater ability of the mixture to distribute applied forces, and greater resistance to creep deformations. The sample containing 2% Rhamnolipid Biosurfactant had the highest Marshall ratio, which indicated that these samples had more hardness and resistance to permanent deformation of asphalt samples [35–46].

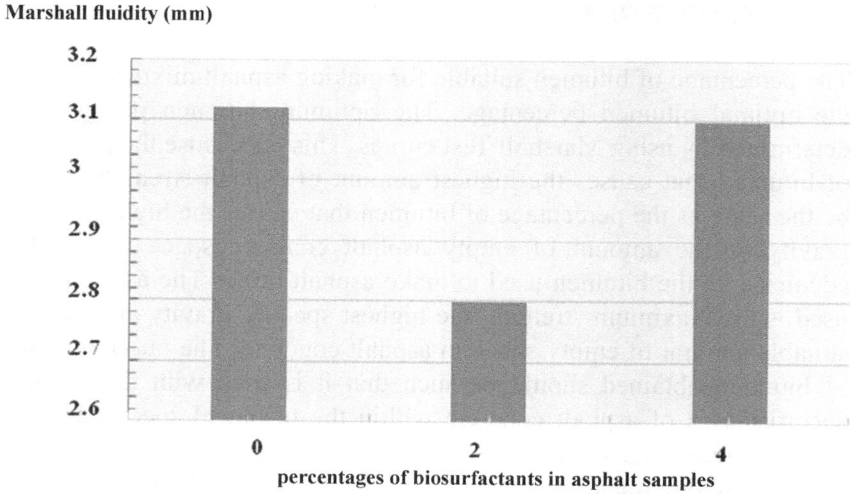

FIGURE 1.3 Results related to Marshall fluidity of asphalt samples containing different percentages of biosurfactants.

1.3.4 TEST RESULTS FOR THE MODULUS OF RESISTANCE

The robustness modulus test is a non-destructive test in which the loads applied to the sample are small. This experiment was performed by the Nottingham test machine. The value of the modulus of resistance is the maximum energy that the unit volume of the asphalt mixture can withstand without any permanent deformation and return to its original state by loading the mixture. The addition of Rhamnolipid Biosurfactant increased the modulus of resistance at all temperatures tested. The samples containing 4% Rhamnolipid Biosurfactant had the highest amount of germination modulus at all temperatures. The sample containing 4% Rhamnolipid Biosurfactant increased by 31% compared to the control sample at 5°C. Also, the sample containing 2% Rhamnolipid Biosurfactant has a 17% increase compared to the control sample. These results indicate that the germination modulus has increased with an increasing amount of Rhamnolipid Biosurfactant. But the remarkable thing is that the amount of resistance modulus decreases too much with increasing temperature.

1.3.5 RESULTS OF PERMANENT DEFORMATION RESISTANCE TESTS

The results at all temperatures showed that samples containing Rhamnolipid Biosurfactant had less final deformation than control samples. Probably the higher strength of samples containing modified bitumen compared to control samples was due to better adhesion of bitumen and aggregates. Also, with increasing temperature, the amount of permanent deformation of the samples had decreased significantly. For example, a sample containing 4% Rhamnolipid Biosurfactant at 40°C to 50°C had a final deformation of 50% to 39%, respectively. Therefore, it can be concluded that bitumen containing Rhamnolipid Biosurfactant also performs better resistance at high temperatures. The results of this work show that the samples with 4% Rhamnolipid Biosurfactant have less permanent deformation.

The results of the creep test are also shown in Figure 1.4. As can be seen from the comparison of the two figures, with increasing stress, the axial strain values decreased for all specimens. The samples containing 2% and 4% of Rhamnolipid Biosurfactant have less axial strain than the control sample. The samples containing 2% and 4% of Rhamnolipid Biosurfactant were reduced by 31% and 39%, respectively, compared to the control sample at 50°C, which ensures the accuracy of the repeated axial load test. According to the results obtained in this section, it can be concluded that a sample containing 4% Rhamnolipid Biosurfactant will have the best performance against permanent deformation.

The description of results for second stage are as follows:

- Based on the GIS and networked sensors, the critical zone with the most needs to the RAP was identified.

The area of pilot includes 8,500 hectares. The roads occupy 27% of the area of pilot. The central area is ranked first among the areas of pilot [34].

In this work, in order to identification of critical zone with the most needs to the pavement, the GIS Ready maps have been prepared by using ArcGIS-ArcMap software based on the as following procedure (Figure 1.5):

- Exchange of graphic information from CAD space to GIS space.
- Remove errors in CAD space.
- Convert graphic information from DWG format to SHP.
- Complete layers of descriptive and spatial information and fix errors in the GIS space (descriptive and spatial).

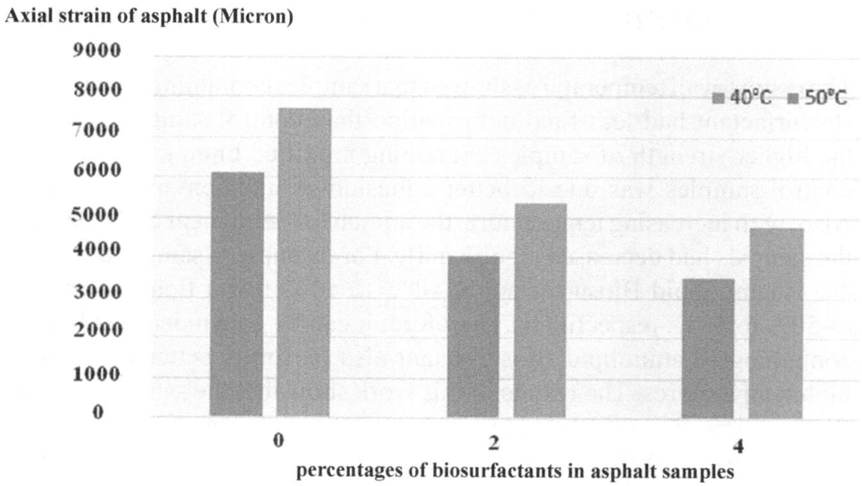

FIGURE 1.4 Axial strain and the creep test results of asphalt samples containing biosurfactant under a stress of 400 kPa.

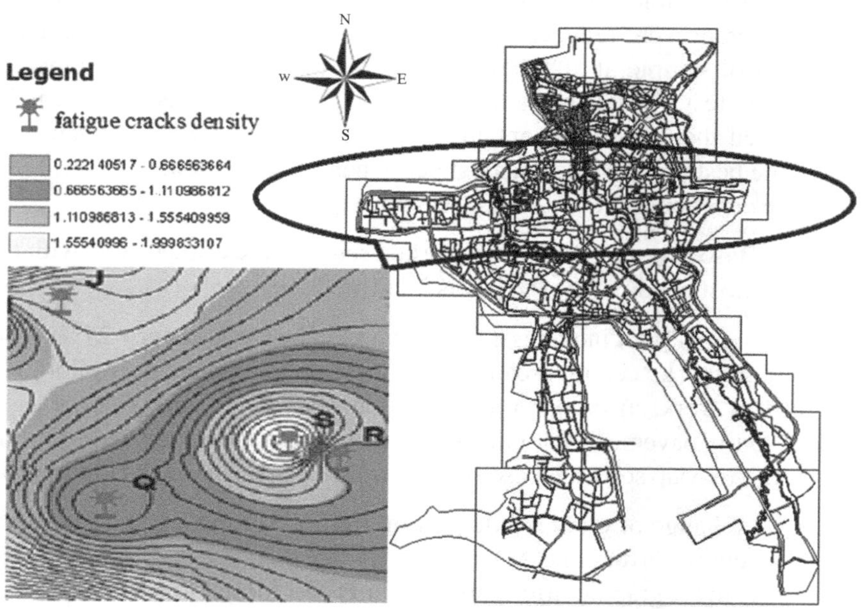

FIGURE 1.5 RAP samples from 23 different locations: Based on the geospatial information system (GIS) and networked sensors.

- Separated effects are snapped together with appropriate tolerance.
- Eliminate toll errors that are in the wrong place.
- Create primary and external keys for the map toll table.
- Create appropriate tolerance and exchange spaghetti space to topology space.
- Preparation of conceptual model for modeling in GIS space.
- Create a suitable terrestrial database of maps.
- Create the ability to track and perform map analysis.

The description of results for the third stage are as follows:

➢ The sample of 4% of Rhamnolipid Biosurfactant had the best fatigue performance according to the results of Step 1.

1.3.6 TEST RESULTS TO DETERMINE FATIGUE LIFE BY INDIRECT TRACTION METHOD

The number of failure cycles decreases sharply with increasing temperature. The samples containing 2% and 4% Rhamnolipid Biosurfactant performed better against the cyclic loads and were broken in more cycles. When repeating the applied loads, cracks were created in the asphalt mixture and these cracks expanded as the loading continues and caused rupture. Rhamnolipid Biosurfactant oil has improved the fatigue performance of asphalt mixtures, which is probably due to the rejuvenation of bitumen and greater flexibility of bitumen against microcracks in bitumen containing oil, as well as greater adhesion of bitumen to stone materials due to the reduced tensile properties of biosurfactant. In the sample containing 2% Rhamnolipid Biosurfactant at 10°C, the number of rupture cycles increased by 120% compared to the control sample and by 213% in the sample containing 4% Rhamnolipid Biosurfactant. Also, the number of rupture cycles of samples containing 2% and 4% of Rhamnolipid Biosurfactant at 20°C compared to the control sample increased by 120% and 249%, respectively. Thus, it can be concluded that samples containing 4% Rhamnolipid Biosurfactant have the best fatigue performance at both temperatures of 10°C and 20°C. The results related to the advanced pavement fatigue cracks criterion are given in Figure 1.6.

Flashpoint, DSR, BBR, and rotational viscosity tests were carried out on modified bitumen samples (Figure 1.7).

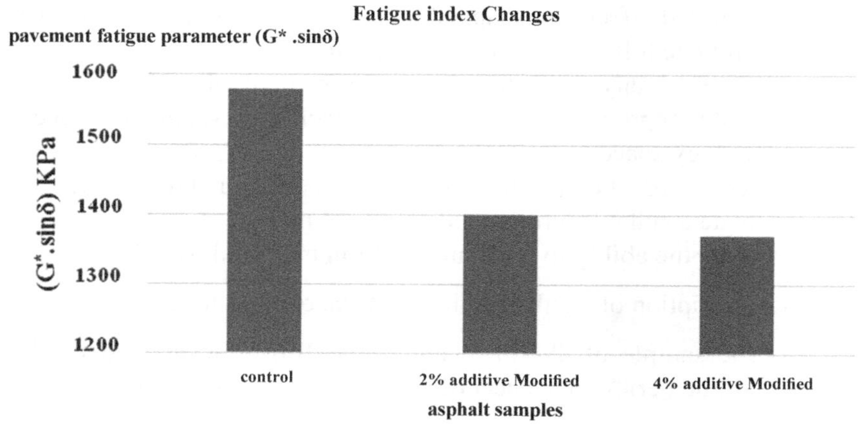

FIGURE 1.6 Fatigue index changes of controlled and modified bitumen for different values of biosurfactant rhamnolipid.

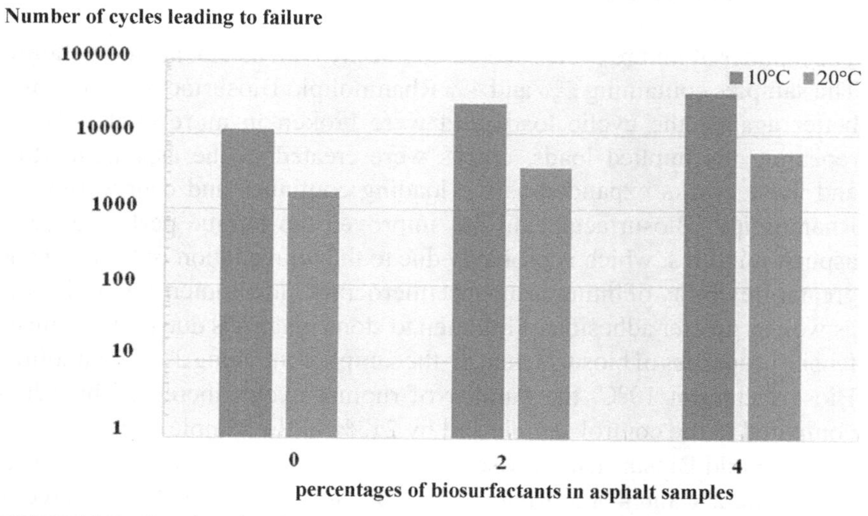

FIGURE 1.7 Results related to the number of cycles leading to failure of asphalt samples containing biosurfactants.

The description of results for fourth stage are as follows:

➢ The RAP include the sample of 4% of Rhamnolipid Biosurfactant was preferred for critical zone based on the laboratory results of Step 1.

Based on the GIS Ready maps the sample of 4% of Rhamnolipid Biosurfactant informed for critical zone with the most needs to the pavement. The GIS Ready maps have been prepared by using ArcGIS-ArcMap software (Figure 1.8).

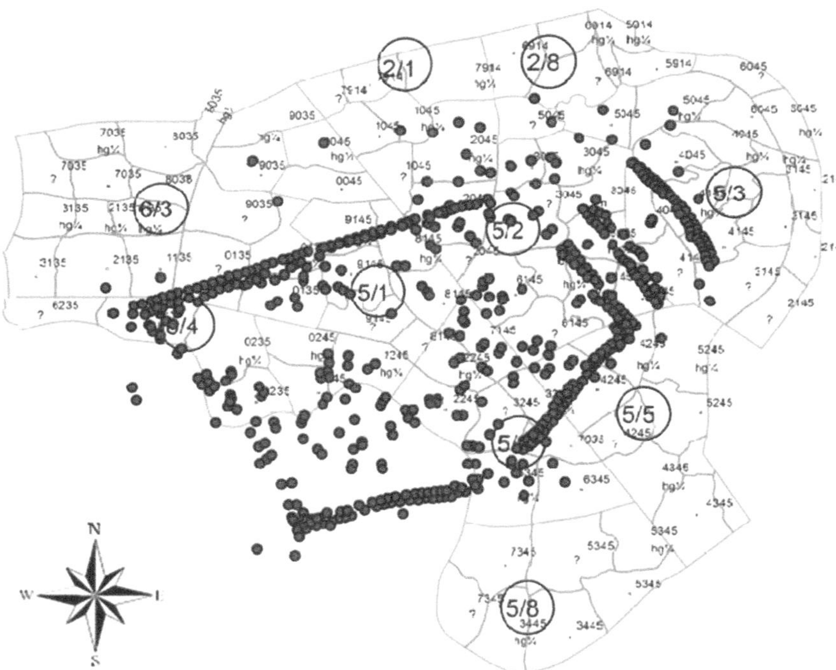

FIGURE 1.8 RAP samples investigation from 250 different locations: Based on the geospatial information system (GIS) and networked sensors.

The description of results for fifth stage are as follows:

- ➢ The regression method investigated the parameters of the fatigue index, MPD and the number of wheel passes for the RAP include the sample of 4% of rhamnolipid biosurfactant during field test at critical zone.

At present, approximately every 7.5 people have a car in pilot. It is predicted for 2032 there will be 1,00,000 cars there. There are 1,250 traffic on the central streets of pilot per hour. It seems that, there are 20,000 number of wheel passes on the central streets of pilot per 24 hours [34].

The Regression method can define the mathematical models for investigation of parameters for Fatigue index, MPD and the number of wheel passes at critical zone according to three cases were showed in (Table 1.2). These dependent and independent variables were applied for the RAP includes 4% of Rhamnolipid Biosurfactant.

During field test, the specimens were collected into the critical zone for 10 times for 10 months (01/03/2017 to 01/12/2017). In order to preselected confidence levels for hypothesis testing in this work the p-value was investigated. Due to the regression analysis and according to the parameter values of specimens, the p-values were calculated. A p-value is a measure of the probability that an observed difference could have occurred just by random chance. The lower the p-value is as similar as the greater the statistical significance of the observed difference. The p-value is a number between 0 and 1. In the most realistic situations, the p-value is not equal to the boundary value (Table 1.2) [16, 34].

TABLE 1.2 Field Tests and Regression Stiffness Model for Asphalt Pavement

Case Number	Date of Test	Fatigue Index for Samples Containing 4% of Rhamnolipid Biosurfactant (kPa)	Mean Profile Depth (mm)*	Number of Wheel Passing per Day**
1.	01/03/2017	1,374	108	20,381
2.	01/04/2017	1,374	118	18,301
3.	01/05/2017	1,373	112	18,368
4.	01/06/2017	1,375	107	19,313
5.	01/07/2017	1,376	108	19,364
6.	01/08/2017	1,379	109	20,387
7.	01/09/2017	1,378	117	18,398
8.	01/10/2017	1,381	113	17,403
9.	01/11/2017	1,383	118	16,342
10.	01/12/2017	1,383	109	16,315

**MPD (independent variable, P value is equal to 0.977).*

***Number of wheels passing (independent variable, P value is equals to 0.962).*

By the regression method it was fitted the function curve and regression analysis. The curve estimation procedure (5–15) allowed quickly estimating regression statistics and producing related plots (Figures 1.9–1.15) for defined model were showed in Table 1.3. The power function had suitable correlation on scatter diagram.

1. **Linear Function:**

$$A = 1358.443 - 33.436x \tag{5}$$

2. **Logarithmic Function:**

$$\log y = \log(1258.832) + (158.682)\log x \tag{6}$$

3. **Inverse Function:**

$$y = 1397.370 + 33.910 f^{-1}(y) \tag{7}$$

4. **Quadratic Function:**

$$y = 730.945 + 1584.44x^{-.0494} \tag{8}$$

5. **Cubic Function:**

$$y = 730.945 + 0x - -049x^2 - 1567.807x^3 \tag{9}$$

6. **Compound Function:**

$$A = 1358.591e^{t} + 1.00 \tag{10}$$

7. **Power Function:**

$$y = 1288.91x^{0.014} \tag{11}$$

8. **S Function:**

$$y = f_0\ (7.242, X, -1.601) \tag{12}$$

9. **Growth Function:**

$$(dA/dt) = 7.214A + 0.0001 \tag{13}$$

10. **Exponential Function:**

$$\mathrm{y} = (1358.591\mathrm{E} - 0.0001)^{x} + g \tag{14}$$

11. **Logistic Function:**

$$y = 0.0007x + 0.999 \tag{15}$$

TABLE 1.3 Curve Fit Modeling for Fatigue Index; Mean Profile Depth (mm)

Equation	Model Summary				Parameter Estimates			
	R Square	F	df1	Sig.	α_0	α_1	α_2	α_3
Linear, (5) $A = 1358.443 - 33.436x$	0.03947	0.32870	8	0.5822	1358.443	33.436	–	–
Logarithmic, (6) $\log y = \log(1258.832) + (158.682) \log x$	0.04013	0.33446	8	0.5790	1285.832	158.682	–	–
Inverse, (7) $y = 1357.370 + 33.910 f^{-1}(y)$	0.04081	0.34037	8	0.5757	1397.370	33.910	–	–
Quadratic, (8) $y = 730.945 + 1584.44x^{-0.0494}$	0.06053	0.22549	7	0.8037	730.945	1584.440	–0.049	–
Cubic, (9) $y = 730.945 + 0x - -049x^2 - 1567.807x^3$	0.06053	0.22549	7	0.8037	730.945	–	–0.049	1567.807
Compound, (10) $A = 1358.519e^t + 1.00$	0.03941	0.32824	8	0.5824	1358.591	1.0001	–	–
Power, (11) $y = 1288.91x^{0.014}$	0.04008	0.33399	8	0.5792	1288.910	0.014	–	–
S, (12) $y = f_0(7.242, X, -1.601)$	0.04075	0.33989	8	0.5760	7.242	–1.601	–	–
Growth, (13) $(dA/dT) = 7.214A + 0.0001$	0.03941	0.32824	8	0.5824	7.214	0.0001	–	–
Exponential, (14) $y = (1358.591E - 0.0001)^x + g$	0.03941	0.32824	8	0.5824	1358.591	0.0001	–	–
Logistic, (15) $y = 0.0007x + 0.999$	0.03941	0.32824	8	0.5824	0.000736	0.999	–	–

Note: F.I.: Fatigue index (kPa) as dependent variable; and t: Mean profile depth (mm) as independent variable.

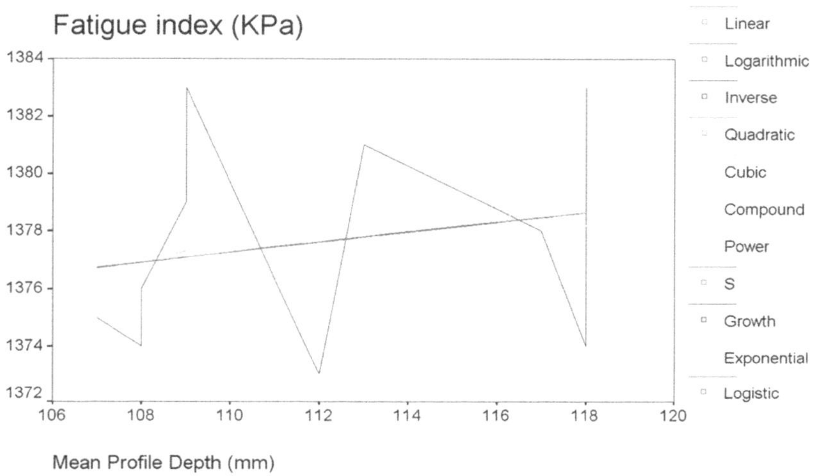

FIGURE 1.9 Curve fit for fatigue index (kPa) and mean profile depth (mm).

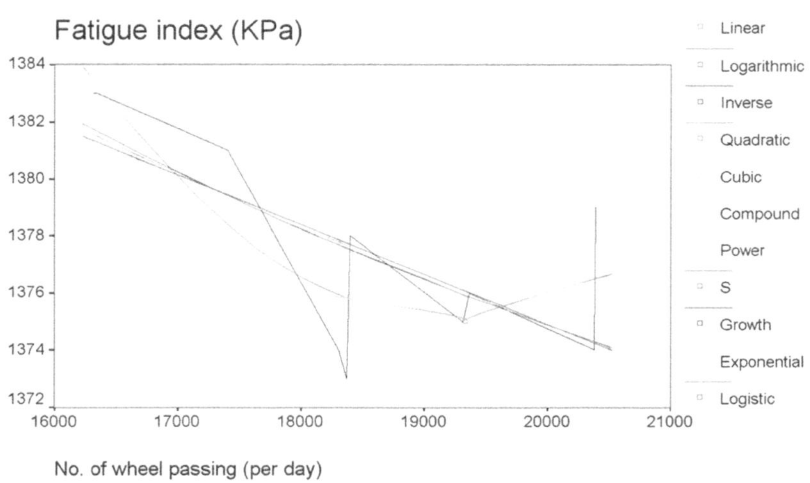

FIGURE 1.10 Curve fit for fatigue index (kPa) and number of wheels passing.

Due to the parameter limitations, the Power function had suitable correlation on scatter diagram and best-fit curve which was used to extrapolate measured data at critical zone according to three cases were showed in Table 1.4. The Fatigue index (kPa) varied with the MPD (mm) and number of wheels passing, as shown in Eqn. (16).

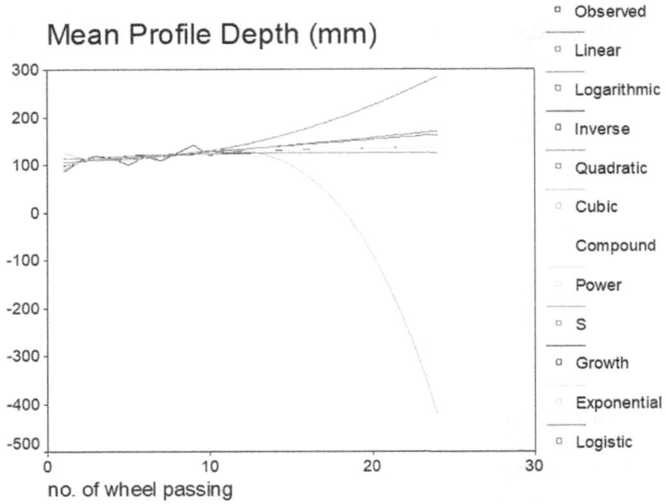

FIGURE 1.11 Curve fit for mean profile depth (mm); number of wheels passing.

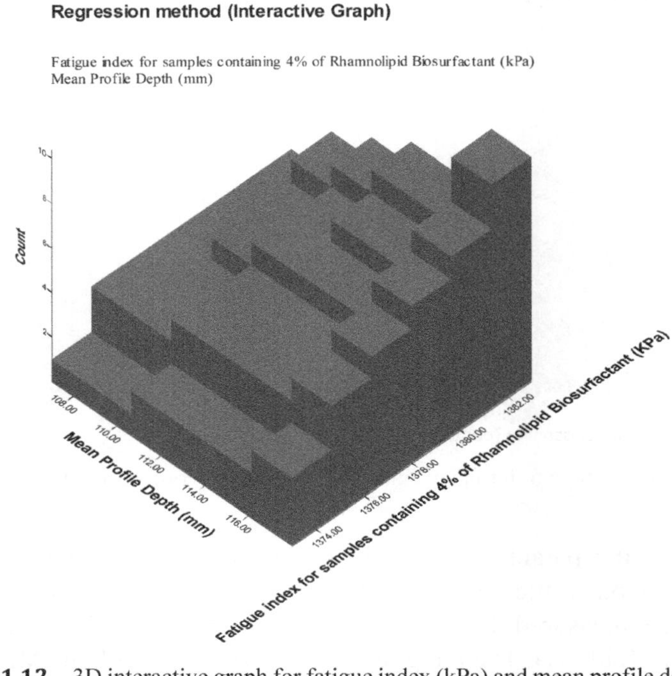

FIGURE 1.12 3D interactive graph for fatigue index (kPa) and mean profile depth (mm).

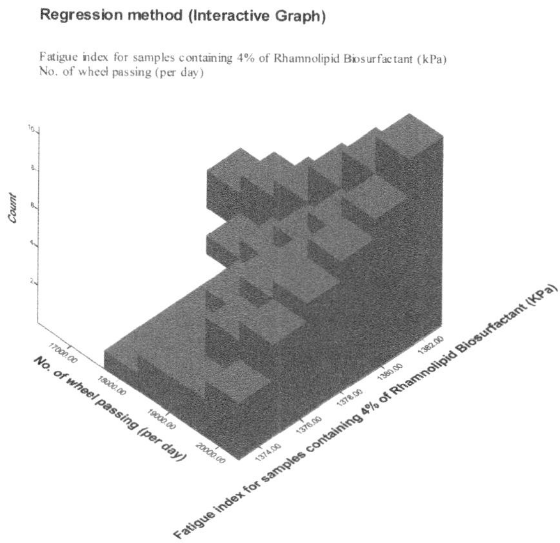

FIGURE 1.13 3D interactive graph for fatigue index (kPa) and number of wheels passing.

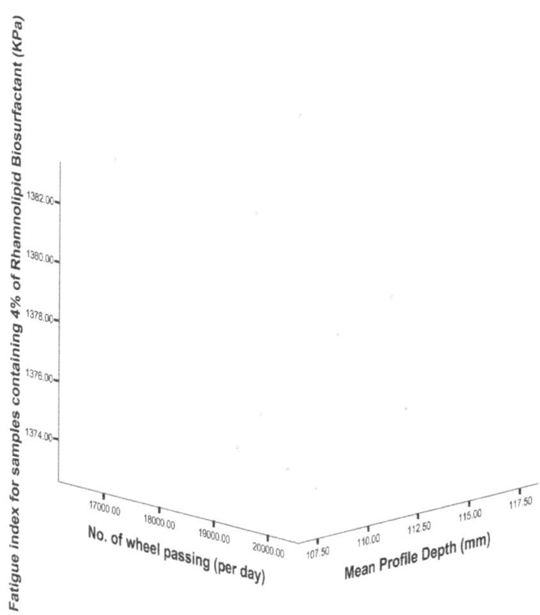

FIGURE 1.14 3D scatter diagram for fatigue index (kPa) as dependent variable; mean profile depth (mm) and number of wheels passing as independent variables.

$$f(x) = kx^a \quad (16)$$

$$F.I. = kt^a$$

where; *F.I.*: Fatigue index (kPa); *t*: MPD (mm); *a* and *k* are parameters determined by the linear least-squares method using data from the 10 parameters tested.

$$y = 1288.91x.^{014}$$

The p-value is the level of marginal significance within a statistical hypothesis test representing the probability of the occurrence of a given event. The p-value is used as an alternative to rejection points to provide the smallest level of significance at which the null hypothesis would be rejected. At this work, the following procedure done for the p-value evaluation:

- Determine the experiment's expected results.
- Determine the experiment's observed results.
- Determine the experiment's degrees of freedom.
- Compare expected results to observed results with chi square.
- Choose a significance level.
- Use a chi square distribution table to approximate the p-value.
- Approximate p value for experiment, it can be decided whether or not to reject the null hypothesis of the experiment.

Generally, based on the hypothesis and the experimental results, If the p value is lower than the significance value, it can be showed that the experimental results would be highly unlikely to occur if there was no acceptable relation between the variables which be manipulated (Table 1.4).

In this work, the p-values for three variables (MPD; daily temperature; number of wheel passes) are as follows (Table 1.4):

1. P-value for fatigue index (kPa):

   Chi 2 =1.200, DF = 7

   The P value equals 0.991.

   By conventional criteria, this difference is considered to be not statistically significant.

2. P Value for MPD:

   Chi 2 = 1.200, DF = 6

   The P value equals 0.977.

By conventional criteria, this difference is considered to be not statistically significant.

3. P Value for number of wheel passes:

$$\text{Chi } 2 = 0.000, \text{ DF} = 9.$$

The two-tailed P value equals 0.962.

TABLE 1.4 Test Statistic and p-Value Calculation

	Fatigue Index (kPa)	Mean Profile Depth (mm)	Number of Wheel Passing (per day)
Chi-square	1.200	1.200	0.000
df	7	6	9
Asymp. Sig.	0.991	0.977	0.962

The description of results for sixth stage are as follows:

- ➢ The laboratory results of Step 1 were compared with the Regression analysis of Step 5 for the RAP include the sample of 4% of Rhamnolipid Biosurfactant at critical zone.

According to the results of DSR test at medium temperature and the study of excellent pavement fatigue parameter (G *.sinδ) and low temperature BBR test (Step 1), it can be expected that the sample of 4% of Rhamnolipid Biosurfactant can improve the performance of asphalt samples against fatigue and heat cracks. The results of Step 5 due to the regression analysis showed the power function had suitable correlation on scatter diagram. The results of Step 1 for Fatigue index due to the sample of 4% Rhamnolipid Biosurfactant as laboratory results was very closely to the Power function solutions (Figure 1.15).

A p-value is a number between 0 and 1, and in most realistic situations, a value at the boundary (especially a value at 0) is impossible. A value of 1 is impossible because when you compute two statistics (Figures 1.16–1.18) from two normally distributions, the probability that those two statistics are exactly equal to 0 and only an exact equality will lead to a p-value of 1.

The p-value is defined as the probability of obtaining a result equal to or "more extreme" than what was actually observed, when the null hypothesis is true. In frequents inference, the p-value is widely used in statistical hypothesis testing, specifically in null hypothesis significance testing. In this work the p-value are as follows:

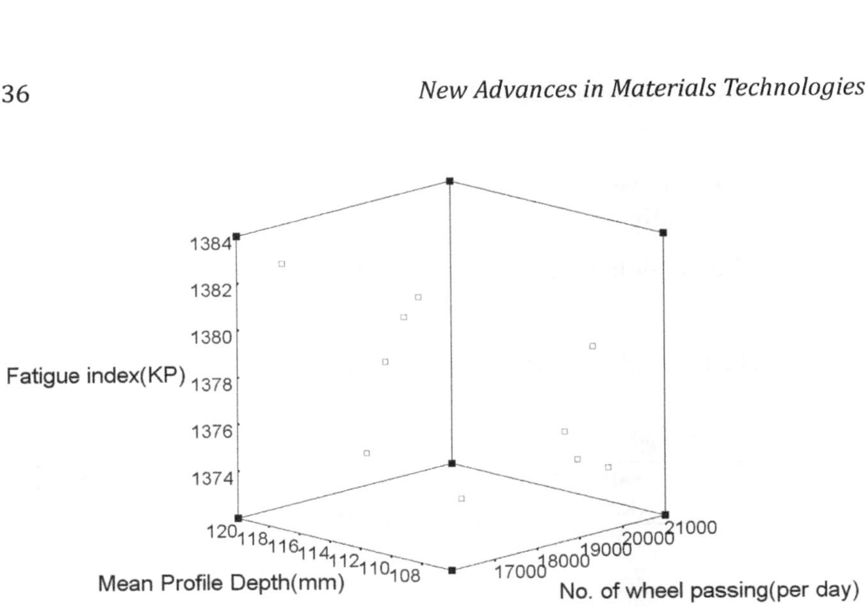

FIGURE 1.15 3D scatter diagram for fatigue index (kPa) as dependent variable; mean profile depth (mm) and number of wheels passing as independent variables.

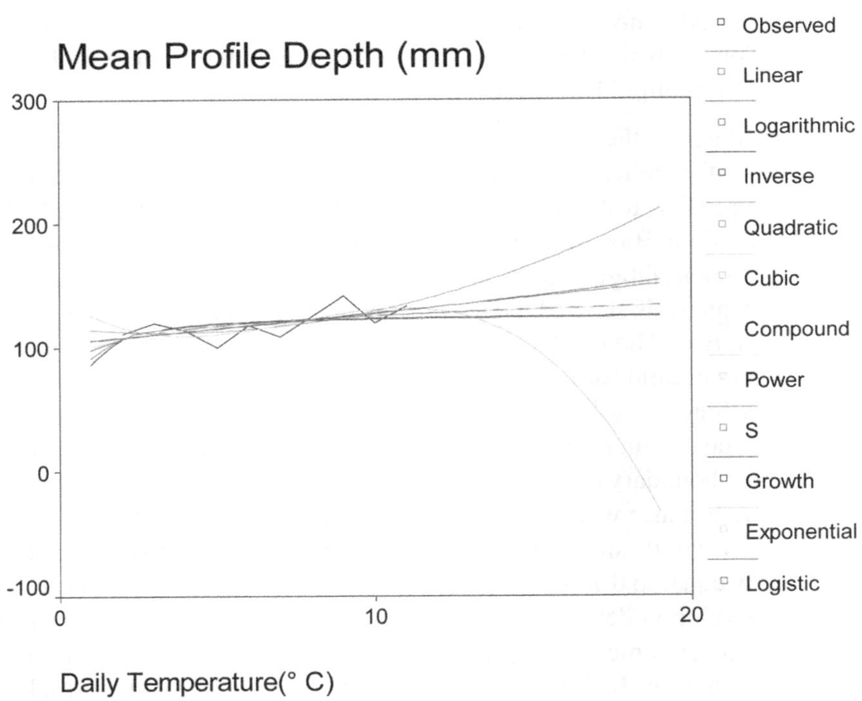

FIGURE 1.16 Curve fit for mean profile depth (mm); daily temperature (°C).

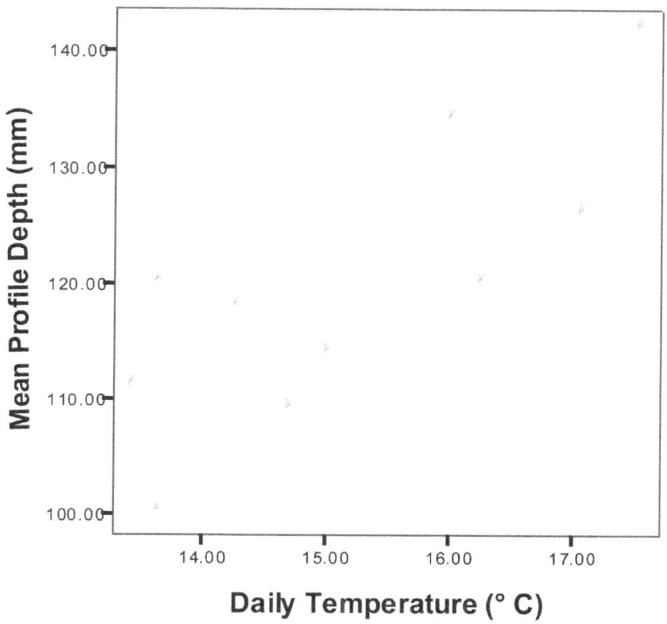

FIGURE 1.17 2D scatter for mean profile depth (mm) and daily temperature (°C).

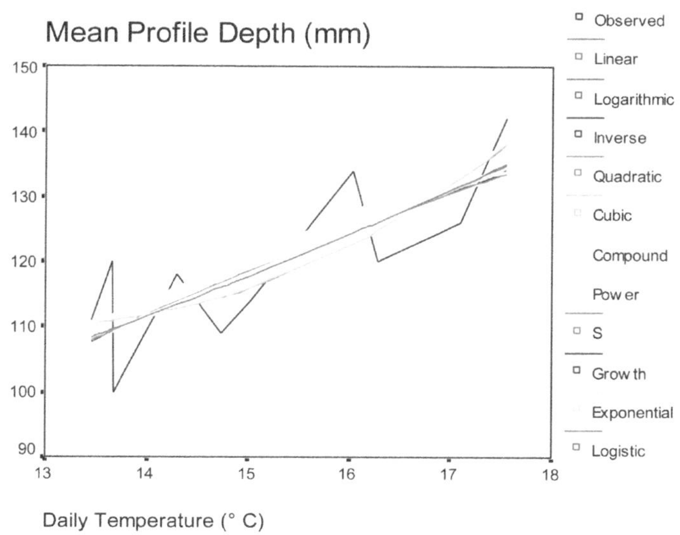

FIGURE 1.18 Curve fit for mean profile depth (mm); daily temperature (°C).

P value results for mean profile depth:

Chi 2 = 0.8, DF = 8

The two-tailed P value equals 0.9992.

By conventional criteria, this difference is considered to be not statistically significant.

P value results for daily temperature.

Chi 2 = 0, DF = 9

The two-tailed P value equals 0.9692.

By conventional criteria, this difference is considered to be not statistically significant.

1.4 DISCUSSION

The Roads are the national assets of any country that provide communication possibility among different points, essentially. Furthermore, the cost of road construction has increased over time, hence, its preservation is a critical job. Thus, the researchers have conducted extensive tests in order to improve maintenance methods, which can be used to achieve further benefits from this national treasure. This work presented the performance of statistical modeling based on the flowing stages:

- Mathematical regression models are developed for first order auto correlated residuals. It provided reliable estimate of both goodness-of-fit measures and significance levels of chosen predictor variables. The Power function had suitable correlation on scatter diagram.
- Present work compares the pavement data for number wheel passes and temperature against MPD. The results include the MPD data as a function of number of wheel passes and temperature for stone matrix asphalt (SMA) of 10 specimens.
- The results showed the good correlation between the parameter: P value results for Depth (mm) equals 0.9992 and P value results for daily temperature (°C) equals 0.9692 and P value results for number of wheel passes equals 0.9909.
- The results of this work will help to energy savings due to optimizing the cost and stiffness of the asphalt pavement.

1.4.1 COMPARISON OF PRESENT WORK WITH OTHER EXPERTS WORK

A more recent study [16] has shown the effect of six rejuvenators on the fatigue behavior of asphalt mixtures was investigated. The six additives including waste vegetable oil, waste vegetable grease, organic oil, industrial wood oil, aromatic extraction oil, and waste engine oil, each with a weight percentage of 12% were combined. The engine oil has the shortest failure cycle and it has the highest probability of fatigue failure compared to other rejuvenating. However, all samples showed longer fatigue life than the base mixture. The fatigue performance improvement for all samples led to a range of 5% to 38%. The organic oil-modified bitumen had the best performance compared to other modified bitumen.

In present work, the performance of the sample containing 4% Rhamnolipid Biosurfactant was increased by 213%. This indicates the excellent performance of bitumen modified with Rhamnolipid Biosurfactant against fatigue of asphalt mixtures. By comparing the results of this work with the discussion by several authors [16], it can be concluded that samples containing Rhamnolipid Biosurfactant had a better behavior against fatigue against to samples made by bitumen modified with other oils. There are some similarities between present work results and more recent work [16] results. But present work also investigated the GIS and networked sensors as a strong tool with the ability of overlapping a lot off queries and fields for precision and rapid identifying the zones with the most needs to RAP. The RS facility is equipped with data loggers and the Internet of Things (IoT) based on GIS and networked sensors, connecting a lot of detectors and sensors at the least time for the identification of zones. The connection between (GIS), and Networked Sensors and (IoT) make possible for modeling in order to prediction of the mechanical behavior due to RAP in a little time up to one second.

1.5 CONCLUSIONS

The purpose of this work was to improve the properties of bitumen. This led to improving the mechanical properties of hot asphalt mixtures. This also led to generating desirable asphalt with the aid of natural, non-toxic, and finally to improve the asphalt pavement quality. Finally, the following subjects can be expected from this work:

- Does the compressive strength of asphalt samples increase by the use of modified bitumen of rhamnolipid biosurfactant?
- Does the tensile strength of asphalt samples increase by the use of modified bitumen of rhamnolipid biosurfactant?
- Does the resilience modulus of asphalt samples increase by the use of modified bitumen of rhamnolipid biosurfactant?

By testing the aggregates and bitumen based on the Marshall method, the optimal bitumen percentage of the mixtures was determined. According to this optimal percentage of bitumen, the necessary samples were evaluated according to the relevant standards. In this work, the following tests were performed on samples of asphalt mixes.

Due to Biosurfactants' environmental compatibility, Biosurfactants decompose easily in nature and are non-toxic agents. In this work, by adding Rhamnolipid Biosurfactant to the bitumen, which is one of the main components of asphalt mixtures, the viscosity and rutting parameter of the superpave has decreased. Also, the fatigue and thermal cracks have been reduced by the influence of the Rhamnolipid Biosurfactant. For the modified asphalt samples, the tests of Marshall stability, flow, resilience modulus, repeated load axial, uniaxial creep, and indirect tensile fatigue were carried out. The mechanical properties of the asphalt mixture were improved by adding more Rhamnolipid Biosurfactant. The Marshall stability and flow evaluation results proved that the adding of more Rhamnolipid Biosurfactant can bring more stability and less flow value. This fact indicates that the modified samples had more resistance to shear stress. The results of resilience modulus tests, deformation, and fatigue tests confirmed that Rhamnolipid Biosurfactant can improve the properties of asphalt mixtures in all situations. The result of this work showed that the Biosurfactants are unique dual molecules that are widely used to remove organic and metallic pollutants from the environment. This work also should be of interest to engineers, contractors, and others involved in the specification and design of asphalt mixtures for flexible pavements. Those involved in promoting the optimal use of rhamnolipid biosurfactant.

- The increasing Rhamnolipid Biosurfactant, caused to decrease in viscosity and rutting parameter, so that this additive reduced the high temperature of bitumen by one degree.
- The presence of Rhamnolipid Biosurfactant in bitumen reduced the susceptibility of fatigue and heat cracks in bitumen.
- The results of Marshall strength test showed that samples containing 2% Rhamnolipid Biosurfactant had the highest Marshall strength, which indicated that this sample was more resistant to static loads.
- According to Marshall test, the sample containing 2% Rhamnolipid Biosurfactant had the best performance.
- The samples containing 2% Rhamnolipid Biosurfactant had the best performance against static loads.
- The resistance modulus test results showed that samples containing 4% rhamnolipid biosurfactant had the highest value at all test temperatures.
- The repeated axial load test results showed that samples containing 4% rhamnolipid biosurfactant had the lowest final strain at all temperatures.
- The results of creep test showed that the sample containing 4% rhamnolipid biosurfactant had the lowest axial strain and the highest resistance to deformation.
- The results of indirect traction fatigue test showed that the sample containing 4% rhamnolipid biosurfactant at both 10°C and 20°C has the highest resistance to fatigue.
- The samples containing 4% rhamnolipid biosurfactant had the best performance against dynamic loads.

At this work, linear regression model by using data includes numbers of wheel passes as independent variable against MPD of asphalt as dependent variable was defined. At the second case, linear regression model by using data includes the daily temperature as independent variable against MPD as dependent variable was defined. The daily temperature of site at pilot had most important effects on regression model. The regression method fitted function curve and provided regression analysis based on the field test data. The results of work emphasize that "regression model" is accurate and an improved model for reclamation of asphalt pavement and reclaimed asphalt shingles

(RAS). This factor can effect on choice of predictors and hence the validity of the model. It provides reliable estimates of both goodness-of-fit measures and significance levels of chosen predictor variables and the Power function had suitable correlation on scatter diagram. Hence it emphasizes on the meaningful relation between numbers of wheel passes against MPD. Also, it emphasizes on the meaningful relation between site temperatures against MPD. The results emphasize on the MPD as a function of numbers of wheel passes and site temperature. It showed the good correlation between the parameters: P value results for depth (mm) equals 0.9992 and P value results for daily temperature (°C) equals 0.9692 and P value results for numbers of wheel passes equals 0.9909. Consequently, the results of this work will help to optimize the asphalt pavement cost and performance.

Present work also investigated the GIS and Networked Sensors as a low cost, high precision and rapid methods for identifying the RAP. Finally, this work showed that GIS can be linked to new techniques, including networked sensors, RS, and IoT, which can be serious subjects for future research in the field of pavement engineering [55–77].

1.6 SUGGESTIONS FOR FUTURE RESEARCH

This work led to raising the following suggestion for improvement of modeling for evaluation of RAP containing rhamnolipid biosurfactant:

- Make connection between networked sensors, GIS, RS, and IoT for rapid data intercommunication in order to zoning the RAP in a little time up to one second.
- Conceptual modeling for prediction of the mechanical behavior due to RAP.
- Improvement of RS facilities equipped with Networked Sensors and data loggers, IoT and GEO-database intercommunication process.
- Evaluate the effect of biosurfactant on moisture sensitivity of hot asphalt samples.
- Evaluate the use of calcareous stone materials and compare it with siliceous stone materials.
- Performing cost-benefit analysis of using Biosurfactants in asphalt pavements for industrial scale application.

KEYWORDS

- **bending beam rheometer**
- **dynamic section rheometer**
- **geospatial information system**
- **networked sensors**
- **Nottingham asphalt testing machine**
- **reclaimed asphalt pavement**
- **regression method**
- **rhamnolipid biosurfactant**

REFERENCES

1. Mrugacz, J., (2020). *The Complete 7-Step Process for Asphalt Pavement Installation.* Sun Prairie: Wolf Paving. https://www.wolfpaving.com/blog/the-complete-7-step-process-for-asphalt-pavement-installation (accessed on 06 September 2023).
2. Hassan, M., Essam, T., Yassin, A. S., & Salama, A., (2016). Optimization of rhamnolipid production by biodegrading bacterial isolates using Plackett–Burman design. *International Journal of Biological Macromolecules, 82*(4), 573–579. doi: 10.1016/j.ijbiomac.2015.09.057.
3. Arabani, M., Yousefpour, V., & Taleghani, M., (2017). Rutting behavior of hot mix asphalt modified by polyvinyl chloride powder. *Journal of Petroleum Science and Technology, 35*(15), 1621–1626. https://doi.org/10.1080/10916466.2017.1336772.
4. Zeng, M., Pan, H., Zhao, Y., & Tian, W., (2016). Evaluation of asphalt binder containing castor oil-based bioasphalt using conventional tests. *Journal of Construction and Building Materials, 126*, 537–543. doi: 10.1016/j.conbuildmat.2016.09.072.
5. Yang, X., & You, Z., (2015). High temperature performance evaluation of bio-oil modified asphalt binders using the DSR and MSCR tests. *Journal of Construction and Building Materials, 76*, 380–387. https://doi.org/10.1016/j.conbuildmat.2014.11.063.
6. Somé, S. C., Pavoine, A., & Chailleux, E., (2016). Evaluation of the potential use of waste sunflower and rapeseed oils-modified natural bitumen as binders for asphalt pavement design. *International Journal of Pavement Research and Technology, 9*(5), 368–375. https://doi.org/10.1016/j.ijprt.2016.09.001.
7. Shirzad, S., Hassan, M. M., Aguirre, M. A., Mohammad, L. N., & Daly, W. H., (2016). Evaluation of sunflower oil as a rejuvenator and its microencapsulation as a healing agent. *Journal of Materials in Civil Engineering, 28*(11), 04016116. https://ascelibrary.org/doi/abs/10.1061/%28ASCE%29MT.1943-5533.0001625 (accessed on 06 September 2023).

8. Fini, E. H., Kalberer, E. W., Shahbazi, A., Basti, M., You, Z., & Ozer, H., (2011). Chemical characterization of biobinder from swine manure: Sustainable modifier for asphalt binder. *Journal of Materials in Civil Engineering, 23*(11), 1506–1513. https://ascelibrary.org/doi/abs/10.1061/%28ASCE%29MT.1943-5533.0000237 (accessed on 06 September 2023).
9. Chen, M., Leng, B., Wu, S., & Sang, Y., (2014). Physical, chemical, and rheological properties of waste edible vegetable oil rejuvenated asphalt binders. *Journal of Construction and Building Materials, 66,* 286–298. https://doi.org/10.1016/j.conbuildmat.2014.05.033.
10. Gardziejczyk, W., (2002). *Texture of Road Surfaces, Methods of Measurement Parameters Evaluation and its Influence on the Tire/Road Noise* (No. 2, pp. 5–29). Drogi Mosty, Poland.
11. Content. A Practical Guide, (2007). Quality Improvement Series, 124 National Asphalt Pavement Association, Lanham, MD, March, 36.
12. Nady, R. M., (1997). *The Quality of Random RAP* (Vol. 2, No. 2, pp. 14–17). Separating Fact from Supposition. National Asphalt Pavement Association, HMAT magazine, Lanham, MD.
13. Newcomb, D. E., Brown, E. R., & Epps, J. A., (2007). Designing HMA Mixtures with High RAP Content. A Practical Guide. *Quality Improvement Series 124.* National Asphalt Pavement Association, Lanham, MD.
14. Kowalski, K. J., McDaniel, R. S., Olek, J., Shah, A., & Li, S., (2008). Development of the international friction index flag value. *10th International Conference on Application of Advanced Technologies in Transportation.* Athens, Greece, Paper No. 705.
15. McDaniel, R. S., & Coree, B. J., (2003). *Identification of Laboratory Techniques to Optimize Superpave HMA Surface Friction Characterization.* Phase I: Draft Final Report. Institute for Safe, Quite and Durable Highways, SQDH 2003–6, West Lafayette, Indiana.
16. Behbahani, H., Ziari, H., Kamboozia, N., Khaki, A. M., & Mirabdolazimi, S., (2015). Evaluation of performance and moisture sensitivity of glasphalt mixtures modified with nanotechnology zycosoil as an anti-stripping additive; some research aspects for reclamation of asphalt pavement. *Journal of Construction and Building Materials, 78*, 60–68. doi: 10.1016/j.conbuildmat.2014.12.053.
17. Li, S., Noureldin, S., & Zhu, K., (2003). *Upgrading the INDOT Pavement Friction Testing Program.* Final Report FHWA/IN/JTRP-2003/23, West Lafayette, Indiana.
18. West, T. R., Choi, J. C., Bruner, D. W., Park, H. J., & Cho, K. H., (2001). Evaluation of dolomite and related aggregates used in bituminous overlays for Indiana pavements. *Transportation Research Record, 1757*, 137–147.
19. Indiana Department of Transportation. Standard Specifications Book. (2009). Section 400, Asphalt Pavements, Indianapolis.
20. Climate.data.org, Climate-Data.org: https://www.yahoo.com/news/weather (accessed on 06 September 2023).
21. Kowalski, K. J., McDaniel, R. S., & Olek, J., (2008). Development of a laboratory procedure to evaluate the influence of aggregate type and mixture proportions on the frictional characteristics of flexible pavements. *Journal of the Association of Asphalt Paving Technologists, 77*, 35–70.

22. Pezzullo, J. C., (2016). *Handbook of Math Functions by Abramowitz*. Adapted from JavaScript written by PhD, Associate Professor, Pharmacology and Biostatistics Georgetown University Medical Center, Washington D.C.
23. *Recycling Hot Mix Asphalt Pavements*, (2007). Information Series 123 (IS-123) (p. 27). National Asphalt Pavement Association, Lanham, MD.
24. Read, J., & Whiteoak, D., (2003). *The Shell Bitumen Handbook*. Thomas Telford.
25. Claudy, P., Letoffe, J., King, G., & Plancke, J., (1992). Characterization of asphalts cements by thermomicroscopy and differential scanning calorimetry: Correlation to classic physical properties. *Fuel Science & Technology International*.
26. Masson, J. F., Leblond, V., & Margeson, J., (2006). Bitumen morphologies by phase-detection atomic force microscopy. *Journal of Microscopy, 221*(1), 17–29.
27. Per Redelius & Hilde Soenen, (2015). Relation between bitumen chemistry and performance. *Fuel*, 140, 34–43, ISSN 0016-2361, https://doi.org/10.1016/j.fuel.2014.09.044. (https://www.sciencedirect.com/science/article/pii/S0016236114009132).
28. Šaban, M., Porter, S., Costello, C., Djuričić, M., & Vitorović, D., (1980). Polar constituents isolated from Aleksinac oil shale. *Physics and Chemistry of the Earth, 12*, 559–566.
29. Hetényi, M., & Sajgó, C., (1990). Hydrocarbon generation potential of some Hungarian low-rank coals. *Organic Geochemistry, 16*(4–6), 907–916.
30. Longstaff, D., Deo, M., Hanson, F., Oblad, A., & Tsai, C., (1992). Hydrotreating the bitumen-derived hydrocarbon liquid produced in a fluidized-bed pyrolysis reactor. *Fuel, 71*(12), 1407–1419.
31. Binet, S., Pfohl-Leszkowicz, A., Brandt, H., Lafontaine, M., & Castegnaro, M., (2002). Bitumen fumes: Review of work on the potential risk to workers and the present knowledge on its origin. *Science of the Total Environment, 300*(1–3), 37–49.
32. Sato, S., Sugimoto, Y., Sakanishi, K., Saito, I., & Yui, S., (2004). Diesel quality and molecular structure of bitumen-derived middle distillates. *Fuel, 83*(14, 15), 1915–1927.
33. Ji, J., Yao, H., Suo, Z., You, Z., Li, H., Xu, S., & Sun, L., (2016). Effectiveness of vegetable oils as rejuvenators for aged asphalt binders. *Journal of Materials in Civil Engineering, 29*(3), D4016003.
34. Asli, H. H., Arabani, M., & Golpour, Y., (2020). Reclaimed asphalt pavement (RAP) BASED on a geospatial information system (GIS). *Slovak Journal of Civil Engineering, 28*(2), 36–42. https://doi.org/10.2478/sjce-2020-0013.
35. Safari, A., Rayaee, M., Abolhassani, A., & Motamedi, H., (2008). *Isolation, Screening, and Identification of Biosurfactant-Producing Bacteria from the Caspian Sea Ecosystem in Order to Biologically Eliminate Environmental Pollution*. The First International Conference on Environmental Change in the Caspian Region, University of Mazandaran, Babolsar.
36. Arabani, M., (2012). *Bitumen and Asphalt Tests* (2nd edn.). University of Guilan.
37. Bagheri, T., Partovi, M., & Bahmani, M., (2012). Production of rhamnolipid biosurfactant by Pseudomonas aeruginosa from vegetable oil processing wastes. *Journal of Cellular-Molecular Biotechnology, 2*(9).
38. Kosaric, N., (2001). Biosurfactants and their application for soil bioremediation. *Food Technology and Biotechnology, 39*(4), 295–304.

39. Nguyen, T. T., & Sabatini, D. A., (2011). Characterization and emulsification properties of rhamnolipid and sophorolipid biosurfactants and their applications. *International Journal of Molecular Sciences*.
40. Azahar, W. N. A. W., Jaya, R. P., Hainin, M. R., Bujang, M., & Ngadi, N., (2016). Chemical modification of waste cooking oil to improve the physical and rheological properties of asphalt binder. *Construction and Building Materials, 126*, 218–226.
41. Ali, A. W., Mehta, Y. A., Nolan, A., Purdy, C., & Bennert, T., (2016). Investigation of the impacts of aging and RAP percentages on effectiveness of asphalt binder rejuvenators. *Construction and Building Materials, 110*, 211–217.
42. Behbahani, H., Ziari, H., Kamboozia, N., Khaki, A. M., & Mirabdolazimi, S., (2015). Evaluation of performance and moisture sensitivity of glasphalt mixtures modified with nanotechnology zycosoil as an anti-stripping additive. *Construction and Building Materials, 78*, 60–68.
43. Zargar, M., Ahmadinia, E., Asli, H., & Karim, M. R., (2012). Investigation of the possibility of using waste cooking oil as a rejuvenating agent for aged bitumen. *Journal of Hazardous Materials, 233*, 254–258.
44. Hansen, J. S., Lemarchand, C. A., Nielsen, E., Dyre, J. C., & Schrøder, T., (2013). Four-component united-atom model of bitumen. *The Journal of Chemical Physics, 138*(9), 094508. arXiv:1307.2468. Bibcode:2013JChPh.138i4508H. doi: 10.1063/1.4792045. ISSN 0021-9606. PMID 23485314. S2CID 17886103.
45. Jones, D., (2011). Roach, P., Setter, J., & Esling, J., (eds.), *Cambridge English Pronouncing Dictionary* (18th edn.). Cambridge University Press. ISBN 978-0-521-15255-6.
46. Abraham, H., (1938). *Asphalts and Allied Substances: Their Occurrence, Modes of Production, Uses in the Arts, and Methods of Testing* (4th edn.). New York: D. Van Nostrand Co., Inc. Full text at Internet Archive (archive.org).
47. Oishimaya, S. N., (2021). *The Unique Pitch Lakes of the World.* World Atlas.
48. Sörensen, A., & Wichert, B., (2009). Asphalt and bitumen. *Ullmann's Encyclopedia of Industrial Chemistry.* Weinheim: Wiley-VCH. doi: 10.1002/14356007.a03_169.pub2.
49. Brown, E. R., Kandhal, P. S., Roberts, F. L., Kim, Y. R., Lee, D. Y., & Kennedy, T. W., (1991). *Hot Mix Asphalt Materials, Mixture Design, and Construction* (3rd edn.). Lanham, Maryland: NAPA Education and Research Foundation. ISBN 978-0914313021.
50. "Oil Sands Glossary" 2007, (2008). *Oil Sands Royalty Guidelines*. Government of Alberta.
51. Chen, S. (2023). The fate of bitumen: an exploratory study of national newspaper coverage of Alberta's bitumen industry during the COVID-19 pandemic. *Humanit Soc Sci Commun 10,* 30, https://doi.org/10.1057/s41599-023-01515-2.
52. *Alberta's Energy Reserves 2014 and Supply/Demand Outlook 2015–2024*. Calgary, Alberta: Alberta Energy Regulator.
53. Polaczyk, P., Han, B., Huang, B., Jia, X., & Shu, X., (2018). Evaluation of the hot mix asphalt compactability utilizing the impact compaction method. *Construction and Building Materials, 187*, 131–137. doi: 10.1016/j.conbuildmat.2018.07.117. ISSN 0950-0618.
54. Zaumanis, M., Mallick, R. B., Poulikakos, L., & Frank, R., (2014). Influence of six rejuvenators on the performance properties of reclaimed asphalt pavement (RAP)

binder and 100% recycled asphalt mixtures. *Construction and Building Materials, 71*, 538–550.

55. Feizbahr, M., Mirhosseini, S. M., Joshaghani, A. H. (2020). *Improving the Performance of Conventional Concrete Using Multi-Walled Carbon Nanotubes Express Nano Letters, 1*(1), 1–9.
56. Kian Hariri Asli & Kaveh Hariri Asli (2022). Isolated pressure zones based on GIS as a solution for water network problems, Water Practice and Technology 2022, wpt2022119. doi: https://doi.org/10.2166/wpt.2022.119, *17*(10), 2125–2140, https://iwaponline.com/wpt/article/doi/10.2166/wpt.2022.119/91338/Isolated-pressure-zones-based-on-GIS-as-a-solution.
57. Kian Hariri Asli & Kaveh Hariri Asli (2023). Minimum night flow (MNF) and corrosion control in compliance with internet of things (IoT) for water systems, *Water Practice and Technology, 18*(3), 608–625, https://doi.org/10.2166/wpt.2023.012, https://iwaponline.com/wpt/article/doi/10.2166/wpt.2023.012/93513/Minimum-night-flow-MNF-and-corrosion-control-in.
58. Meissam Nazeryan, & Mahdi Feizbahr, (2022). Seismic Evaluation of the Cheng and Chen Modified Model Using Shear Keys in Steel Beam-to-Concrete Column Connections, *Transactions of Civil and Environmental Engineering, 8*(2) Article ID: 2787, 1–6.
59. Asli, K. H., & Asli, K. H. (2023). Smart Water System and Internet of Things. *J Mod Ind Manuf, 2*, 5. DOI: 10.53964/jmim.2023005.
60. Amritanshu Shukla, Kian Hariri Asli, Neha Kanwar Rawat, Ann Rose Abraham, & Haghi, A. K. (2024). *Technological Advancement in Clean Energy Production* (Apple Academic Press, Hard ISBN: 9781774915585), https://www.appleacademicpress.com/title.php?id=1416.
61. Ali Pourhashemi, Kian Hariri Asli, Ann Rose Abraham, & Haghi, A. K. (2024). *Sustainable Water Engineering; Smart and Emerging Technologies,* (Apple Academic Press, Hard ISBN: 9781774915714), https://appleacademicpress.com/sustainable-water-engineering-smart-and-emerging-technologies/9781774915714.
62. Ann Rose Abraham, Heru Susanto, Haghi, A. K. & Kian Hariri Asli (2024). "*Sustainability in Energy and Environment Engineered Materials and Smart Computational Techniques,*" (Apple Academic Press, Hard ISBN: 9781774916209), https://www.appleacademicpress.com/sustainability-in-energy-and-environment-engineered-materials-and-smart-computational-techniques/9781774916209#bios.
63. Sonia Khanna, Ann Rose Abraham, Kian Hariri Asli, & Haghi, A. K. (2024). "*Sustainable Environmental Engineering Water Security, Energy Conservation, and Green Processes*" (Apple Academic Press, Hard ISBN: 9781774916902), https://www.appleacademicpress.com/sustainable-environmental-engineering-water-security-energy-conservation-and-green-processes/9781774916902.
64. Kian Hariri Asli (2024). Technological Advancement in HVAC&R System Energy Efficiency, Technological Advancement in Clean Energy Production, (Apple Academic Press, Hard ISBN: 9781774915585)https://www.appleacademicpress.com/title.php?id=1416.
65. Mahdi Feizbahr & Morteza Jamshidi, Jayaprakash, (2013). *Review on Various Types and Failures of Fibre Reinforcement Polymer, Middle-East Journal of Scientific Research 13*(10), 1312–1318.

66. Kian Hariri Asli (2024). *IoT-Based Smart Water Leak Detection System for a Sustainable Future: A Case Study, Sustainable Water Engineering; Smart and Emerging Technologies*, (Apple Academic Press, Hard ISBN: 9781774915714). https://appleacademicpress.com/sustainable-water-engineering-smart-and-emerging-technologies/9781774915714.
67. Mahdi Feizbahr, & Shova, R. Bhattarai, (2022). Some Aspects of Research for Reclamation of Asphalt Based on Remote Sensing (RS), *CRPASE: Transactions of Civil and Environmental Engineering, 8*(3), Article ID: 2810, 1–6.
68. Kian Hariri Asli (2024). Measurement, Modeling, and Sustainable Control of Environmental Contaminants Using Smart Computational Technologies: A Case Study, Sustainability in Energy and Environment; *Engineered Materials and Smart Computational* Techniques, (Apple Academic Press, Hard ISBN: 9781774916209, https://www.appleacademicpress.com/sustainability-in-energy-and-environment-engineered-materials-and-smart-computational-techniques/9781774916209#bios.
69. Mohammad Nikookar, Nicholas A. Brake, Mubarak Adesina, Ashiqur Rahman, Thinesh Selvaratnam, Haley A. Snyder, & Ozge Günaydın-Sen, (2022). Reutilization of oil and gas produced water in cement composite manufacturing, Journal of Cleaner Production, Volume 381, Part 1, 135113, ISSN 0959–6526, https://doi.org/10.1016/j.jclepro.2022.135113.
70. Kian Hariri Asli (2024). *Fuzzy Logic for Smart Water and Wastewater Systems, Sustainable Environmental Engineering Water Security, Energy Conservation, and Green Processes*, (Apple Academic Press, Hard ISBN: 9781774916902), https://www.appleacademicpress.com/sustainable-environmental-engineering-water-security-energy-conservation-and-green-processes/9781774916902.
71. Oruji, S., Brake, N. A., Hosseini, S., Adesina, M., Nikookar, M., (2023). Enhancing Recycled Aggregate Concrete Using a Three-Stage Mixed Coal Bottom Ash Slurry Coating, *Journal of Materials in Civil Engineering, 35*(5), https://doi.org/10.1061/(ASCE)MT.1943–5533.0004723.

CHAPTER 2

Mathematical Modeling of Rogue Waves in a Multi-Ion Cometary Plasma to Study the Effect of Heavier Ions and Dust

G. SREEKALA, ASHWINI S. PILLAI, and NAJIYA NAZAR

Department of Plant Science, Bharathidasan University, Tiruchirappalli, India

ABSTRACT

We have studied the formation of rogue waves in a multi-ion cometary plasma by deriving the Korteweg de-Vries (KdV) equation and nonlinear Schrodinger equation (NLSE) by reductive perturbation method. The superthermal kappa distribution which provides an unambiguous replacement for a Maxwellian distribution in space plasmas is connected with non-extensive Statistical Mechanics and provides a continuous spectrum of energy. Hence, we model electrons by kappa distribution. We considered plasma composed of pair-ions (positively and negatively charged hydrogen ions), dust (positively and negatively charged) and electrons (cold and super thermal) which well satisfied with the coma of comet Halley and studied the effect of each component especially super thermal electrons and charged dust grains in the formation of solitary and rogue structures. For numerical study, the parameters observed in the coma of comet Halley are used.

New Advances in Materials Technologies. Hossein Hariri Asli, Ali Pourhashemi, Ann Rose Abraham, & A. K. Haghi (Eds.)

2.1 INTRODUCTION

A plasma is a gas of charged particles which consists of equal number of free positive and negative charge carriers [1]. The plasma medium has its properties of anisotropy, inhomogeneity, loss free and dispersive nature. The anisotropy is the quality of exhibiting properties with different values when measured in different directions.

The dispersive means that the properties of the medium depend on frequency of the wave. The ion acoustic waves are low frequency waves found in collision less plasma and terms fundamental plasma modes. 'Rog wave' is a scientific term, marked its energy in the science of ocean waves. These are the "freak waves" with large amplitude which are short lived violent phenomenon occurring spontaneously in coastal water. The name denotes that the wave that appear from nowhere and disappear without a trace [2]. Indeed, rogue waves can be found in various nonlinear physical environment such as optical system, B.E condensates, superfluid helium, atmosphere [3].

Tolba et al. [33] studied the evolution of rogue waves related to the dynamics of positively charged dust grains that have the interaction with the streaming electrons and ions. It is found that the existence the region of rogue waves depends on the acoustic speed and the streaming densities of the ions and electrons. Williams [4] had discovered three dimensional rogue waves in a dusty plasma system, which represents the wave particle interaction view on their information.

Ruderman [5] analyzed the presence of rogue waves in laboratory and space plasmas in which two existing plasma modes are considered. Moslem [6] analyzed Langmuir rogue waves in collision less electron positron plasmas. This study was helpful in understanding the excitation of nonlinear rogue pulses in astrophysical environments. Sabry et al. [7] provide an investigation for the generation of rogue waves in dusty plasma. El-Tantawy et al. [8] investigated IA super rogue waves in ultracold neutral plasmas with non-thermal electrons. The generation of nonlinear ion acoustic waves in plasma with non-extensive electrons was also studied. Stenflo & Marklund [9] indicates the possibility for the existence of atmospheric rogue waves and the advent of rogue waves is widely known in oceanographics, optics, and cold matter systems.

A possible mechanism for the formation of rogue waves is the modulation instability (MI) [10], a universal phenomenon in many physical systems, such as hydrodynamics [11], biology [12], optics [13], etc. A MI is one in which a plane wave solution is unstable against longitudinal perturbations. An experimental observation of the MI of the monochromatic ion acoustic wave was reported by Watanabe [14]. Also, there are many more studies on rogue waves in plasma [6, 9, 15–19], etc. Pierre [20] investigated the appearance of rogue waves in turbulence and it currently attracted much interest in neutral and conducting fluids. This experiment is carried out in a new toroidal unmagnetized plasma device.

Pathak Sharma, & Bailing [21] found the evolution of super rogue waves having greater amplitude of 5 times the background wave in multi-component plasma. Veldes et al. [22] found the occurrence of rogue waves associated with the interaction of electromagnetic pulse propagation with plasma. The implications of non-Maxwell electron distributions were examined on rogue waves in dusty warm plasma [17].

Chabchoub et al. [23] reported about the mathematical analysis and the properties acquired by nonlinear waves. He also explained peregrine waves which serves as the solutions of the Nonlinear Schrodinger equation. The superposition of rogue waves results in multi-rogue wave solutions. Abdel-wahed et al. [24] analyzed the properties of rogue waves in ion pair plasma with super thermal electrons, which is applied in Earth's ionosphere plasmas. Chan [34] studied the features of rogue waves exhibited only nonlinear Schrödinger equation with MI background. El-Tantawy [25] derived the modified Korteweg-de-Vries equation and studied the properties of rogue waves and dark soliton and its collision in complex plasma. These rogue waves were investigated in the three-component degenerative relativistic quantum plasma system, which is of extra ordinarily high density and low temperature. The rogue waves triplets were described by the interaction of polarization force and further transform into super rogue waves on its superposition and studied the characteristics of dust acoustic waves in an electron depleted dusty plasma. There are so many latest studies on rogue waves [8, 25, 35] in plasma which inspired us to model the rogue waves in a five-component plasma. Wael et al. [26] investigated the role of electrons superthermality at the rogue waves in a weakly relativistic warm plasma.

2.2 THEORETICAL MODEL

We studied the effect of rogue waves in a multi-ion dusty plasma with super thermal electrons. We considered interaction of solar wind with the cometary plasma with separate quasi neutrality conditions. Our system is a collision less magnetized plasma consisting of pair ions, positively and negatively charged dust grains, cold electrons and super thermal hot electrons. The dust particles considered is said to be massive point charge but smaller than Debye length. The positive dust grain arises due to photoelectric effect caused by presence of electromagnetic radiation and negative dust grain charges due to absence of electromagnetic radiation. The total electron density, of the plasma is $n_{ce} = n_{ce} + n_{he}$ where n_{ce} represents colder electron component where temperature is of the order of temperature of ions present. The hot electron component whose temperature is much higher than ions present. The hot electrons are described by kappa distribution. The kappa function is said to be a convenient tool to describe plasma systems out of thermodynamic equilibrium. For modeling we used equation of continuity and equation of motion of cold electrons, pair ions and dust along with Poisson's equation.

For cold electrons:

$$\frac{\partial n_c}{\partial t} + \nabla.(n_c u_c) = 0 \tag{1a}$$

$$\frac{\partial u_c}{\partial t} + (u_c.\nabla)u_c - p\nabla\phi - p\omega_{c+}(u_c \times \hat{z}) = 0 \tag{1b}$$

For positive ions:

$$\frac{\partial n_+}{\partial t} + \nabla\cdot(n_+ u_+) = 0 \tag{2a}$$

$$\frac{\partial u_+}{\partial t} + (u_+\nabla)u_+ + \nabla\phi + \omega_{c+}(u_+ \times \hat{z}) = 0 \tag{2b}$$

For negative ion:

$$\frac{\partial n_-}{\partial t} + \nabla\cdot(n_- u_-) = 0 \tag{3a}$$

$$\frac{\partial u_-}{\partial t} + \nabla\cdot(u_-\nabla)u_- - q\nabla\phi - q\omega_{c+}(u_- \times \hat{z}) = 0 \tag{3b}$$

For dust grains:

$$\frac{\partial n_d}{\partial t} + \nabla\cdot(n_d u_d) = 0 \tag{4a}$$

$$\frac{\partial u_d}{\partial t} + (u_d \cdot \nabla)u_d + \delta z_d r(u_d \times \hat{z}) = 0 \tag{4b}$$

For hot electrons, kappa distributions is used:

$$\mathrm{n}_{he} = \left(1 - \frac{\Delta\varphi}{\kappa - \frac{3}{2}}\right) \quad \text{with } \Delta = \frac{T_+}{T_{he}} \tag{5}$$

and κ is the spectral index.

Poisson's equation:

$$\nabla^2\phi = \left[\alpha n_- - n_+ + \beta n_{ce} + \gamma n_{he} - \sigma\delta z_d n_d\right] \tag{6}$$

$$\text{with } p = \frac{m_+}{m_e}, q = \frac{m_+}{m_-}, \ \mathrm{r} = \frac{m_+}{m_d} \tag{7a}$$

$$\alpha = \frac{n_{-0}}{n_{+0}}, \ \beta = \frac{n_{c0}}{n_{+0}}, \gamma = \frac{n_{h0}}{n_{+0}}, \ \sigma = \frac{n_{d0}}{n_{+0}} \tag{7b}$$

where; in the above equations n_j (j = ce, he, +, –, d) represents the number density of the jth species normalized by its equilibrium value n_{j0} (j = ce, he, +, –, d), u_j represents fluid velocity of the jth species, z_d is the charge number of dust particle and δ is +1 for the positively charged dust and –1 for negatively charged dust. T_+ is the temperature of positive ion; and T_h is the temperature of the hot electron.

2.3 SOLITARY STRUCTURE

The independent variables are scaled by the following stretched coordinates:

$$\xi = \varepsilon^{1/2}\left(l_x x + l_y y + l_z z - \lambda t\right) \tag{8}$$

$$\tau = \varepsilon^{3/2} t$$

where; ϵ is the smallness parameter; λ is the phase velocity; and l_x, l_y, and l_z are the direction cosines of the wave vector along x, y, and z-axis, respectively with:

$$l_x^2 + l_y^2 + l_z^2 = 1$$

The power series expansion:

$$n_j = 1 + \varepsilon n_{j1} + \varepsilon^2 n_{j2} \ldots\ldots \quad (9a)$$

$$u_{jx} = \varepsilon u_{jy1} + \varepsilon^2 u_{jx2} \ldots\ldots \quad (9b)$$

$$u_{jy} = \varepsilon u_{jy1} + \varepsilon^2 u_{jy2} \ldots\ldots \quad (9c)$$

$$u_{jz} = \varepsilon u_{jz1} + \varepsilon^2 u_{jz2} \ldots\ldots \quad (9d)$$

$$\phi = \varepsilon \phi_1 + \varepsilon^2 \phi_2 \ldots\ldots \quad (9e)$$

We get the following expressions by equating the lowest order of ε, from the equation of continuity and equation of motion of the cold electron:

$$n_{ce1} = \frac{-pl_z^2}{\lambda^2}\varphi_1;\ u_{cex1} = \frac{-l_y}{\omega_c}\frac{\partial \varphi_1}{\partial \xi};\ u_{cex2} = -\frac{l_x\lambda}{p\omega_{c+}^2}\frac{\partial^2 \varphi_1}{\partial \xi^2};$$
$$u_{cey1} = \frac{l_x}{\omega_c}\frac{\partial \varphi_1}{\partial \xi};\ u_{cey2} = \frac{-l_y\lambda}{p\omega_{c+}^2}\frac{\partial^2 \varphi_1}{\partial \xi^2} \text{ and } u_{cez1} = \frac{-pl_z}{\lambda}\phi_1 \quad (10a)$$

The corresponding ones for positive and negative pair ions are:

$$n_{+1} = \frac{l_z^2}{\lambda^2}\varphi_1;\ u_{+x1} = \frac{-l_y}{w_{c+}}\frac{\partial \varphi_1}{\partial \xi};\ u_{+x2} = \frac{l_x\lambda}{\omega_{c+}^2}\frac{\partial^2 \varphi_1}{\partial \xi^2}; \quad (10b)$$

$$u_{+y1} = \frac{l_x}{\omega_{c+}}\frac{\partial \phi_1}{\partial \xi};\ u_{+y2} = \frac{l_y\lambda}{\omega_{c+}^2}\frac{\partial^2 \phi_1}{\partial \xi^2} \text{ and } u_{+z1} = \frac{l_z}{\lambda}\phi_1$$

$$n_{-1} = \frac{ql_z^2\phi_1}{\lambda^2};\ u_{-x1} = -\frac{l_y}{\omega_{c+}}\frac{\partial \phi_1}{\partial \xi};\ u_{-x2} = \frac{-l_x\lambda}{q\omega_{c+}^2}\frac{\partial^2 \phi_1}{\partial \xi^2} \quad (10c)$$

$$u_{-y1} = \frac{l_x}{\omega_{c+}}\frac{\partial \phi_1}{\partial \xi};\ u_{-y2} = \frac{-l_y\lambda}{q\omega_{c+}^2}\frac{\partial^2 \phi_1}{\partial \xi^2} \text{ and } u_{-z1} = \frac{-ql_z}{\lambda}\phi_1$$

For the dust, these quantities are:

$$n_{d1} = \frac{\delta z_d r l_z^2}{\lambda^2}\phi_1;\ u_{dx1} = \frac{-l_y}{\omega_{c+}}\frac{\partial \phi_1}{\partial \xi};\ u_{dx2} = \frac{l_x\lambda}{\delta z_d r\omega_{c+}^2}\frac{\partial^2 \phi_1}{\partial \xi^2};$$
$$u_{dy1} = \frac{l_x}{\omega_{c+}}\frac{\partial \phi_1}{\partial \xi};\ u_{dy2} = \frac{l_y\lambda}{\delta z_d r\omega_{c+}^2}\frac{\partial^2 \phi_1}{\partial \xi^2} \text{ and } u_{dz1} = \frac{\partial z_d r l_z}{\lambda}\phi_1 \quad (10d)$$

The compatibility condition is obtained from Poisson's equation:

$$\lambda = \frac{l_z}{\gamma}\sqrt{\left(\left(\frac{2\kappa-1}{2\kappa+1}\right)\times\left[1+\alpha q+\beta p+\delta^2 z_d^2\sigma r\right]\right)} \quad (11)$$

Using Eqns. (8) and (9) in (1)–(6) and applying reductive perturbation method, Korteweg de Vries equation is derived. It has the term:

$$\frac{\partial \phi_1}{\partial \tau} + A\phi_1 \frac{\partial \phi_1}{\partial \xi} + B\frac{\partial^3 \phi_1}{\partial \xi^3} = 0 \tag{12}$$

where; the coefficients are:

$$\text{A}= -\frac{\lambda}{2}\left\{\frac{(2\kappa+3)}{(2\kappa-1)} - 3\gamma\frac{(2\kappa+1)}{(2\kappa-1)} \times \left[\frac{1-\alpha q^2 - \beta p^2 + \delta^3 z_d^3 \sigma r^2}{\left[\left[1+\alpha q+\beta p+\delta^2 z_d^2 \sigma r\right]^2\right]}\right]\right\} \tag{13}$$

$$B = \frac{\lambda}{2\gamma}\frac{(2\kappa-1)}{(2\kappa+1)}\left\{1+\left(\frac{1-l_z^2}{w_{c+}^2}\right)\times\left(1+\frac{\alpha}{q}+\frac{\beta}{p}+\frac{\sigma_1+\sigma_2}{r}\right)\right\} \tag{14}$$

The solution of Eqn. (12) is:

$$\phi_1 = \phi_m \operatorname{sech}^2\left(\chi/\omega\right) \tag{15}$$

where; $\phi_m = \frac{3u}{A}$ and $\omega = 2\sqrt{B/u}$

To obtain Eqn. (15), we introduce the transformation $\lambda = \xi - u\tau$ where u is a dimensionless real constant representing the deviation from linear phase speed λ. We now study the propagation of dust ion acoustic rogue waves that arise due to modulational instability. For that we derived the NLSE by considering the periodic solution:

$$\phi = \sum_{m=1}^{\infty} \varepsilon^m \sum_{l=-\infty}^{\infty} \phi_1^{(m)}(\zeta,\eta)\exp\left[il(k\xi - \omega\tau)\right] \tag{16}$$

where; k and ω represents the carrier wave number and the frequency of the given IA wave, respectively. Using the scaling:

$$\zeta = \varepsilon\left[\xi + V_g\tau\right] \text{ and } \eta = \varepsilon^2\tau \tag{17}$$

We derived NLSE which has the form:

$$i\frac{\partial \psi}{\partial \eta} + \frac{1}{2}\frac{\partial^2 \psi}{\partial \xi^2} + Q\psi|\psi|^2 = 0 \tag{18}$$

NLSE admits a localized rational solution given by Ankiewicz et al. [32]:

$$\psi = \frac{1}{\sqrt{Q}}\left[\frac{4(1+2i\eta)}{1+4\eta^2+\left(4/P\right)\zeta^2}\right]\exp(i\eta) \tag{19}$$

This solution concentrates the energy of the background into a small region due to the nonlinearity property of the medium. We use the arbitrary value of φ of the form:

$$|\psi| = \psi\psi^* = \frac{3kB}{A^2}\left[\left(\frac{9kB - 12kB\eta^2 - 4\xi^2}{9kB + 12kB\eta^2 + 4\xi^2}\right) + \left(\frac{24kB\eta^2}{3kB + 12kB\eta^2 + 4\xi^2}\right)\right] \quad (20)$$

2.4 RESULTS

The parameters used in this study was observed at comet Halley [27–30]. The density of positive ion fluid and the density of the negative ion fluid was taken as 4.9 cm^{-3} and 0.4 cm^{-3}, respectively; the temperature of these ions was $T_+ = 8 \times 10^4$ K. The densities of hydrogen ion found at comet Halley was 4.95 cm^{-3}. The temperature of super thermal electron is $T_{he} = 2 \times 10^5$ K. The negatively charged ions in the mass peaks of 7–19, 22–65, and 85–110 amu found in inner coma of comet Halley and it indicates the presence of negatively charged oxygen ions unambiguously [31]. The positively and negatively charged oxygen ion was taken as the positively and negatively charged dust with equilibrium densities $n_{10} = 0.05$ cm^{-3} (negative dust) and $n_{20} = 0.5$ cm^{-3} (positive dust) respectively. The magnetic field was taken as $B_0 = 8 \times 10^{-5}$ G. The rogue wave structure plotted using these parameters is shown in Figure 2.1.

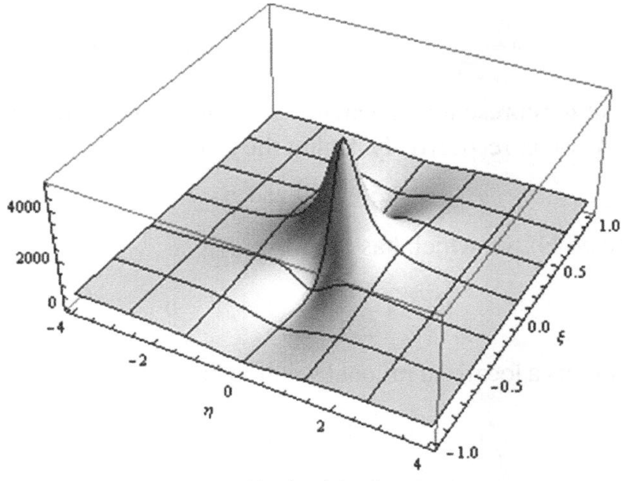

FIGURE 2.1 Rogue wave structure obtained for the observe parameters.

2.5 CONCLUSION

We modeled rogue waves in a multi-ion cometary plasma consisting of pair ions, charged dust particles and hot and colder electron components. Method of reductive perturbation is used to derive the K-dV equation and NLSE is derived from K-dV equation by the method of modulational instability.

KEYWORDS

- **dusty plasma**
- **kappa distribution**
- **Korteweg de-Vries (KdV) equation**
- **modulational instability**
- **multi-ion cometary plasma**
- **nonlinear Schrodinger equation (NLSE)**
- **nonlinearity**
- **rogue waves**

REFERENCES

1. Baumjohann, W., & Treumann, R. A., (2012). *Basic Space Plasma Physics Revised Edition.* Imperial College Press.
2. Kharif, C., Pelinovsky, E., & Slunyaev, A., (2009). *Rogue Waves in the Ocean.* Springer Verlag, Berlin.
3. Shalini, & Saini, N. S., (2015). *J. Plasma Physics, 1*, 17.
4. Williams, J., (2016). *Nature Physics, 12*, 529.
5. Ruderman, M. S., (2010). *Eur. Phys. J. Spec. Top., 185*, 57.
6. Moslem, W. M., (2011). *Phys. Plasmas, 18*, 032301.
7. Sabry, R., Moslem, W. M., & Shukla, P. K., (2012). *Phys. Plasmas, 19*, 122903.
8. El-Tantawy, S. A., El-Bedwehy, N. A., & El-Labany, S. K., (2013). *Physics of Plasmas, 20*, 072102.
9. Stenflo, L., & Marklund, M., (2010). *J. Plasma Phys., 76*, 293.
10. Onorato, M., Osborne, A. R., & Serio, M., (2010). *Phys. Rev. Lett., 96*, 014503.
11. Benjamin, T. B., & Feir, J. E., (1967). *J. Fluid Mech., 27*, 417.
12. Turing, A. M., (1952). *Philos. Trans. R. Soc. London B., 237*, 37.

13. Bespalov, V. I., & Talanov, V. I., (1966). *JETP Lett., 3*, 307.
14. Watanabe, S., (1977). *J. Plasma Phys., 17*, 487.
15. Shats, M., Punzmann, H., & Xia, H., (2010). *Phys. Rev. Lett., 104*, 104503.
16. Marklund, M., & Stenflo, L., (2009). *Physics, 2*, 86.
17. El-Awady, E. I., & Moslem, W. M., (2011). *Phys. Plasmas, 18*, 082306.
18. Moslem, W. M., Shukla, P. K., & Eliasson, B., (2011). *Europhys. Lett., 96*, 25002.
19. Onorato, M., Residori, S., Bortolozzo, U., Montina, A., & Arecchi, F. T., (2013). *Phys. Rep., 528*, 47.
20. Pierre, T., (2018). *APS Division of Plasmas Physics Meeting Abstracts, 63,* CP11.055.
21. Pathak, S. K., Sharma, P., & Bailing, H., (2015). *Proceedings of the Tenth Asia Plasma and Fusion Association Conference: Book of Abstracts, 47.*
22. Veldes, G. P., Borhanian, J., McKerr, M., Saxena, V., Frantzeskakis, D. J., & Kourakis, I., (2013). *J. Opt., 15*, 064003.
23. Chabchoub, A., Slunyaev, A., Hoffmann, N., Dias, F., Kibler, B., Genty, G., Dudley, J. M., & Akhmediev, N., (2021). *Frontiers of Physics, 9*, 633549.
24. Abdelwahed, H. G., El-Shewy, E. K., Zahran, M. A., & Elwakil, S. A., (2016). *Phys. Plasmas, 23*, 22102.
25. Abdikian, A., & Sultana, S. (2023). Stability of dust-acoustic solitary waves in magnetized dusty plasmas: effect of polarization force and degenerate electron temperature. *Phys. Scr. 98*, 055603.
26. Wael, F., El-Taibany, El-Bedwehy, N. A., El-Shafeay, N. A., & El-Labany, S. K., (2021). *Galaxies, 9*(3), 48.
27. Brinca, A. L., & Tsurutani, B. T., (1987). *Astron. Astrophys., 187*, 311.
28. Brinca, A. L., & Tsurutani, B. T., (1989). *J. Geophys. Res., 94*, 3.
29. Brinca, A. L., & Tsurutani, B. T., (1990). *J. Geophys. Res., 95*, 8291.
30. Brinca, A. L., Tsurutani, B. T., & Scarf, F. L., (1989). *J. Geophys. Res., 94*, 60.
31. Chaizy, P., Reme, H., Sauvaud, J. A., D'uston, C., Lin, R. P., et al., (1991). *Nature, 349*, 393.
32. Ankiewicz, A., Devine, N., & Akhmediev, N., (2009). *Phys. Lett. A, 373*, 3997.
33. Tolba, R., Moslem, W., Elbedwehy, N., & El-Labany, S., (2015). Evolution of rogue waves in dusty plasmas. *Physics of Plasmas. 22*. 043707. 10.1063/1.4918706.
34. Chan, H. N., Ding, E., Kedziora, D. J., Grimshaw, R., & Chow, K. W., (2016), Rogue waves for a long wave-short wave resonance model with multiple short waves, *Nonlinear Dynamics, 85*(4), 2827–2841.
35. Rahman, M. H., Chowdhury, N. A., Mannan, A., & Mamun, A.A. (2021). Dust-Acoustic Rogue Waves in an Electron-Positron-Ion-Dust Plasma Medium. *Galaxies, 9*, 31. https://doi.org/10.3390/galaxies9020031.

CHAPTER 3

Modeling in Fluid Mechanics: Atmospheric Modeling Using Navier Stokes Equation

M. S. SREERAJ[1], S. L. SRUTHI[2], and S. G. SUMOD[1]

[1]*Space Science Group, Department of Physics, Sacred Heart College, Thevara, Kochi, Ernakulam, Kerala, India*

[2]*Department of Basic Sciences and Humanities, Rajagiri School of Engineering and Technology, Rajagiri Valley, Kakkanad, Kochi, Ernakulam, Kerala, India*

ABSTRACT

In this chapter, we derive the partial differential equations governing the motion of viscous fluids, the Navier-Stokes equations, in the general form. Along the way, the forces acting on a cubic element of the fluid is described and expressions for the resultant forces are formulated. To get an idea on the viscous force, the viscous stress tensor, which describes the viscous stresses acting on the fluid element is discussed in detail. The Navier-Stokes equation has a wide range of applications, one of them being in the weather forecast. The Lorenz weather model, one of the most important weather models, is described here as an application of Navier-Stokes equation. Lorenz equations are derived from the simplification of Saltzmann model using Boussinesq approximation and Fourier–Galerkin procedure. In addition, Lorenz attractor is simulated for particular values of the control parameters and their changes amid the changes in initial conditions are also simulated.

New Advances in Materials Technologies. Hossein Hariri Asli, Ali Pourhashemi, Ann Rose Abraham, & A. K. Haghi (Eds.)

3.1 INTRODUCTION

We all like to see the calm, beautiful flow of a river. Only little things give as much pleasure as that. But have we ever wondered about the complexity of the movement of water? On the outside, it seems smooth and regular. We might think that the path would be somewhat linear. But not, it's not that simple. On the inside, the currents and the obstacles break the path of water lines differentially, which makes the flow unpredictable. Each molecule of water will behave differently and follow a random path in three-dimensions, which makes it impossible to track each of them. Still imagine, what if we could predict this unpredictability? What if, even the random motion of air around us can be predicted? This incredible complexity associated with the movement of all fluids are regulated by one of the seven *Millennium Prize Problems*, the Navier Stokes equation.

As described by Hosch [1], Navier-Stokes equation is a partial differential equation that describes the motion of incompressible fluids, named after French engineer and physicist Claude Louis Navier and Anglo-Irish physicist and mathematician George Gabriel Stokes. They derived the equations independently in the early 18th century. The equations were extension of Euler equations, which relates the velocity, pressure, and density of a moving fluid, constructed by Leonhard Euler to describe the flow of incompressible, frictionless fluids. Effects of viscosity on the flow, which was not considered by Euler, were included to make it Navier-Stokes equation, the generalized form.

Even though Navier Stokes equation describes the fluid motion, a complete solution has not yet evolved. The nonlinearity of acceleration and the second derivative viscous force terms make the equations too difficult to solve in a given flow problem. Here, in the following section, the general form of Navier-Stokes equation is derived by considering the forces acting on a fluid element, especially viscous forces.

3.2 DERIVATION OF NAVIER STOKES EQUATION

The inviscid and unsteady flows of compressible or incompressible fluids is described with the help of Euler equation of motion. Euler related the change in velocity of a fluid particle to the presence of a force, which makes it a form of Newton's second law.

$$\frac{\partial \vec{v}}{\partial t} + \left(\vec{v}.\vec{\nabla}\right)\vec{v} + \frac{1}{\rho}\vec{\nabla}p = \vec{g} \tag{1}$$

where; ρ and $\vec{v}$ are density of the fluid and its velocity of flow, both of which are position dependent. $\vec{\nabla}p$ being the pressure gradient at a point and $\vec{g}$ is the gravitational acceleration. As the fluid flows, viscosity leads to frictional forces between the layers. These frictional forces are accounted to the Euler equation to reach Navier Stokes equation. Similar approach has been adopted in many of the earlier works also [2].

Consider the flow of an incompressible fluid of density ρ from left to right through a bottleneck as shown in Figure 3.1. To start with, we consider an infinitesimally small cubic fluid element and the forces acting on it. We take x and z axes to be horizontal and vertical, respectively on the plane of the paper and y axis perpendicular to the plane. Consider the point (x, y, x) on the cube of length dx height dz and width dy. Only the motion along x direction of the fluid is taken into consideration. The same mathematics is applicable to y and z direction. As the fluid flows, the element flows through it steadily. Consider three regions 1, 2, and 3 along the flow.

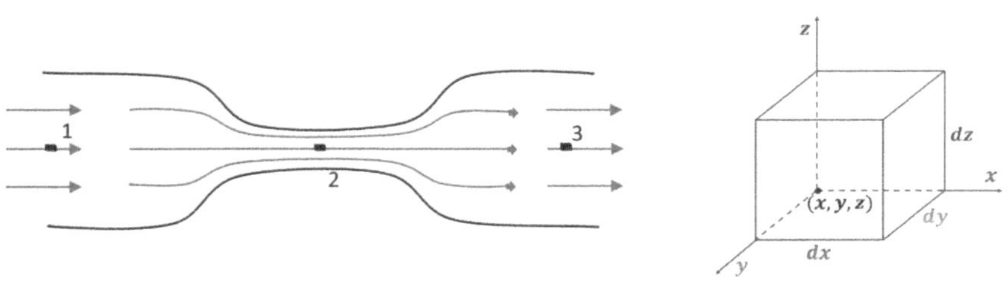

FIGURE 3.1 On the left, the flow of fluid through the bottleneck is shown. The three positions – 1, 2, and 3 are marked. On the right, the considered cubic element is given.

The element is flowing steadily from left to right slowly in region 1. As it approaches the 2nd region (constriction) it begins to accelerate. When it emerges to the 3rd region it slows down and recovers to a steady velocity as it moves towards the exit.

3.2.1 FORCES ACTING ON THE CUBIC ELEMENT

The element is acted upon by three forces which we are concerned of – the force due to gravity, forces due to differences in pressure on the faces of the element and the forces originate due to viscosity of the fluid. Each of these forces is discussed in detail in subsections. The method followed by Faro [3] is adopted here.

3.2.1.1 GRAVITATIONAL FORCE

In the considered case, gravity acts along the z direction, i.e., vertically downward. But the flow need not be horizontal. In that case, which is more general, the component of gravitational force along the direction of motion is only relevant. Let g_x be the gravitational acceleration component in the x direction. Then, gravitational force on the element is:

$$F_{gx} = dmg_x$$

where; $dm = \rho.dV$ is the mass of the cubic element; and $dV = dxdydz$ is the volume of the cubic element.

Thus,

$$F_{gx} = \rho g_x.dV \tag{2}$$

The gravitational force acts on the entire fluid element. So, it is considered as a body force.

3.2.1.2 NORMAL FORCE

The motion of this cubic element (considering it from left to right) is affected by the pressure forces acting on the front and back faces. These forces try to displace the face of the element in the x direction as detailed in earlier works [4, 5]. The pressure points out towards the right on the front face and towards left on the back face. To illustrate, this pressure forces are represented using the letter σ. As the pressure along the direction of flow generally decreases, this normal stress acting on both front and back faces are different. This gives rise to a gradient in force with respect to the position. If σ_x is the normal stress acting at the position x, the stress gradient across the length x is $\left(\frac{\partial \sigma_x}{\partial x}\right)$. Through the distance dx, the normal stress changes by an amount $\left(\frac{\partial \sigma_x}{\partial x}.dx\right)$. Pressure or stress is the force acting per unit area. So, area times the stress gives the total force, i.e.,

Normal force at point x is, $F_x = \sigma_x.dydz$

Similarly, normal force at $x + dx$ is, $F_{x+dx} = \left(\sigma_x + \frac{\partial \sigma_x}{\partial x}dx\right).dydz$

Since the forces act along different directions, the resulting normal force is:

$$F_{\sigma x} = F_x - F_{x+dx}$$

which gives:

$$F_{\sigma_x} = -\frac{\partial \sigma_x}{\partial x}.dV \tag{3}$$

3.2.1.3 SHEAR FORCE

Due to the relative motion, shear forces act on lateral (front–back and top–bottom) faces of the cubic element. The resulting lateral force on front–back faces is to be calculated. The same method can be applied to obtain the resulting forces on top–bottom face. Unlike normal stress, shear stress acts parallel to the surfaces and tries to shear or deform the fluid element. Bistafa [4] and Coleman [6] explained that the resultant shear stress can be found by applying Newton's law of fluid motion, which states that "The shear stress (τ) between two adjacent layers of a fluid is directly proportional to the velocity gradient $\left(\frac{\partial vj}{\partial i}\right)$ between the two layers," i.e.,

$$\tau_{ij} = \eta.\left(\frac{\partial v_j}{\partial i}\right) \tag{4}$$

where; η represents the viscosity. Here, indices i and j denote the direction for which the stress is determined and the shear stress acts, respectively. So, in this case, $j = x$ and $i = y, z$. For the front–back faces, the change is along the y direction. Then $i = y$.

Let τ_{yx} denotes the shear stress on the back face, i.e., at y.

Along the width dy, there exists a velocity gradient which produces a difference in stress acting on front and back faces. Let the shear stress gradient corresponding to the velocity gradient be $\left(\frac{\partial \tau_{yx}}{\partial y}\right)$, which when multiplied by the distance gives the shear stress at (y+dy). Then, shear stress on the front face, i.e., at:

$$(y+dy)=\tau_{yx}+\frac{\partial\tau_{yx}}{\partial y}\,.\,dy$$

According to Newton, if the shear stress gradient is positive, velocity gradient increases along that direction. Hence, the flow velocity increases along the distance *dy*, i.e., the flow velocity at location ($y + dy$) is greater than that at location *y*. So, along the positive *x* direction (direction of flow) fluid flow is slower on the left (back face) side and faster on the right (front face) side of the fluid. This implies that shear stress points in the negative *x* direction on the back face and positive *x* direction on the front face (Figure 3.2). The resulting shear stress on the faces, represented by τ_{L1}, is the difference in shear stresses and is:

$$\tau_{L1}=\tau_{yx}+\frac{\partial\tau_{yx}}{\partial y}\,.\,dy-\tau_{yx}=\frac{\partial\tau_{yx}}{\partial y}\,.\,dy \tag{5}$$

This is the force per unit area. In order to get the total force, τ_{L1} has to be multiplied with the area.

Then the lateral force on the front–back face is:

$$F_{L1}=\frac{\partial\tau_{yx}}{\partial y}\,.\,dy\,.\,dxdz=\frac{\partial\tau_{yx}}{\partial y}\,.\,dV \tag{6}$$

Analogously, lateral force on the top–bottom face is then:

$$F_{L2}=\frac{\partial\tau_{zx}}{\partial z}\,.\,dV \tag{7}$$

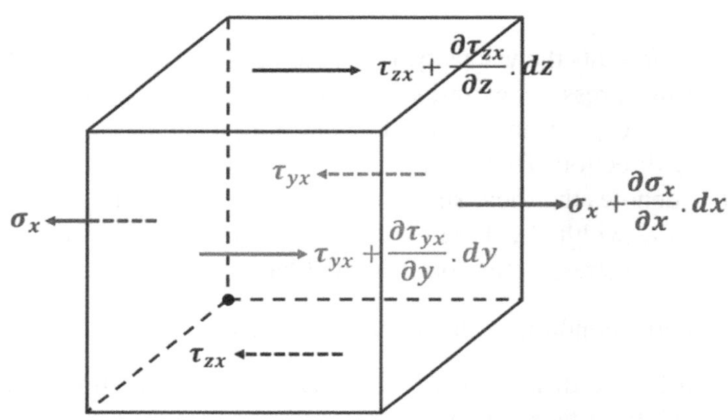

FIGURE 3.2 The shear and normal stresses acting on each face of the cubic element.

3.2.2 ACCELERATION

The sum of gravitational force F_{gx}, normal force F_{σ_x}, and shear forces F_{L1}, F_{L2} gives the resultant force, F_{res}, acting on the fluid element in the x direction.

And so:

$$F_{res} = F_{gx} + F_{\sigma_x} + F_{L1} + F_{L2}$$

$$= \rho g_x \, . \, dV + \left(-\frac{\partial \sigma_x}{\partial x} \, . \, dV \right) + \frac{\partial \tau_{yx}}{\partial y} \, . \, dV + \frac{\partial \tau_{zx}}{\partial z} \, . \, dV$$

As Newton proposed in the second law of motion, a non-zero force leads to acceleration. The resultant force then leads to an acceleration on the fluid element towards the x direction, which is given by:

$$a_x = \frac{F_{res}}{\rho dV} = \frac{\rho g_x \, . \, dV + \left(-\frac{\partial \sigma_x}{\partial x} \, . \, dV \right) + \frac{\partial \tau_{yx}}{\partial y} \, . \, dV + \frac{\partial \tau_{zx}}{\partial z} \, . \, dV}{\rho dV}$$

$$= \frac{\rho g_x - \frac{\partial \sigma_x}{\partial x} + \frac{\partial \tau_{yx}}{\partial y} + \frac{\partial \tau_{zx}}{\partial z}}{\rho}$$

This is the observable acceleration of the fluid element and is called substantial acceleration. The equation can be written as:

$$a_x \rho = \rho g_x - \frac{\partial \sigma_x}{\partial x} + \frac{\partial \tau_{yx}}{\partial y} + \frac{\partial \tau_{zx}}{\partial z} \tag{8}$$

Now, let us consider velocity of the fluid along the x direction, v_x. Since there exists a velocity gradient all over the element, this velocity is a function of position and time, i.e., $v_x = v_x(x, y, z, t)$. Then the acceleration is:

$$a_x = \frac{dv_x}{dt}$$

Using chain rule:

$$a_x = \frac{\partial v_x}{\partial x}\frac{dx}{dt} + \frac{\partial v_x}{\partial y}\frac{dy}{dt} + \frac{\partial v_x}{\partial z}\frac{dz}{dt} + \frac{\partial v_x}{\partial t}$$

where; $\frac{dx}{dt} = v_x, \frac{dy}{dt} = v_y, \frac{dz}{dt} = v_z$ are the components of fluid velocity along the three directions, which makes acceleration:

$$a_x = v_x \frac{\partial v_x}{\partial x} + v_y \frac{\partial v_x}{\partial y} + v_z \frac{\partial v_x}{\partial z} + \frac{\partial v_x}{\partial t} \tag{9}$$

The first three terms in Eqn. (9) are called convective acceleration and the last term is called local acceleration. The substantial acceleration, which is the observable acceleration of the fluid element, is not only due to the change in velocity with time but also accounts for change in velocity with space. The fluid element changes its velocity from time to time at a fixed location. Local acceleration is thus a consequence of change in flow of a fluid over time at a fixed location. One would think that if the flow is steady, there is no acceleration to the fluid elements. But this is not the scenario. Even though the flow is steady, the fluid elements generally change their velocity while flowing. As mentioned in the beginning, fluid flows at greater speeds in a constricted region and slower where the cross section is larger. Thus, a fluid element is accelerated when it changes the location. This acceleration is called the convective acceleration. Both these acceleration components together result in the observable acceleration.

Substituting Eqns. (9) in (8), we get:

$$\left(v_x \frac{\partial v_x}{\partial x} + v_y \frac{\partial v_x}{\partial y} + v_z \frac{\partial v_x}{\partial z} + \frac{\partial v_x}{\partial t} \right) \rho = \rho g_x - \frac{\partial \sigma_x}{\partial x} + \frac{\partial \tau_{yx}}{\partial y} + \frac{\partial \tau_{zx}}{\partial z} \tag{10}$$

For a fluid at rest, there are only normal components of stress on a surface. But a moving fluid additionally develops components of stress due to viscosity, as explained by Kundu [7]. The stress or force created due to viscosity can be represented as a tensor called the viscous stress tensor and is explained in the following section. As the flows in physical world are three-dimensional, so are the tensors involved.

3.2.3 VISCOUS STRESS

When sufficient shear stress acts on a material, it undergoes deformation. This is exactly what happens during the three-dimensional flow of a fluid. As we discussed earlier, due to viscosity, shear forces act on a fluid element as it flows with the fluid. This will create velocity gradient in all three directions. Stresses which originate from viscosity are described by the stress tensor, the *viscous stress tensor* ($\bar{\bar{\tau}}$).

$$\bar{\bar{\tau}} = \left[\tau_{ij}\right] = \begin{bmatrix} \tau_{xx} & \tau_{xy} & \tau_{xz} \\ \tau_{yx} & \tau_{yy} & \tau_{yz} \\ \tau_{zx} & \tau_{zy} & \tau_{zz} \end{bmatrix} \tag{11}$$

Each element of the matrix τ_{ij} represents the particular stress component that affects a plane perpendicular to the *i* axis and in the direction of *j* axis. Blazek [5] gave an idea about the elements of the tensor. The diagonal elements represent the normal stresses and the off-diagonal elements stand for the shear stresses. Papanastasiou, Georgiou, & Alexandrou [8] detailed that the viscous normal stress is caused by accelerating/decelerating motions towards the element surfaces and is proportional to viscosity of the medium and velocity gradient along the direction of streamline. Accordingly, the viscous shear stress is due to shearing motion of element surface adjacent to the surface. This is proportional to the viscosity and velocity gradient perpendicular to the streamlines (Figure 3.3).

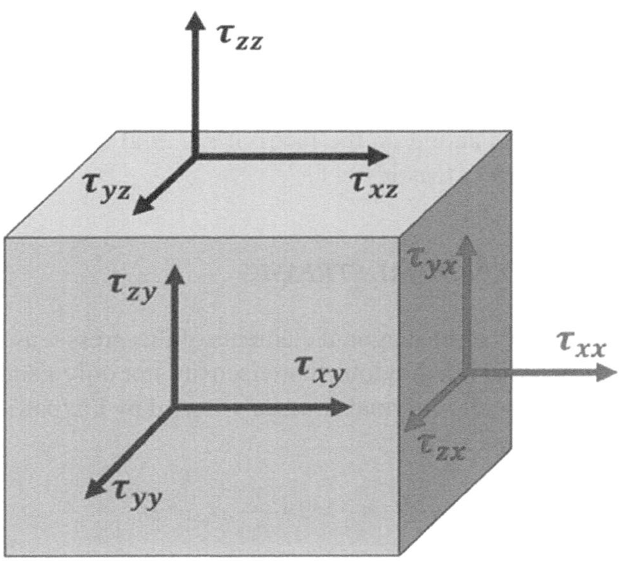

FIGURE 3.3 The stress forces acting on the different faces of the cube along different directions.

The components of stress can be found from the assumption that for a Newtonian fluid, the stress acting on any fluid element is directly proportional to the rate of deformation, which can be related to the change in velocity in the direction of stress, i.e.,

$$\tau_{ij} = \eta\left(\frac{\partial v_i}{\partial j}\right)$$

To apply this and to get Navier-Stokes equation, Stokes made some assumptions. He found that in three-dimensional flow of a fluid, the shear stress depends on a further velocity gradient. Such that the stress becomes:

$$\tau_{ij} = \eta\left(\frac{\partial v_i}{\partial j} + \frac{\partial v_j}{\partial i}\right) \tag{12}$$

The stress tensor then can be expanded as:

$$\tau = \eta.\begin{bmatrix} \left(2\frac{\partial v_x}{\partial x}\right) & \left(\frac{\partial v_x}{\partial y} + \frac{\partial v_y}{\partial x}\right) & \left(\frac{\partial v_x}{\partial z} + \frac{\partial v_z}{\partial x}\right) \\ \left(\frac{\partial v_y}{\partial x} + \frac{\partial v_x}{\partial y}\right) & \left(2\frac{\partial v_y}{\partial y}\right) & \left(\frac{\partial v_y}{\partial z} + \frac{\partial v_z}{\partial y}\right) \\ \left(\frac{\partial v_z}{\partial x} + \frac{\partial v_x}{\partial z}\right) & \left(\frac{\partial v_z}{\partial y} + \frac{\partial v_y}{\partial z}\right) & \left(2\frac{\partial v_z}{\partial z}\right) \end{bmatrix} \tag{13}$$

The normal stress acting on the faces of the fluid element can now be deduced from the stress tensor.

3.2.4 RESULTANT NORMAL STRESSES

We have discussed that the diagonal elements of the stress tensor represent the normal stress. So, it is obvious that viscosity not only generates shear stresses but also generates normal stresses as stated by Fitzpatric [9]. Then, $\tau_{xx} = \sigma_{\eta x}, \tau_{yy} = \sigma_{\eta y}, \tau_{zz} = \sigma_{\eta z}$.

$$\sigma_{\eta x} = 2\eta\frac{\partial v_x}{\partial x}, \sigma_{\eta y} = 2\eta\frac{\partial v_y}{\partial y}, \sigma_{\eta z} = 2\eta\frac{\partial v_z}{\partial z}$$

Normal viscous stress acts perpendicularly outward from the surface of the fluid element. In addition to this, for an incompressible media, static pressure also acts perpendicular to the surface but is against the normal force, i.e., into the surface. The resultant normal stress therefore is the result of this static pressure and normal viscous stress [10]. If p is the static pressure, then along the x direction:

Normal stress, $\sigma_x = p - \sigma_{\eta x} = p - 2\eta \frac{\partial v_x}{\partial x}$ (14)

Combining all the above, we can now formulate the Navier-Stokes equation for an incompressible fluid.

3.2.5 NAVIER-STOKES EQUATION OF INCOMPRESSIBLE FLUIDS

Taking the values $\tau_{yx} = \left(\frac{\partial v_y}{\partial x} + \frac{\partial v_x}{\partial y}\right)$ and $\tau_{zx} = \left(\frac{\partial v_z}{\partial x} + \frac{\partial v_x}{\partial z}\right)$ from Eqn. (13) and $\sigma_x = p - 2\eta \frac{\partial v_x}{\partial x}$ from Eqn. (14) and substituting to the acceleration Eqn. (10):

$$\left(v_x \frac{\partial v_x}{\partial x} + v_y \frac{\partial v_x}{\partial y} + v_z \frac{\partial v_x}{\partial z} + \frac{\partial v_x}{\partial t}\right)\rho = \rho g_x - \frac{\partial \sigma_x}{\partial x} + \frac{\partial \tau_{yx}}{\partial y} + \frac{\partial \tau_{zx}}{\partial z}$$

$$= \rho g_x - \frac{\partial}{\partial x}\left(p - 2\eta \frac{\partial v_x}{\partial x}\right) + \frac{\partial}{\partial y}\left(\eta\left\{\frac{\partial v_y}{\partial x} + \frac{\partial v_x}{\partial y}\right\}\right) + \frac{\partial}{\partial z}\left(\eta\left\{\frac{\partial v_z}{\partial x} + \frac{\partial v_x}{\partial z}\right\}\right)$$

$$= \rho g_x - \frac{\partial p}{\partial x} + 2\eta\left(\frac{\partial^2 v_x}{\partial x^2}\right) + \eta \frac{\partial}{\partial y}\left(\frac{\partial v_y}{\partial x}\right) + \eta\left(\frac{\partial^2 v_x}{\partial y^2}\right) + \eta \frac{\partial}{\partial z}\left(\frac{\partial v_z}{\partial x}\right) + \eta\left(\frac{\partial^2 v_x}{\partial z^2}\right)$$

Splitting the third term:

$$\left(v_x \frac{\partial v_x}{\partial x} + v_y \frac{\partial v_x}{\partial y} + v_z \frac{\partial v_x}{\partial z} + \frac{\partial v_x}{\partial t}\right)\rho = \rho g_x - \frac{\partial p}{\partial x} + \eta \frac{\partial}{\partial x}\left(\frac{\partial v_x}{\partial x}\right) + \eta\left(\frac{\partial^2 v_x}{\partial x^2}\right)$$
$$+ \eta \frac{\partial}{\partial y}\left(\frac{\partial v_y}{\partial x}\right) + \eta\left(\frac{\partial^2 v_x}{\partial y^2}\right) + \eta \frac{\partial}{\partial z}\left(\frac{\partial v_z}{\partial x}\right) + \eta\left(\frac{\partial^2 v_x}{\partial z^2}\right) \quad (15)$$

Now, we can rearrange the right-hand side of the equation as:

$$\left(v_x \frac{\partial v_x}{\partial x} + v_y \frac{\partial v_x}{\partial y} + v_z \frac{\partial v_x}{\partial z} + \frac{\partial v_x}{\partial t}\right)\rho = \rho g_x - \frac{\partial p}{\partial x} + \eta$$
$$\left[\frac{\partial}{\partial x}\left(\frac{\partial v_x}{\partial x}\right) + \frac{\partial}{\partial y}\left(\frac{\partial v_y}{\partial x}\right) + \frac{\partial}{\partial z}\left(\frac{\partial v_z}{\partial x}\right)\right] + \eta\left[\left(\frac{\partial^2 v_x}{\partial x^2}\right) + \left(\frac{\partial^2 v_x}{\partial y^2}\right) + \left(\frac{\partial^2 v_x}{\partial z^2}\right)\right] \quad (16)$$

Since the order of differentiation of the partial derivative terms inside the first bracket does not make any change, we can swap the coordinates. This gives:

$$\left(v_x\frac{\partial v_x}{\partial x}+v_y\frac{\partial v_x}{\partial y}+v_z\frac{\partial v_x}{\partial z}+\frac{\partial v_x}{\partial t}\right)\rho=\rho g_x-\frac{\partial p}{\partial x}+\eta$$

$$\left[\frac{\partial}{\partial x}\left(\frac{\partial v_x}{\partial x}\right)+\frac{\partial}{\partial x}\left(\frac{\partial v_y}{\partial y}\right)+\frac{\partial}{\partial x}\left(\frac{\partial v_z}{\partial z}\right)\right]$$

$$+\eta\left[\left(\frac{\partial^2 v_x}{\partial x^2}\right)+\left(\frac{\partial^2 v_x}{\partial y^2}\right)+\left(\frac{\partial^2 v_x}{\partial z^2}\right)\right]$$

$$\left(v_x\frac{\partial v_x}{\partial x}+v_y\frac{\partial v_x}{\partial y}+v_z\frac{\partial v_x}{\partial z}+\frac{\partial v_x}{\partial t}\right)\rho=\rho g_x-\frac{\partial p}{\partial x}+\eta\frac{\partial}{\partial x}$$

$$\left[\left(\frac{\partial v_x}{\partial x}\right)+\left(\frac{\partial v_y}{\partial y}\right)+\left(\frac{\partial v_z}{\partial z}\right)\right]+\eta\left[\left(\frac{\partial^2 v_x}{\partial x^2}\right)+\left(\frac{\partial^2 v_x}{\partial y^2}\right)+\left(\frac{\partial^2 v_x}{\partial z^2}\right)\right] \tag{17}$$

The third term on the right-hand side, i.e., divergence of the velocity of flow, $(\nabla\cdot\vec{v})$, is zero for incompressible fluids. Eqn. (17) thus reduces to:

$$\left(v_x\frac{\partial v_x}{\partial x}+v_y\frac{\partial v_x}{\partial y}+v_z\frac{\partial v_x}{\partial z}+\frac{\partial v_x}{\partial t}\right)$$

$$\rho=\rho g_x-\frac{\partial p}{\partial x}+\eta\left[\left(\frac{\partial^2 v_x}{\partial x^2}\right)+\left(\frac{\partial^2 v_x}{\partial y^2}\right)+\left(\frac{\partial^2 v_x}{\partial z^2}\right)\right] \tag{18}$$

This equation describes the motion of an incompressible fluid along the x direction. The equations along y and z can be formulated analogously as:

$$\left(v_x\frac{\partial v_y}{\partial x}+v_y\frac{\partial v_y}{\partial y}+v_z\frac{\partial v_y}{\partial z}+\frac{\partial v_y}{\partial t}\right)$$

$$\rho=\rho g_y-\frac{\partial p}{\partial y}+\eta\left[\left(\frac{\partial^2 v_y}{\partial x^2}\right)+\left(\frac{\partial^2 v_y}{\partial y^2}\right)+\left(\frac{\partial^2 v_y}{\partial z^2}\right)\right] \tag{19}$$

and

$$\left(v_x\frac{\partial v_z}{\partial x}+v_y\frac{\partial v_z}{\partial y}+v_z\frac{\partial v_z}{\partial z}+\frac{\partial v_z}{\partial t}\right)$$

$$\rho=\rho g_z-\frac{\partial p}{\partial z}+\eta\left[\left(\frac{\partial^2 v_z}{\partial x^2}\right)+\left(\frac{\partial^2 v_z}{\partial y^2}\right)+\left(\frac{\partial^2 v_z}{\partial z^2}\right)\right] \tag{20}$$

respectively.

Combining Eqns. (18)–(20), the Navier-Stokes equation for incompressible fluids can be written in general form as:

$$\left[\frac{\partial\vec{v}}{\partial t}+\left(\vec{v}.\vec{\nabla}\right)\vec{v}\right]\rho=-\vec{\nabla}\,p+\eta\left(\vec{\nabla}^{2}\vec{v}\right)+\rho\,\vec{g}$$

The bracketed term on the left-hand side represents the substantial acceleration, with first term being local acceleration and second term the convective acceleration. On right-hand side first term is normal force (or pressure force) followed by shear force (or frictional force) and gravitational force (or body force).

The Navier Stokes equation is one of the pillar stones of fluid mechanics. But these equations are considered to be one of the hardest equations to solve. Nevertheless, limited solutions can be obtained subject to specified initial and boundary conditions. Such a complete solution to the equation can be utilized to understand the unsteady flow of three-dimensional viscous fluids. This very aspect has a wide range of applications in the field of engineering and science. One of the many such aspects is atmospheric modeling, especially for weather forecast. The Navier Stokes equations allow us to predict the temperature, pressure, and wind velocity at some time in future. However, the non-linearity of the equation and lack of precise initial conditions of the variables make it difficult to predict the weather for more than a few days in advance. In this context, the relevance of the equation in view of the simplest atmospheric weather model, the Lorenz weather model, is discussed briefly in succeeding sections.

3.3 ATMOSPHERIC MODELING

The atmosphere is a fluid in constant motion and the Navier stokes equation can be utilized to describe the flow of atmospheric gas, allowing us to predict and model it. Thus, Navier Stokes equation becomes fundamental to every climate model and weather forecast. The basic idea of weather forecasting is to sample the state of atmosphere at a particular time and using fluid equations coupled with thermodynamics to predict the future state. The simplest weather model was introduced by Edward Lorenz in 1963 based on convective flow of fluids. The model picturizes weather related variables in an atmosphere when it is heated from below.

The Lorenz weather model is based on two-dimensional thermal convection of fluid under the condition of Rayleigh-Bénard flow as explained by Hilborn [11]. Rayleigh-Bénard convection occurs during the two-dimensional flow of a fluid between two thermally conducting plates, when there exists a temperature gradient between the plates. Considering that the plates are heated from below; the lower plate will be with a higher temperature T_l than that of the upper plate with a temperature T_u ($T_l > T_u$). The temperature gradient between the plates is denoted as ΔT. L is the vertical distance between the plates. The existing temperature gradient, $\Delta T = T_l - T_u$, makes the fluid density non uniform, i.e., a density gradient, $\Delta \rho$, is created with higher density near the upper plate (Figure 3.4).

FIGURE 3.4 The upper and lower plates with plane in the horizontal (x) direction. The distance L is measured in (z) direction.

The buoyancy of fluid and its viscous drag are the factors that are responsible for the convection of fluid between the plates. When the temperature gradient established becomes sufficiently large, a small layer of fluid near the lower plate experiences an upward buoyant force. As a result, this layer moves towards the upper region. Since the gravitational force will be higher in the upper region, the denser upper layer is pulled downward to the bottom. These upward and downward motion of the layers set in a convective roll and sort out into regular patterns. Thus, heat energy is transferred into the upper plate (Figure 3.5).

In modeling the atmosphere, the Rayleigh-Bénard convection can be explained by considering layers close to the earth surface. As the Sun heats the surface, the temperature on the surface will be greater there and gradually decreases as we move upward. So, if we consider two atmospheric layers, one close to the surface and the other at some height, then there exists a temperature gradient. And hence convective rolls will be formed between these layers. Figure 3.6 shows the convective rolls formed between two such layers.

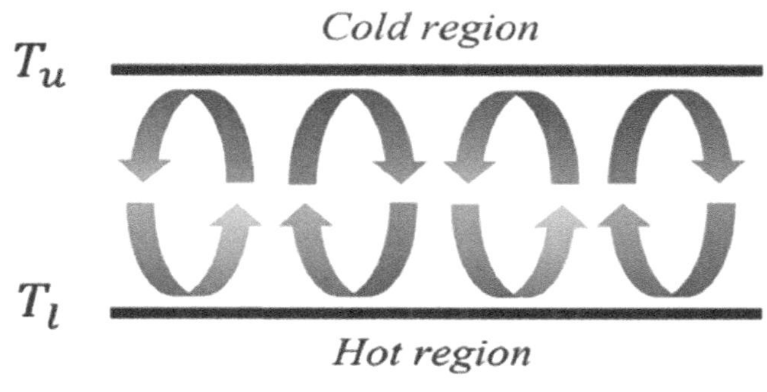

FIGURE 3.5 The convective rolls formed between the plates. Both clockwise and anti-clockwise rolls can be formed.

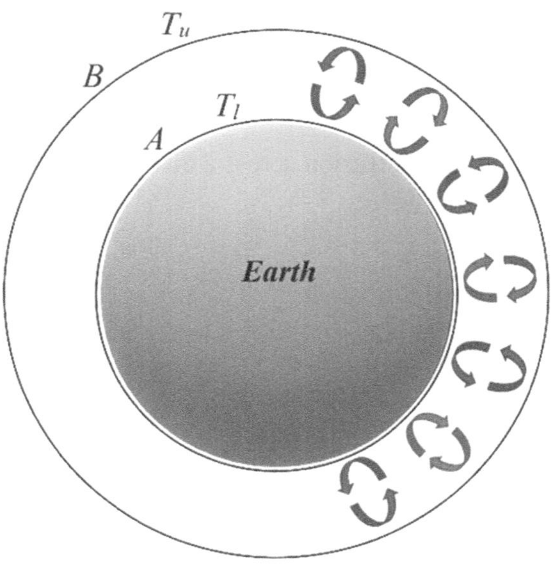

FIGURE 3.6 Convective rolls formed between layers A and B on the Earth's atmosphere. The convective rolls formed in the sunlit hemisphere is shown.

The process involved in the heat transfer is not only the convection but also conduction. The transition from conductive motion to convective motion can be related using Rayleigh number, which is the critical parameter for Rayleigh-Bénard convection. The Rayleigh number is defined as:

$$R = \frac{thermal\ diffusion\ time}{displacement\ time\ for\ convection} = \frac{t_{td}}{t_c} \tag{21}$$

A fluid of density gradient $\Delta\rho$, which is due to the temperature difference, between the plates separated by the length L experiences a gravitational force of the order of $\Delta\rho L^3 g$ (assuming that the fluid volume between the plates have size L in all three-dimensions). The viscous drag of fluid is proportional to $\eta L v$, where η is the viscosity of the fluid and v is the velocity with which the fluid moves. When the viscous drag balances the gravitational pull, the fluid moves with uniform velocity. When these forces are equated, the velocity of the fluid can be approximated as:

$$v = \frac{\Delta\rho L^2 g}{\eta} \tag{22}$$

Thus, the displacement time via convection is:

$$t_c = \frac{L}{v} = \frac{\eta}{\Delta\rho L g} \tag{23}$$

The time for thermal diffusion across a distance L is:

$$t_{td} = \frac{L^2}{D} \tag{24}$$

where; D is the thermal diffusivity.

Then, the Rayleigh number:

$$R = \frac{t_{td}}{t_c} = \frac{\Delta\rho L^2 g}{\eta D} \tag{25}$$

We know that the thermal expansion coefficient is defined as the relative change in density per unit temperature change, i.e.,

$$\alpha = \frac{\Delta\rho / \rho_0}{\Delta T} \tag{26}$$

where; ρ_0 is the original density of the fluid.

which converts Eqn. (25) as:

$$R = \frac{\alpha\rho_0 L^3 g \Delta T}{\eta D} \tag{27}$$

If the thermal diffusion time is longer, fluid will continue to experience an upward buoyant force and the convection begins. Thus, the value of the Rayleigh number decides whether the convection will initiate or not.

Before deriving the Lorenz equation, we have to consider the Boussinesq approximation, which is important in coupling the Navier-Stokes equation with thermal convection. Following Giga & Novotný [2], here we also have accounted for the two-dimensional flow of fluid (z and x direction). The corresponding Navier Stokes equation for x and z components of fluid velocity are given by:

$$\left[\frac{\partial v_x}{\partial t}+\left(v_x.\vec{\nabla}\right)\vec{v}\right]\rho=-\frac{\partial p}{\partial x}+\eta\left(\vec{\nabla}^2 v_x\right) \quad (28)$$

$$\left[\frac{\partial v_z}{\partial t}+\left(v_z.\vec{\nabla}\right)\vec{v}\right]\rho=-\frac{\partial p}{\partial z}+\eta\left(\vec{\nabla}^2 v_z\right)-\rho g \quad (29)$$

Note that the gravity term is neglected in x component equation here, as it is assumed that the gravity has no effect along the x direction. Also, the gravity is acting vertically downward, i.e., in the negative z direction. So, the gravity term is taken as negative in the z component equation.

The temperature T of the fluid is modeled by the thermal diffusion equation:

$$\frac{\partial T}{\partial t}+\vec{v}.\nabla T = D\nabla^2 T \quad (30)$$

In the non-convecting state, the temperature varies linearly between the plates from bottom to top, i.e.,

$$T(x,z,t)=T_l-\frac{z}{L}\Delta T$$

Let us consider a function $\Phi(x,z,t)$. This function, which helps to describe the deviation of temperature from its linear behavior, is defined as:

$$\Phi(x,z,t)=T(x,z,t)-T_l+\frac{z}{L}\Delta T \quad (31)$$

Since the fluid convection is initiated due to decrease in density with temperature, let us now consider the variation in density of fluid with temperature. The fluid density at a particular temperature T can be expanded in power series. By neglecting the higher terms, we get:

$$\rho(T) = \rho_0 + \frac{\partial \rho}{\partial T}(T - T_l) \tag{32}$$

where; ρ_0 is the fluid density evaluated at temperature T_l. In terms of thermal expansion coefficient, Eqn. (32) can rewrite as:

$$\rho(T) = \rho_0 - \alpha\rho_0\left[\Phi(x,z,t) - \frac{z}{L}\Delta T\right]$$

From Eqn. (32), the thermal diffusion Eqn. (30) can be written as:

$$\frac{\partial \Phi}{\partial t} + \vec{v}\,.\,\nabla\Phi - v_z\frac{\Delta T}{L} = D\nabla^2\Phi$$

The convection problems that involve non-isothermal flow can be solved using Boussinesq approximation. So, the above-discussed scenario can be simplified as in subsections.

3.3.1 BOUSSINESQ APPROXIMATION

The Boussinesq approximation is applied to fluid motions when there is a temperature variation in fluid causing the fluid to flow. The approximation states that we can ignore the density difference except in terms which involves gravity. Thus, the Navier Stokes equation for Eqn. (28) reduces to:

$$\frac{\partial v_x}{\partial t} + \vec{v}\,.\,\nabla v_x = -\frac{1}{\rho_0}\frac{\partial P}{\partial x} + \vartheta\,\nabla^2 v_x \tag{33}$$

And Eqn. (29) becomes:

$$\frac{\partial v_z}{\partial t} + \vec{v}.\nabla v_z = -\frac{1}{\rho_0}\frac{\partial P}{\partial z} + \alpha\Phi g + \vartheta\nabla^2 v_z \tag{34}$$

As given by Batchelor [22], here $P = p + \rho_0 gz + \left[(\alpha g\rho_0).\left(\frac{z^2\Delta T}{2L}\right)\right]$ and $\vartheta = \eta/\rho_0$ is called the kinematic viscosity.

Before further proceeding the derivation, a new parameter is introduced to represent the velocity called the stream function.

3.3.2 STREAM FUNCTION

The stream function $\Psi(x, z, t)$ can be used to specify the velocity of a fluid at a particular location (x, z) in a two-dimensional fluid flow motion. Stream function carries all the information about the fluid flow. By taking partial derivative of stream function, the velocity components can be represented as:

$$v_x = -\frac{\partial \Psi(x,z,t)}{\partial z}, \; v_z = \frac{\partial \Psi(x,z,t)}{\partial x} \quad (35)$$

To express the Navier Stokes equation in terms of Ψ and Φ, we introduce Eqn. (35) into Eqns. (33) and (34) then differentiating them with respect to z and x respectively. Thes x component equation is subtracted from the z component equation to give:

$$\frac{\partial}{\partial t}(\nabla^2\Psi) - \frac{\partial \Psi}{\partial z}\frac{\partial}{\partial x}(\nabla^2\Psi) + \frac{\partial \Psi}{\partial x}\frac{\partial}{\partial z}(\nabla^2\Psi) - g\alpha\frac{\partial \Phi}{\partial x} - \vartheta\nabla^4\Psi = 0 \quad (36)$$

and the thermal diffusion equation from the above section becomes:

$$\frac{\partial \Phi}{\partial t} - \frac{\Delta T}{L}\frac{\partial \Psi}{\partial x} - \frac{\partial \Psi}{\partial z}\frac{\partial \Phi}{\partial x} + \frac{\partial \Psi}{\partial x}\frac{\partial \Phi}{\partial z} - D\nabla^2\Phi = 0 \quad (37)$$

Eqns. (36) and (37) provide the base for Lorenz model. However, as these equations are not tractable, they need to be simplified as described in the next section.

3.3.3 DIMENSIONLESS VARIABLES

For the development of the Lorenz model, we have to express Eqns. (36) and (37) in terms of dimensionless variables. The method was followed from Saltzman [12].

We introduce the dimensionless time variable as, $t^* = \frac{D}{L^2}t$ (38)

In a similar manner, the dimensionless distance, dimensionless temperature function and non-dimensional stream function are defined as:

$$x^* = \frac{x}{L}, \; z^* = \frac{z}{L}, \; \Phi^* = \frac{\Phi}{\Delta T}, \; \Psi^* = \frac{\Psi}{D} \quad (39)$$

The Laplacian operator can also be expressed in terms of new variables as:

$$\nabla^{*2} = L^2\nabla^2 \tag{40}$$

We use the new dimensionless variables in Eqn. (36) and multiply the resulting equation throughout with $\frac{L^4}{\vartheta D}$ to get:

$$\frac{D}{\vartheta}\left[\frac{\partial}{\partial t^*}\left(\nabla^{*2}\Psi\right)-\frac{\partial\Psi^*}{\partial z^*}\frac{\partial}{\partial x^*}\left(\nabla^{*2}\Psi^*\right)+\frac{\partial\Psi^*}{\partial x^*}\frac{\partial}{\partial z^*}\left(\nabla^{*2}\Psi^*\right)\right]$$
$$-\frac{g\alpha\Delta TL^3}{\vartheta D}\frac{\partial\Phi^*}{\partial x^*}-\nabla^{*4}\Psi^* = 0 \tag{41}$$

The thermal diffusion Eqn. (37) reduces to:

$$\frac{\partial\Phi^*}{\partial t^*}-\frac{\partial\Psi^*}{\partial x^*}-\frac{\partial\Psi^*}{\partial z^*}\frac{\partial\Phi^*}{\partial x^*}+\frac{\partial\Psi^*}{\partial x^*}\frac{\partial\Phi^*}{\partial z^*}-\nabla^{*2}\Phi^* = 0 \tag{42}$$

The above equation contains only dimensionless ratios. Now we can compare the rate of energy loss due to viscosity and the rate of energy loss due to thermal conduction from a small layer of fluid. The parameter Prandtl number (σ) which is defined as the ratio of kinematic viscosity to the thermal diffusion coefficient is used for this purpose:

$$\sigma = \frac{\vartheta}{D} \tag{43}$$

Along with this, the temperature gradient between the top and bottom plates can be measured using another parameter, the Rayleigh number (R). It is a dimensionless number that characterizes the heat transfer in case of convection, that is in the buoyancy-driven flow.

Rayleigh number is defined as:

$$R = \frac{\alpha gL^3}{\vartheta D}\Delta T \tag{44}$$

Substituting Prandtl number and Rayleigh number in Eqn. (41):

$$\frac{1}{\sigma}\left[\frac{\partial}{\partial t^*}\left(\nabla^{*2}\Psi\right)-\frac{\partial\Psi^*}{\partial z^*}\frac{\partial}{\partial x^*}\left(\nabla^{*2}\Psi^*\right)+\frac{\partial\Psi^*}{\partial x^*}\frac{\partial}{\partial z^*}\left(\nabla^{*2}\Psi^*\right)\right]$$
$$-R\frac{\partial\Phi^*}{\partial x^*}-\nabla^{*4}\Psi^* = 0 \tag{45}$$

The partial differential Eqns. (42) and (45) contains all the information about the fluid flow. We can solve these equations to obtain the flow of a fluid subject to initial conditions by the following procedure.

3.4 FOURIER–GALERKIN PROCEDURE AND BOUNDARY CONDITIONS

Solving the partial differential equation that describes the Lorenz model is a tedious task. So, on the path to solve the equations, we use truncation process called Galerkin procedure (after Hateley [13]). Let us consider the Fourier–Galerkin expansion in complex form for stream function and temperature function:

$$\Psi(x^*, z^*, t^*) = \sum_{m,n\epsilon z} \Psi\left(m, n, t^*\right).exp\left(in\pi z^*\right).exp(imx^*)$$

$$\Phi(x^*, z^*, t^*) = \sum_{m,n\epsilon z} \Phi\left(m, n, t^*\right).exp\left(in\pi z^*\right).exp(imx^*)$$

As mentioned by Rafiq et al. [14], to solve the Lorenz model, we choose only limited sine and cosine terms which will satisfy the boundary conditions that are satisfied by temperature function and stream function.

Let us look into the boundary condition for temperature function Φ^* which represents the deviation from linear temperature behavior. Since the temperatures at the top and bottom plates are fixed, $\Phi^* = 0$ at both the plates. In the case of stream function, we consider the boundary conditions of velocity components. We first assume that the vertical component of velocity v_z is zero. Further we assume that boundaries are stress free and so we neglect the tangential stress, which is proportional to tangential velocity component, therefore $\frac{\partial v_x}{\partial z} = 0$ at the plates. So, for Lorenz model the stream function and temperature function that satisfies these conditions can be represented as:

$$\Psi(x^*, z^*, t^*) = \Psi(t^*).sin(\pi z^*).sin(ax^*)$$
$$\Phi(x^*, z^*, t^*) = \Phi_1(t^*).sin(\pi z^*).cos(ax^*) - \Phi_2(t^*).sin(2\pi z^*)$$

Here $\Psi(t^*)$ represents the convective rolls which is observed when fluid between the plates begins to convect. Also $\Phi_1(t^*)$ gives the difference in temperature between the upward and downward moving parts of a convective roll. $\Phi_2(t^*)$ represents the variation of temperature linearity as a function of vertical positions.

To find the solution of Lorenz model accounting to the above-mentioned boundary condition, we have to substitute the forms of stream function and temperature function in Eqns. (42) and (45). Also, we have:

$$\nabla^{*2}\Psi^* = -\left(a^2+\pi^2\right)\Psi \text{ and } \nabla^{*4}\Psi^* = \left(a^2+\pi^2\right)^2\Psi$$

Eqn. (45) the reduces to:

$$-\frac{d\Psi\left(t^*\right)}{dt^*}\left(a^2+\pi^2\right).sin\left(\pi z^*\right).sin\left(ax^*\right)$$

$$= -\sigma R\Phi_1\left(t^*\right).sin\left(\pi z^*\right).sin\left(ax^*\right)+\sigma\left(a^2+\pi^2\right)^2.\Psi\left(t^*\right).sin\left(\pi z^*\right).sin(ax^*)$$

The common terms cancel out and finally:

$$\frac{d\Psi}{dt^*} = \left[\frac{\sigma R}{\left(\pi^2+a^2\right)}\Phi_1\left(t^*\right)\right] - \left[\sigma\left(a^2+\pi^2\right).\Psi(t^*)\right]$$

The temperature function takes the form:

$$\dot{\Phi}_1.sin\left(\pi z^*\right).cos\left(ax^*\right) - \dot{\Phi}_2.sin\ (2\pi z^*) + \left(\pi^2+a^2\right)\Phi_1.sin\ (\pi z^*).cos$$

$$(ax^*) - 4\pi^2\Phi_2.sin\ (2\pi z^*) - a\Psi.sin\ (\pi z^*).cos\ (ax^*) = -\left[\pi\Psi.\cos(\pi z^*).\sin(ax^*)\right]$$

$$[a\Phi_1.sin\left(\pi z^*\right).sin\left(ax^*\right)] - \left[a\Psi.sin(\pi z^*).cos\left(ax^*\right)\right]\ [\pi\Phi_1.cos\left(\pi z^*\right).cos\ (ax^*)]$$

$$+\left[a\Psi.sin\left(\pi z^*\right).cos\left(ax^*\right)\right]\ [2\pi\Phi_2.cos(2\pi z^*)] \tag{46}$$

$sin(\pi z^*).cos(2\pi z^*)$ in the last term on the right-hand side is expanded using trigonometric identity $sin(x).cos(y) = \left[\frac{1}{2}sin(x-y)+\frac{1}{2}sin(x+y)\right]$ as:

$$sin\left(\pi z^*\right).cos\left(2\pi z^*\right) = -\frac{1}{2}sin\left(\pi z^*\right)+\frac{1}{2}sin\left(3\pi z^*\right)$$

The term $sin(3\pi z^*)$ can be neglected, as it has more spatial dependence. Equating the coefficients of $sin(\pi z^*).cos(ax^*)$ we obtain:

$$\dot{\Phi}_1 = a\Psi - \left(\pi^2+a^2\right)\Phi_1 - \pi a\Psi\ \Phi_2$$

Taking the first two terms on the right-hand side of Eqn. (46).

$$-\left[\pi\Psi.\cos(\pi z^*).\sin(ax^*)\right]\left[a\Phi_1.sin\left(\pi z^*\right).sin\left(ax^*\right)\right]$$
$$-\left[a\Psi.sin(\pi z^*).cos\left(ax^*\right)\right]\left[\pi\Phi_1.cos\left(\pi z^*\right).cos\left(ax^*\right)\right]$$
$$=-\pi a\Psi\Phi_1.sin(\pi z^*)\left[cos(\pi z^*).sin^2(ax^*)+cos\left(\pi z^*\right).cos^2\left(ax^*\right)\right]$$

i.e.,

$$-\left[\pi\Psi.\cos(\pi z^*).\sin(ax^*)\right]\left[a\Phi_1.sin\left(\pi z^*\right).sin\left(ax^*\right)\right]$$
$$-\left[a\Psi.\sin(\pi z^*).cos\left(ax^*\right)\right]\left[\pi\Phi_1.cos\left(\pi z^*\right).cos\left(ax^*\right)\right]$$
$$=-\pi a\Psi\Phi_1.sin(\pi z^*).\cos\left(\pi z^*\right)=-\pi\, a\Psi\;\Phi_1.\frac{1}{2}sin(2\pi z^*)$$

Substituting this in Eqn. (46) and equating the coefficients of $sin(2\pi z^*)$ gives $\dot{\Phi}_2$:

$$\dot{\Phi}_2=\frac{\pi a}{2}\;\Psi\;\Phi_1-4\pi^2\,\Phi_2$$

Let us define the new time variable as $t' = (\pi^2 + a^2)t^*$. Lorenz introduced three functions X, Y and Z such that:

$$X(t')=\frac{a\pi}{\left(\pi^2+a^2\right)\sqrt{2}}\Psi(t')$$

$$Y(t')=\frac{r\pi}{\sqrt{2}}\Phi_1(t')$$

$$Z(t')=\pi\, r\,\Phi_2(t')$$

Here, X represents the rate of convection, Y the ascending/descending in temperature of the fluid and Z the deviation from temperature with respect to vertical position.

The new notation r is the reduced Rayleigh number formulated as:

$$r=\frac{a^2}{(a^2+\pi^2)^3}R$$

As the next step, a new parameter b is introduced which is defined by:

$$b=\frac{4\pi^2}{a^2+\pi^2}$$

With all these substitutions we finally arrive at the Lorenz equations;

$$\frac{dX}{dt'} = \sigma\,(Y - X)$$

$$\frac{dY}{dt'} = r\,X - X\,Z - Y$$

$$\frac{dZ}{dt'} = X\,Y - b\,Z$$

The Lorenz model is a simplified mathematical model for atmosphere in which *X, Y* and *Z* represent the state of the atmosphere. For example, temperature, humidity, pressure, etc., can be represented by these equations. Lorenz [15] treated the atmosphere as a dynamic chaotic system and so, this will give a way for weather and climate predictability. Slingo & Palmer [16] also pointed that when we follow the trajectory, which is the solution of above equation, with sufficient initial conditions and parameter value, we observe the evolution of weather.

Every dynamical system is studied using phase space or state space, a finite dimensional Euclidean space spanned by variables characterizing the system (e.g., temperature, pressure, humidity, etc., in case of atmosphere). Each point in the state space represents the instantaneous state of a system. The variation in the state is represented by the trajectory in phase space. Here we simulate the Lorenz equation using Runge-Kutta 4th order method by following Razali & Ahmad [17]. The value of parameters used are $\sigma = 10$, $r = 28$ and $b = 8/3$. The state space occupied by the solution of Lorenz equation which exhibits the complicated geometric structure, although deterministic, is called Lorenz attractor (Figure 3.7).

As the Lorenz equation contains nonlinear terms, the evolution of trajectory changes if there is a small change in the initial condition. This implies that, after a particular time the trajectory will not be deterministic but probabilistic. Therefore, weather forecast is considered to be rather probabilistic than deterministic. The predictability of system depends on the flow of trajectory, i.e., some weather patterns are unpredictable but some show considerable predictability. Figure 3.8 shows the change in trajectory with some change in initial conditions.

Thus, it is evident that ambiguities in initial condition lead to uncertainty in weather forecast. For example, small temperature changes may cause significant effects on the pattern of trajectory that the atmosphere follows.

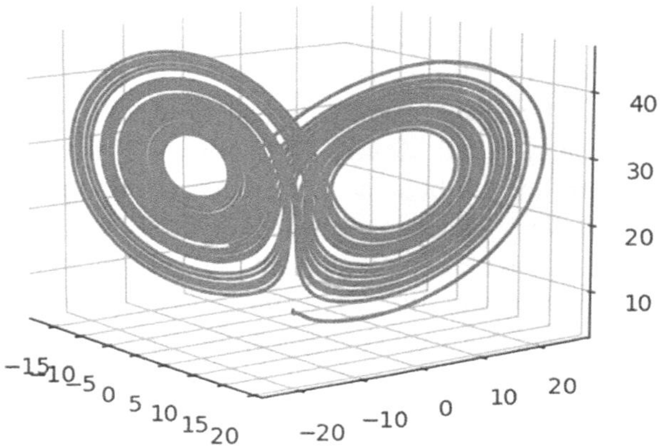

FIGURE 3.7 Lorenz attractor. The solution of Lorenz equation when $\sigma = 10$, $r = 28$, and $b = 8/3$.

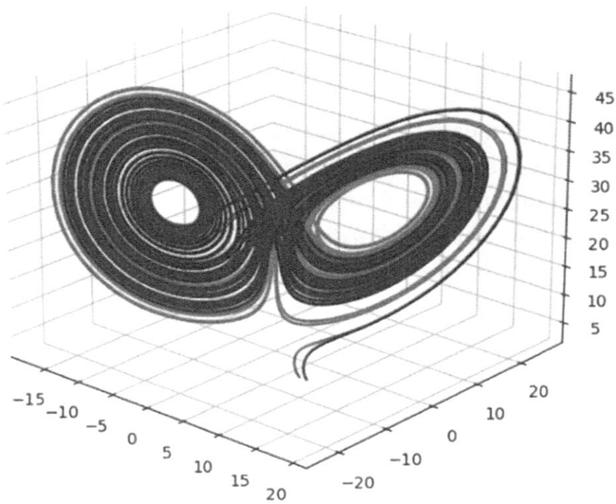

FIGURE 3.8 Lorenz attractor subject to change in initial conditions; two trajectories – blue and red – starting at two pints are given. At the beginning they differ only by a small value in the x-coordinate and also seem to coincide. But after some time, they deviate.

3.5 SUMMARY AND CONCLUSION

The Navier-Stokes equations are thought to govern the motion of all fluids and are very essential in continuum mechanics. For Newtonian fluids, they express the conservation of mass and momentum. In their simplified form, Navier Stokes equations have immense applications in aerodynamics [18, 24], unsteady fluid flows [19] and analysis of pollution [20]. In fact, Marmanis [21] has shown the analogy between Navier-Stokes equation and Maxwell's equations. This coupling between electromagnetism and turbulence are is very important, particularly in modeling studies governing magnetohydrodynamics. It must be mentioned that comprehensive understanding of this equation is highly appreciated particularly in fluid dynamics, cosmology, nuclear fusion, electrodynamics, and aerospace engineering. Therefore, an attempt is made in this work to derive Navier-Stokes equation for an incompressible fluid in the general form, considering the forces acting on a cubic element of a fluid. Gravitational force, shearing force and normal force equations are formulated by incorporating viscous forces acting on the fluid element.

Furthermore, we have explained atmospheric modeling as an immediate application of Navier-Stokes equation. With the help of Boussinesq approximation and Fourier–Galerkin procedure, the set of Lorenz weather model equations are obtained. As is known, prediction of weather and climate are rather difficult mainly due to random and complex motion of atmospheric fluids. Therefore, understanding the features in the transient behavior of Lorenz model have important implications in atmospheric weather forecast and climate modeling. Nevertheless, as suggested by Lorenz himself, predicting the future state of atmosphere is only probabilistic and the evolution of weather patterns are highly sensitive to initial conditions.

KEYWORDS

- **atmospheric modeling**
- **Boussinesq approximation**
- **Lorenz attractor**
- **Lorenz model**
- **Navier-Stokes equation**
- **normal and shear forces**
- **stream function**
- **viscous stress tensor**

REFERENCES

1. Hosch, W. L., (2020). *Navier–Stokes Equation.* Encyclopedia Britannica. https://www.britannica.com/science/Navier-Stokes-equation (accessed on 06 September 2023).
2. Giga, M. H., Kirshtein, A., & Liu, C., (2018). Variational modeling and complex fluids. In: Giga, Y., & Novotný, A., (eds.), *Handbook of Mathematical Analysis in Mechanics of Viscous Fluids*. Springer, Cham. doi: 10.1007/978-3-319-13344-7_2.
3. Faro, A., (2020). *Navier-Stokes Equation (An Overview and the Simplification).* 10.13140/RG.2.2.17406.00323.
4. Bistafa, S. R., (2018). On the development of the Navier-stokes equation by Navier. *Revista Brasileira de Ensino de Física, 40.*
5. Blazek, J., (2015). Governing equations contents. In: *Computational Fluid Dynamics: Principles and Applications*. Butterworth-Heinemann.
6. Coleman, N., (2010). A derivation of the Navier–Stokes Equations. *Mathematics Exchange*, *7*(1), 20–26.
7. Kundu, P. K., & Cohen, I. M., (1990). *Fluid Mechanics*. Academic press. Philadelphia, Pennsylvania.
8. Papanastasiou, T., Georgiou, G., & Alexandrou, A., (2000). *Viscous Fluid Flow*. CRC press. 10.1201/9780367802424.
9. Fitzpatrick, R., (2017). Mathematical models of fluid motion. In: *Theoretical Fluid Mechanics* (pp. 1–32). IOP Publishing. doi: 10.1088/978-0-7503-1554-8ch1.
10. Deissler, R. G., (1976). *American Journal of Physics, 44*(11), 1128–1130. Derivation of the Navier–Stokes equation.
11. Hilborn, R. C., (2000). *Chaos and Nonlinear Dynamics: An Introduction for Scientists and Engineers*. Oxford University Press on Demand.
12. Saltzman, B., (1962). Finite amplitude free convection as an initial value problem—I. *Journal of Atmospheric Sciences*, *19*(4), 329–341.
13. Hateley, J., (2021). *The Lorenz System*. Lecture Notes. https://web.math.ucsb.edu/~jhateley/paper/lorenz.pdf (accessed on 06 September 2023).
14. Rafiq, T., Qamar, A., Mirza, A. M., & Murtaza, G., (2000). Chaotic behavior of ion temperature–gradient driven drift–dissipative modes. *Physics of Plasmas, 7*(11), 4499–4505.
15. Lorenz, E. N., (1963). Deterministic nonperiodic flow. *Journal of Atmospheric Sciences, 20*(2), 130–141.
16. Slingo, J., & Palmer, T., (2011). Uncertainty in weather and climate prediction. *Philosophical Transactions of the Royal Society A: Mathematical, Physical and Engineering Sciences, 369*(1956), 4751–4767.
17. Razali, N., & Ahmad, R., (2009). *Solving Lorenz System by Using Runge-Kutta Method, 32*, 241–251.
18. Gacherieu, C., Collercandy, R., Larrieu, P., Soumillon, S., Tourrette, L., & Viala, S., (2000). *Navier-Stokes Calculations at Aerospatiale Matra Airbus for Aircraft Design.* In: ICAS 2000 Congress.
19. Varley, E., & Seymour, B., (1994). Applications of exact solutions to the Navier–Stokes equations: Free shear layers. *Journal of Fluid Mechanics, 274*, 267–291. doi: 10.1017/S0022112094002120.

20. Adair, D., & Jaeger, M., (2015). Reynolds-averaged Navier–Stokes modeling of air pollution at the local urban scale. *Eng. Appl. Comput. Fluid Dyn., 4*, 119–136.
21. Marmanis, H., (1998). Analogy between the Navier–Stokes equations and Maxwell's equations: Application to turbulence. *Physics of Fluids, 10*(6), 1428–1437.
22. Batchelor, G., (2000). Flow of a uniform incompressible viscous fluid. In: *An Introduction to Fluid Dynamics* (pp. 174–263). Cambridge Mathematical Library. Cambridge: Cambridge University Press. doi: 10.1017/CBO9780511800955.006.
23. Jameson, A., Martinelli, L., & Pierce, N., (1998). Optimum aerodynamic design using the Navier–stokes equations. *Theoretical and Computational Fluid Dynamics, 10*, 213–237. 10.1007/s001620050060.
24. Jameson, Antony & Martinelli, Luigi & Pierce, Niles. (1998). Optimum Aerodynamic Design Using the Navier–Stokes Equations. *Theoretical and Computational Fluid Dynamics. 10*, 213–237. 10.1007/s001620050060.

CHAPTER 4

Deformations in a Nonlocal Isotropic Thermoelastic Material with Two Temperatures Due to Ramp Type Heat Source Using Memory Dependent Derivatives

SUKHVEER SINGH[1], PARVEEN LATA[2], and SATYA BIR SINGH[2]

[1]*Punjabi University APS Neighborhood Campus, Dehla Seehan, Punjab, India*

[2]*Department of Mathematics, Punjabi University, Patiala, Punjab, India*

ABSTRACT

This investigation is related to the two-dimensional deformations occurring in a homogeneous nonlocal isotropic thermoelastic solid with two temperatures due to ramp type source using memory dependent derivatives. Laplace and Fourier transforms have been used to find the analytical expressions of displacement components, stress components and conductive temperature in the transformed domain. Numerical inversion technique has been used to obtain the results in the physical domain. The effects of nonlocal parameter on the components of displacements, stresses, and conductive temperature along with some special cases have been deduced graphically.

New Advances in Materials Technologies. Hossein Hariri Asli, Ali Pourhashemi, Ann Rose Abraham, & A. K. Haghi (Eds.)

4.1 INTRODUCTION

Thermoelasticity is the branch of applied mechanics which deals with the study of elastic properties of a material due to heat transfer, i.e., it deals with the theory of strains and stresses due to heat flow. It covers a broad field of growth associated with science and technology. The nonlocal theory of thermoelasticity is a very important theory in the field of solid mechanics. It states that the stresses at any point are dependent upon strains at all points and not just due to strain at that point only. If we neglect the effects of strains at points other than the point under consideration then classical theory is recovered. The nonlocal theory was mainly given due to the contribution of Refs. [1–3]. They proved the existence of nonlocality in elastic bodies and obtained the constitutive equations for the theory. Artan [4] proved the superiority of the nonlocal theory as he compared various results to differentiate between the stress distributions in the case of the local and the nonlocal elasticity theories. Eringen & Wegner [5] developed nonlocal continuum field theories and thus developed basic field equations for nonlocal continuum field theories.

Thermoelasticity with two temperature is an important theory of thermoelastic materials which was developed by Chen & Gurtin [6]. They proved that the deformations in a thermoelastic material are dependent upon two distinct temperatures known as the thermodynamic temperature and the conductive temperature. Youssef [7] proposed the two-temperature generalized thermoelasticity by extending the concept of Chen and Gurtin. He obtained the uniqueness theorem for the said theory. Youssef & Al-Lehaibi [8] proved the importance of the two-temperature generalized thermoelasticity in describing the state of an elastic body in comparison to one temperature theory. The memory-dependent derivative is an integral form of a common derivative with a kernel function on a slip in the interval. It is playing an important part in the study of various thermoelastic properties in place of use of fractional calculus. Yu et al. [9] established a new generalized model based on the concept of memory-dependent derivatives. Ezzat, El-Karamany, & El-Bary [10] proposed a new model of magneto-thermoelasticity theory in the context of heat conduction with memory-dependent derivatives. Ezzat et al. [11] extended the results to establish a new mathematical model of generalized thermoelasticity with memory-dependent derivatives for the dual-phase-lag heat conduction law subjected to ramp-type heat source.

Peddieson, Buchanan, & McNitt [12] solved some problems to illustrate the magnitude of nonlocal effects by developing a nonlocal Euler beam model. Khurana & Tomar [13] studied the reflection of plane waves from the stress-free boundary of a nonlocal isotropic micropolar solid. Kumar, Sharma, & Lata [14] studied the thermomechanical disturbances in a transversely isotropic thermoelastic rotating medium with two temperatures under the combined effects of Hall current and magnetic field. Khurana & Tomar [15] derived frequency equation and studied the propagation of Rayleigh type surface waves in nonlocal micropolar elastic solid. Kaur, Singh, & Tomar [16] derived dispersion relations and investigated the propagation of surface waves in an isotropic nonlocal elastic solid with voids. Sarkar, Ghosh, & Lahiri [17] studied the magneto-thermoelastic interaction in an isotropic medium using a new two-temperature generalized thermoelasticity theory with memory-dependent derivative. Sarkar, De, & Sarkar [18] used generalized thermoelasticity model with memory-dependent derivative to study the reflection phenomenon of the magneto-thermoelastic plane waves from a stress-free surface of an isotropic conducting solid. Mondal [19] proposed a novel mathematical model of generalized thermoelasticity to investigate the influence of magnetic field and moving heat source in a rod in the context of Eringen's nonlocal elasticity. Saeed & Abbas [20] used Eringen's nonlocal continuum to obtain a novel nonlocal model without energy dissipation and studied the impacts of the nonlocal thermoelastic parameters in a nanoscale material.

Singh, Lata, & Singh [21] studied and discussed the recent advancements in the nonlocal theory of elasticity and thermoelasticity. Lata & Singh [22] discussed the thermomechanical interactions in a nonlocal magneto-thermoelastic medium under the effect of two temperature and Hall current with memory dependence. Lata & Singh [23] studied the deformation in an isotropic nonlocal magnetothermoelastic solid with two temperatures under the effects of inclined load at different inclinations. Lata & Singh [24] discussed the combined effect of nonlocal parameter and Hall current in a magneto-thermoelastic rotating medium with fractional order heat transfer due to normal load. Lata & Singh [25] used Green-Naghdi theory without energy dissipations to study the axisymmetric deformations in a nonlocal isotropic thermoelastic solid. Lata & Singh [26] studied the propagation of Rayleigh wave in a nonlocal rotating magneto-thermoelastic solid under the effect of hall current subjected to multi-dual-phase lag heat transfer. Lata & Singh [27] used the theory of generalized thermoelasticity to discuss the disturbances in a homogeneous nonlocal thermoelastic solid due to normal ramp type heat with two temperatures.

In the present work, we aim at investigating the deformations in a nonlocal isotropic thermoelastic material with two temperatures due to ramp type heating using memory dependent derivatives. The study might prove to be beneficial for the researchers working in the field of nonlocal thermoelastic materials.

4.2 BASIC EQUATIONS

Following Refs. [7, 5, 17], the equations of motion, heat conduction with memory dependent derivatives and constitutive relations in a homogeneous nonlocal isotropic thermoelastic solid with two temperatures are given by:

$$(\lambda+2\mu)\nabla(\nabla.u)-\mu(\nabla\times\nabla\times u)-\beta\nabla\theta=\left(1-\epsilon^2\nabla^2\right)\rho\frac{\partial^2 u}{\partial t^2}, \tag{1}$$

$$k\nabla^2\varphi=\rho C^*\frac{\partial\theta}{\partial t}+\beta\theta_0\frac{\partial}{\partial t}(\nabla.u)+\int_{t-\tau}^{t}K(t-\xi)(\rho C^*\frac{\partial\theta}{\partial t}+\beta\theta_0\frac{\partial}{\partial t}(\nabla.u))d\xi, \tag{2}$$

where;

$$\theta=\left(1-a\nabla^2\right)\varphi \tag{3}$$

$$\left(1-\epsilon^2\nabla^2\right)t_{ij}=\lambda u_{k,k}\delta_{ij}+\mu\left(u_{i,j}+u_{j,i}\right)-\beta\theta\delta_{ij} \tag{4}$$

where; K $(t-\xi)$ is the kernel function, which can be chosen freely as [10, 11]:

$$K(t-\xi)=1-\frac{2b}{\omega}(t-\xi)+\frac{a^2}{\omega^2}(t-\xi)^2=\begin{cases}1\,if\ a=0,b=0\\ 1-\frac{(t-\xi)}{\omega}if\ a=0,b=\frac{1}{2}\\ 1-(t-\xi)if\ a=0,b=\frac{\omega}{2}\\ \left(1-\frac{t-\xi}{\omega}\right)^2 if\ a=b=1\end{cases} \tag{5}$$

where; *a*, *b* are constants; λ, μ are material constants; ϵ is the nonlocal parameter; ρ is the mass density; $u=(u_1,0,u_3)$ is the displacement vector; φ

is the conductive temperature; a is two temperature parameter; θ is absolute temperature; and θ_0 is reference temperature; k is the coefficient of the thermal conductivity; C^* the specific heat at constant strain; $\beta = (3\lambda + 2\mu)\alpha$ is a material constant characteristic where α is coefficient of liner thermal expansion; δ_{ij} is the Kronecker delta; t_{ij} are the components of stress tensor.

4.3 FORMULATION OF THE PROBLEM

We consider a homogeneous nonlocal isotropic thermoelastic body in an initially undeformed state at temperature θ_0. We take a rectangular Cartesian co-ordinate system (x,y,z). We restrict our analysis to two-dimensional problem with:

$$u = (u,\ 0,\ w) \tag{6}$$

Using Eqn. (6) in Eqns. (1)–(2), yields:

$$(\lambda+2\mu)\frac{\partial^2 u}{\partial x^2}+(\lambda+\mu)\frac{\partial^2 w}{\partial x\partial z}+\mu\frac{\partial^2 u}{\partial z^2}-\beta\frac{\partial\theta}{\partial x}=\left(1-\in^2\nabla^2\right)\rho\frac{\partial^2 u}{\partial t^2} \tag{7}$$

$$(\lambda+2\mu)\frac{\partial^2 w}{\partial z^2}+(\lambda+\mu)\frac{\partial^2 u}{\partial x\partial z}+\mu\frac{\partial^2 w}{\partial x^2}-\beta\frac{\partial\theta}{\partial y}=\left(1-\in^2\nabla^2\right)\rho\frac{\partial^2 u}{\partial t^2} \tag{8}$$

$$k\nabla^2\varphi=\rho C^*\frac{\partial\theta}{\partial t}+\beta\theta_0\frac{\partial e}{\partial t}+\int_{t-\tau}^{t}K(t-\xi)(\rho C^*\frac{\partial\theta}{\partial t}+\beta\theta_0\frac{\partial e}{\partial t})d\xi \tag{9}$$

where; $e=\frac{\partial u_1}{\partial x_1}+\frac{\partial u_3}{\partial x_3},\ \nabla^2=\frac{\partial^2}{\partial x_1^2}+\frac{\partial^2}{\partial x_3^2}.$

we define the following dimensionless quantities:

$$(x',z',u',w')=c_1\xi(x,z,u,w),t'_{ij}=\frac{t_{ij}}{\rho c_1^2},t'=c_1^2\xi t,a'=c_1^2\xi^2 a,(\theta',\varphi')$$
$$=\frac{\beta}{\rho c_1^2}(\theta,\varphi-\theta_0) \tag{10}$$

where; $c_1^2=\frac{\lambda+2\mu}{\rho}$ and $\xi=\frac{\rho C^*}{K}.$

Upon introducing the quantities defined by Eqn. (10) in Eqns. (7)–(9), suppressing the primes and using the potential functions defined by:

$$u=\frac{\partial q}{\partial x}-\frac{\partial\psi}{\partial z},\ w=\frac{\partial q}{\partial z}+\frac{\partial\psi}{\partial x},\ \text{where } e=\nabla^2 q \tag{11}$$

where; $q(x_1, x_3, t)$, and $\psi(x_1, x_3, t)$, are scalar potential functions, on solving we get:

$$\left[\nabla^2 - \left(1 - \in^2 \nabla^2\right)\frac{\partial^2}{\partial t^2}\right]q - \left(1 - a\nabla^2\right)\varphi = 0, \tag{12}$$

$$\left[\nabla^2 - a_1\left(1 - \in^2 \nabla^2\right)\frac{\partial^2}{\partial t^2}\right]\psi = 0, \tag{13}$$

$$\nabla^2\varphi = \left\{\left[1 + a\frac{\partial}{\partial t}\left(1 + \omega D_\omega\right)\right]\nabla^2 - \frac{\partial}{\partial t}\left(1 + \omega D_\omega\right)\right\}\varphi - a_2\frac{\partial}{\partial t}\left(1 + \omega D_\omega\right)\nabla^2 q = 0 \tag{14}$$

where; $a_1 = \dfrac{\lambda + 2\mu}{\mu}$, $a_2 = \dfrac{\beta^2\theta_0}{\rho C^*(\lambda + 2\mu)}$

where;

$$G(s) = \left(1 - e^{-s\omega}\right)\left(1 - \frac{2b}{s\omega} + \frac{2a^2}{s^2\omega^2}\right) - e^{-s\omega}\left(a^2 - 2b + \frac{2a^2}{s\omega}\right) \tag{15}$$

and a, b are constants such that:

$$L\left(\omega D_\omega f(t)\right) = \begin{cases} 1 - e^{-s\omega} \; if\ a = 0, b = 0 \\ 1 - \dfrac{\left(1 - e^{-s\omega}\right)}{s\omega} if\ a = 0, b = \dfrac{1}{2} \\ \left(1 - e^{-s\omega}\right) - \dfrac{1}{s}\left(1 - e^{-s\omega}\right) + \omega e^{-s\omega} \; if\ a = 0, b = \dfrac{\omega}{2} \\ \left(1 - \dfrac{2}{s\omega}\right) + \dfrac{2(1 - e^{-s\omega})}{s^2\omega^2} if\ a = b = 1 \end{cases} \tag{16}$$

The initial and regularity conditions are given by:

$$u(x, z, 0) = 0 = \dot{u}(x, z, 0)$$

$$w(x, z, 0) = 0 = \dot{w}(x, z, 0)$$

$$\varphi(x, z, 0) = 0 = \dot{\varphi}(x, z, 0) \text{ for } z \geq 0, -\infty < x < \infty$$

$$u(x, z, t) = w(x, z, t) = \varphi(x, z, t) = 0 \text{ for } t > 0 \text{ when } z \to \infty$$

Applying Laplace and Fourier Transform defined by:

$$\overline{f}(\mathrm{x}, \mathrm{z}, \mathrm{s}) = \int_0^\infty f(x, z, t)e^{-st}dt \tag{17}$$

$$\overline{f^*}(\varsigma, z, s) = \int_{-\infty}^\infty \overline{f}(x, z, s)e^{i\varsigma x}dx \tag{18}$$

In Eqns. (12)–(14), we obtain a system of equations:

$$\left[\zeta_1 D^2-\zeta_2\right]\overline{q^*}+[aD^2-\zeta_6]\overline{\varphi^*}=0, \tag{19}$$

$$\left[\zeta_4 D^2-\zeta_5\right]\overline{\psi^*}=0, \tag{20}$$

$$\left[\left[\zeta_3 D^2-\zeta_3\varsigma^2\right]\right]\overline{q^*}+\left[\zeta_7 D^2+\zeta_8\right]\overline{\varphi^*}=0. \tag{21}$$

From Eqns. (19)–(21), we get a set of homogeneous equations having a nontrivial solution if the determinant of coefficient $[\overline{q^*},\overline{\psi^*},\overline{\varphi^*}]^T$ vanishes so as to give a characteristic equation as:

$$\left[D^6+QD^4+RD^2+S\right](\overline{q^*},\overline{\psi^*},\overline{\varphi^*})=0. \tag{22}$$

where;

$$Q=\frac{1}{P}\{(\zeta_1\zeta_8+\zeta_3\zeta_6)\zeta_4-(\zeta_1\zeta_5+\zeta_2\zeta_4)\zeta_7+a(\zeta_5+\zeta_4\varsigma^2)\zeta_3\}$$

$$R=\frac{1}{P}\{(\zeta_2\zeta_7-\zeta_1\zeta_8-\zeta_3\zeta_6)\zeta_5-\zeta_2\zeta_4\zeta_8-(a\zeta_5+\zeta_4\zeta_6)\zeta_3\varsigma^2\}$$

$$S=\frac{1}{P}\{(\zeta_2\zeta_8+\zeta_3\zeta_6\varsigma^2)\zeta_5\}$$

$$P=(\zeta_1\zeta_7-a\zeta_3)\zeta_4$$

where; $D=\frac{d}{dz}$, $\zeta_1=1+\epsilon^2 s^2$, $\zeta_2=\varsigma^2 s^2(1+\epsilon^2\varsigma^2)$, $\zeta_3=a_2 s(1+G)$, $\zeta_4=1+a_1+\epsilon^2 s^2$, $\zeta_6=1+a\varsigma^2$, $\zeta_7=1+as(1+G)$, $\zeta_8=\varsigma^2+s(1+a\varsigma^2)(1+G)$.

The roots of Eqn. (22) are $\pm\lambda_i$ $(i=1,2,3)$ satisfying the radiation condition that $\overline{q^*},\overline{\psi^*},\overline{\varphi^*}\rightarrow 0$ as $z\rightarrow\infty$, the solutions of equation can be written as:

$$\overline{q^*}=A_1e^{-\lambda_1 z}+A_2e^{-\lambda_2 z}+A_3e^{-\lambda_3 z}, \tag{23}$$

$$\overline{\psi^*}=d_1A_1e^{-\lambda_1 z}+d_2A_2e^{-\lambda_2 z}+d_3A_3e^{-\lambda_3 z}, \tag{24}$$

$$\overline{\varphi^*}=l_1A_1e^{-\lambda_1 z}+l_2A_2e^{-\lambda_2 z}+l_3A_3e^{-\lambda_3 z}. \tag{25}$$

where;

$$d_i=\frac{P^*\lambda_i^4+Q^*\lambda_i^2+R^*}{S^*\lambda_i^4+T^*\lambda_i^2+U^*}\, i=1,2,3 \tag{26}$$

$$l_i=\frac{P^{**}\lambda_i^4+Q^{**}\lambda_i^2+R^{**}}{S^*\lambda_i^4+T^*\lambda_i^2+U^*}\, i=1,2,3 \tag{27}$$

where,

$$P^* = \zeta_1\zeta_7 - a\zeta_3$$
$$Q^* = (\zeta_1\zeta_8 - \zeta_2\zeta_7) + (a\varsigma^2 + \zeta_6)\zeta_3$$
$$R^* = \zeta_2\zeta_8 + \zeta_3\zeta_6\varsigma^2$$
$$S^* = \zeta_4\zeta_7$$
$$T^* = \zeta_4\zeta_8 - \zeta_5\zeta_7$$
$$U^* = -\zeta_5\zeta_8$$
$$P^{**} = \zeta_1\zeta_4$$
$$Q^{**} = -\zeta_1\zeta_5 - \zeta_2\zeta_4$$
$$R^{**} = \zeta_2\zeta_5$$

4.4 APPLICATIONS

For a ramp type heat source, the boundary conditions are given by:

1. $t_{zz}(x, z, t) = 0,$ (28)
2. $t_{zx}(x, z, t) = 0,$ (29)
3. The boundary condition of the half-space depends upon time t and coordinate x and is of the form of:

$$\varphi(x, 0, t) = G(t)\delta(x) \tag{30}$$

where; $\delta(x)$ is Dirac delta function of x while $G(t)$ is a function defined as follows:

$$G(t) = \begin{cases} 0, t \le 0 \\ T_1 \dfrac{t}{t_0}, 0 < t \le t_0, \\ T_1, t > t_0 \end{cases} \tag{31}$$

where; t_0 corresponds to the length of the time required to raise the heat while T_1 is a constant. It means that the boundary of the half space is at a fixed temperature t_0 and at rest initially. Then it is suddenly raised to a temperature which is equal to a function $G(t)\delta(x_1)$ and then is maintained at this temperature.

Laplace and Fourier transforms are applied to both sides of Eqn. (31) to obtain:

$$\overline{\varphi^*}(x,0,t) = \overline{G^*}(s), \text{ where } \overline{G^*}(s) = T_1 \frac{(1-e^{-st_0})}{\Delta t_0 s^2} \tag{32}$$

Using the dimensionless quantities defined by Eqn. (10) and using Eqns. (3), (4), (11), (17), and (18) in Eqns. (28)–(30) then substituting values of $\overline{q^*}$, $\overline{\psi^*}$ and $\overline{\varphi^*}$ from Eqns. (23)–(25), and solving, we obtain the components of displacement, normal stress, tangential stress and conductive temperature as follows:

$$\overline{u^*} = T_1 \frac{(1-e^{-st_0})}{\Delta t_0 s^2} \sum_{i=1}^{3} K_i \Delta_i e^{-\lambda_i z} \tag{33}$$

$$\overline{w^*} = T_1 \frac{(1-e^{-st_0})}{\Delta t_0 s^2} \sum_{i=1}^{3} L_i \Delta_i e^{-\lambda_i z} \tag{34}$$

$$\overline{\varphi^*} = T_1 \frac{(1-e^{-st_0})}{\Delta t_0 s^2} \sum_{i=1}^{3} l_i \text{Ä}_i e^{-\lambda_i z} \tag{35}$$

$$\overline{t_{xx}^*} = T_1 \frac{(1-e^{-st_0})}{\Delta t_0 s^2} \sum_{i=1}^{3} (1+\in^2 \lambda_i^2 - \in^2 \varsigma^2) P_i \Delta_i e^{-\lambda_i z} \tag{36}$$

$$\overline{t_{zx}^*} = T_1 \frac{(1-e^{-st_0})}{\Delta t_0 s^2} \sum_{i=1}^{3} (1+\in^2 \lambda_i^2 - \in^2 \varsigma^2) N_i \Delta_i e^{-\lambda_i z} \tag{37}$$

$$\overline{t_{zz}^*} = T_1 \frac{(1-e^{-st_0})}{\Delta t_0 s^2} \sum_{i=1}^{3} (1+\in^2 \lambda_i^2 - \in^2 \varsigma^2) M_i \Delta_i e^{-\lambda_i z} \tag{38}$$

$$\Delta = M_1 \Delta_{23} + M_2 \Delta_{31} + M_3 \Delta_{12} \tag{39}$$

where;

$$\Delta_{12} = \mathrm{N}_1 \mathrm{l}_2 - \mathrm{N}_2 \mathrm{l}_1, \Delta_{23} = \mathrm{N}_2 \mathrm{l}_3 - \mathrm{N}_3 \mathrm{l}_2, \Delta_{31} = \mathrm{N}_3 \mathrm{l}_1 - \mathrm{N}_1 \mathrm{l}_3.$$

$$\Delta_1 = \mathrm{M}_2 \mathrm{N}_3 - \mathrm{M}_3 \mathrm{N}_2, \Delta_2 = \mathrm{M}_3 \mathrm{N}_1 - \mathrm{M}_1 \mathrm{N}_3, \Delta_3 = \mathrm{M}_1 \mathrm{N}_2 - \mathrm{M}_2 \mathrm{N}_1.$$

$$\mathrm{M}_i = \lambda_i^2 - \left(1 - \frac{2}{a_1}\right)\varsigma^2 - \frac{\iota\varsigma 2}{a_1}\lambda_i d_i - l_i(1 + a\varsigma^2 - a\lambda_i^2), \ \mathrm{N}_i = \left(\lambda_i^2 + \varsigma^2\right)$$

$$d_i + \iota\varsigma 2\lambda_i, \ \mathrm{K}_i = \iota\varsigma + \lambda_i d_i,$$

$$\mathrm{L}_i = \iota\varsigma d_i - \lambda_i, \ \mathrm{P}_i = \iota\varsigma K_i - \left(1 - \frac{2}{a_1}\right)\lambda_i L_i - l_i(1 + a\varsigma^2 - a\lambda_i^2); \ j = 1,2,3$$

4.5 PARTICULAR CASES

- If $a = 0$, then from Eqns. (33)–(38), the corresponding expressions for displacements, stresses, and conductive temperature for nonlocal isotropic solid without two temperatures are obtained.
- If $\in = 0$, then from Eqns. (33)–(38), the corresponding expressions for displacements, stresses, and conductive temperature for isotropic solid without nonlocal effects and with two temperatures are obtained.
- If $\in = a = 0$, then from Eqns. (33)–(38), the corresponding expressions for displacements, stresses, and conductive temperature for isotropic solid without nonlocal effects and two temperatures are obtained.

4.6 INVERSION OF THE TRANSFORMATION

To obtain the solution of the problem in physical domain, we must invert the transforms in Eqns. (33)–(38). Here the displacement components, normal, and tangential stress and conductive temperature are functions of z and the parameters of Laplace and Fourier transforms s and ξ respectively and hence of the form $f(\xi, z, s)$. To obtain the function $f(x, z, t)$ in the physical domain, we invert the Fourier transform using:

$$\tilde{f}(x,z,s) = \frac{1}{2\pi}\int_{-\infty}^{\infty} e^{-i\xi x}\hat{f}(\xi,z,s)\,d\xi = \frac{1}{2\pi}\int_{-\infty}^{\infty}\left|\cos(\xi x) f_e - i\sin(\xi x) f_0\right| d\xi \quad (40)$$

where; f_e and f_0 are, respectively the even and odd parts of $\hat{f}(\xi, z, s)$. Thus the expression Eqn. (40) gives the Laplace transform $\tilde{f}(\xi, z, s)$ of the function $f(x, z, t)$. Following Res. [28], the Laplace transform function $\tilde{f}(x, z, s)$ can be inverted to $f(x, z, t)$.

The Last step is to calculate the integral in Eqn. (40). The method for evaluating this integral is described in Ref. [29]. It involves the use of Romberg's integration with adaptive step size. This also uses the results from successive refinements of the extended trapezoidal rule followed by extrapolation of the results to the limit when the step size tends to zero.

4.7 NUMERICAL RESULTS AND DISCUSSION

Magnesium material is chosen for the purpose of numerical calculation which is isotropic and according to Ref. [30], physical data for which is given as:

$$\lambda = 9.4\times10^{10}\,Nm^{-2}, \mu = 3.278\times10^{10}\,Nm^{-2}, K^{*}$$
$$= 1.7\times10^{2}\,Wm^{-1}K^{-1}, \rho = 1.74\times10^{3}\,Kgm^{-3}, \theta_0 = 298\,K, C^{*}$$
$$= 10.4\times10^{2}\,JKg^{-1}deg^{-1}, \omega_1 = 3.58, a = 0.05$$

A comparison of values of displacement components u and w, stress components t_{zz}, t_{xx} and t_{zx} and conductive temperature φ for an isotropic nonlocal thermoelastic solid with respect to distance x has been made for the local parameter ($\in = 0$) and nonlocal parameter ($\in = 0.2$) and time-delay parameter ($\omega = 0.3$) and $\omega = 0.5$.

1. The dotted black colored line with center symbol square corresponds to local parameter ($\in = 0$) and time delay parameter $\omega = 0.3$.
2. The big dashed red colored line with center symbol circle corresponds to local parameter ($\in = 0$) and time delay parameter $\omega = 0.5$.
3. The small dashed blue colored line with center symbol upward triangle corresponds to nonlocal parameter ($\in = 0.2$) and time delay parameter $\omega = 0.3$.
4. The big dashed and small dashed purple colored line with center symbol downward triangle corresponds to nonlocal parameter ($\in = 0.2$) and time delay parameter $\omega = 0.5$.

Figure 4.1 shows the variations in values of displacement component$u$ with respect to displacement X. It is clear that the values of u follow an increasing pattern in magnitude which becomes steady during the end. For local parameter, the variations are less in magnitude as compared to nonlocal parameter. The trend is similar for the time-delay parameter, i.e., it increases with increase in value. But the trend is more during the first half which is less during the second half. Figure 4.2 depicts the variation of values of displacement component w. The pattern is identical to the behavior of displacement component u, i.e., it follows increasing behavior. But during the first half nonlocal parameter lags behind local parameter while it is reversed during the second half for time delay parameter 0.3 and almost the opposite for 0.5. Figure 4.3 describes the variations of stress component t_{xx} with respect to displacement. For both local and nonlocal parameters, the behavior is oscillatory but the local parameter is seen to be different for time-delay parameter 0.3 but for all else the difference is very small comparatively. Also, the magnitude of oscillations is less during the second half. Figure 4.4 shows the variations of stress component t_{zx} with respect to displacement. The magnitude of oscillations is more for local parameter while the being less

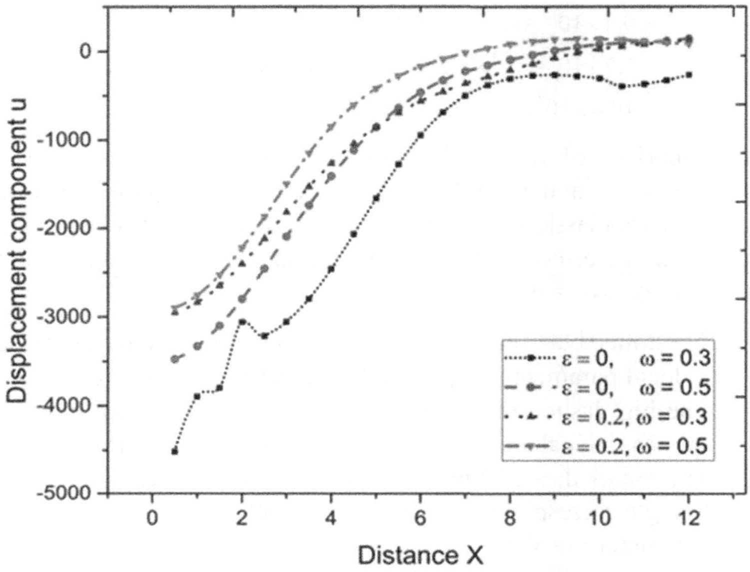

FIGURE 4.1 Variation of displacement component *u* with respect to displacement x.

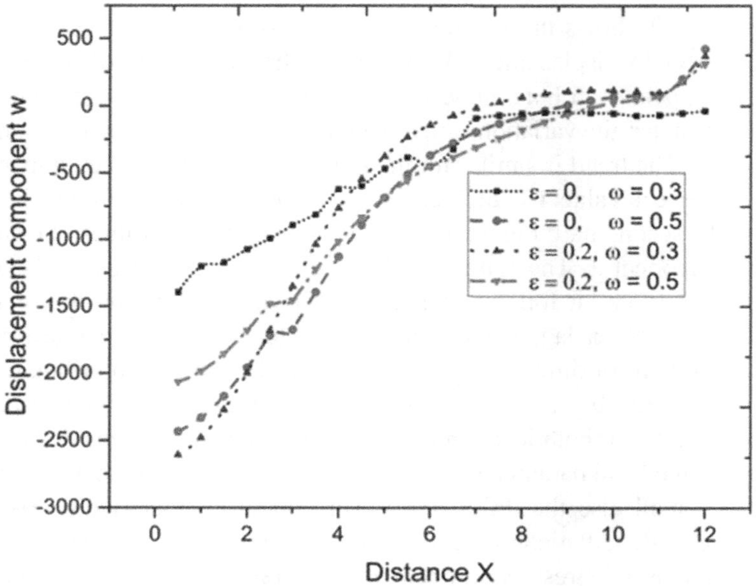

FIGURE 4.2 Variation of displacement component *w* with respect to displacement x.

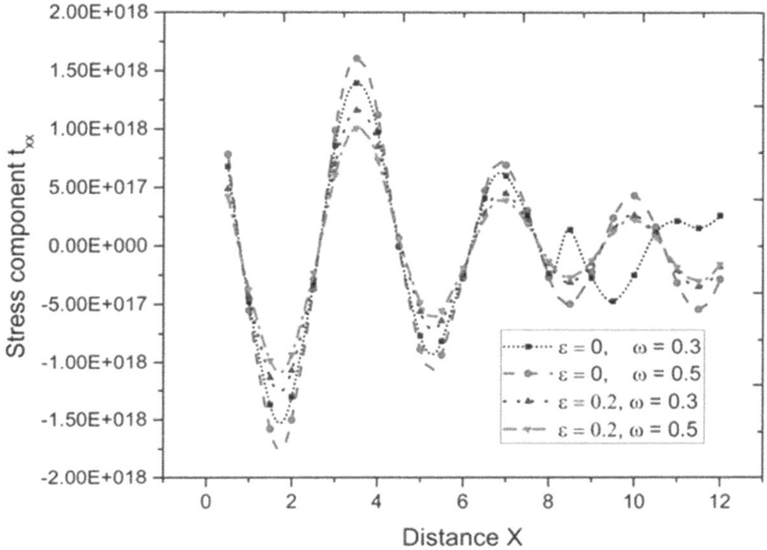

FIGURE 4.3 Variation of stress component t_{xx} with respect to displacement x.

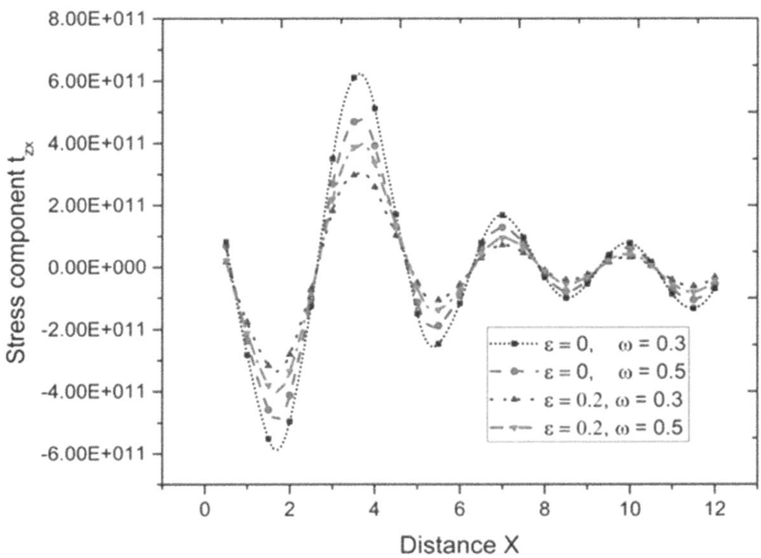

FIGURE 4.4 Variation of stress component t_{zx} with respect to displacement x.

for the nonlocal parameter. But the trend is different for time-delay parameter as while it is decreasing for local parameter but it is opposite in case of nonlocality. Figure 4.5 shows the variation of stress component t_{zz}. Here too the behavior followed is oscillatory with trend almost similar to the behavior for stress component t_{zx}. The only difference being that the trend is similar for both local as well as nonlocal parameters that the magnitude is increasing with increase in time-delay parameter. Figure 4.6 illustrates the variation of conductive temperature φ. The behavior followed is oscillatory with a big dip for $2 \leq x \leq 4$ and while the magnitude is more for nonlocal parameter during the first half, it increases for local parameter during the last part of the graph. But overall, the difference is less as compared to other quantities.

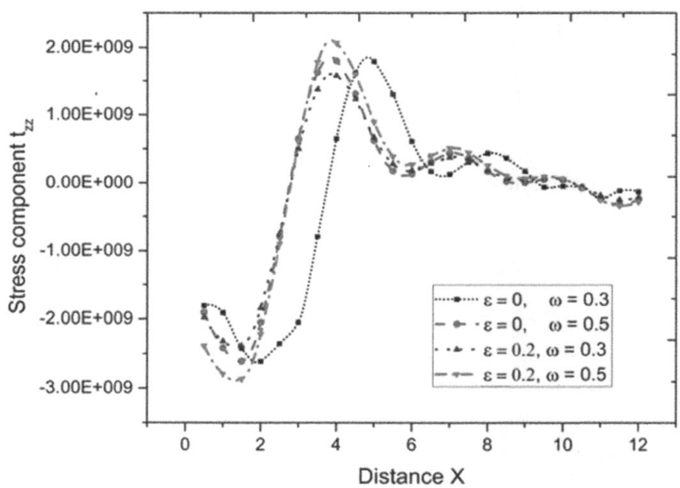

FIGURE 4.5 Variation of stress component t_{zz} with respect to displacement x.

4.8 CONCLUSION

In the present discussion the numerical results have been depicted graphically showing the effects of nonlocal parameter and the time-delay parameter on the components of displacements, stresses, and conductive temperature. From above it is observed that both the nonlocal parameter and the time-delay parameter are playing a significant effect on the variations in displacements, stresses, and conductive temperature. It is observed from Figures 4.1–4.6 that the trends in the variations are significant enough for all the components with exception of conductive temperature as there the effect is less visible as compared to other quantities. But overall, we conclude that nonlocality as well as time-delay cannot be ignored and they play their part for further studies. The results of this investigation can be helpful for the researchers working in the field of nonlocal materials, geophysics, material engineering, acoustics, nanoparticles (NPs), etc.

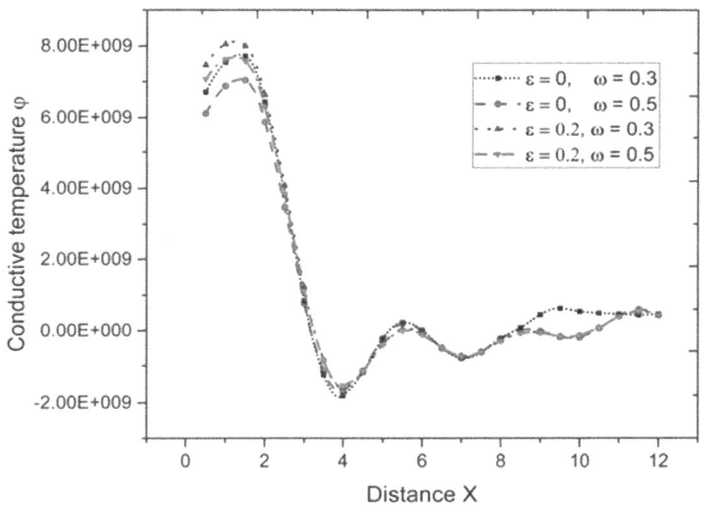

FIGURE 4.6 Variation of conductive temperature φ with respect to displacement x.

KEYWORDS

- **Eringen model of nonlocal theories**
- **geophysics**
- **material engineering**
- **memory dependent derivative**
- **nonlocality**
- **ramp type heat source**
- **time-delay parameter**
- **two temperatures**

REFERENCES

1. Edelen, D. G. B., & Laws, N., (1971). On the thermodynamics of systems with nonlocality. *Arch. Ration Mech. Anal., 43*, 24–35. https://doi.org/10.1007/BF00251543.
2. Edelen, D. G. B., Green, A. E., & Laws, N., (1971). Nonlocal continuum mechanics. *Arch. Ration Mech. Anal., 43*, 36–44. https://doi.org/10.1007/BF00251544.
3. Eringen, A. C., & Edelen, D. G. B., (1972). On nonlocal elasticity. *Int. J. Eng. Sci., 10*, 233–248. https://doi.org/10.1016/0020-7225(72)90039-0.
4. Artan, R., (1996). Elastic half plane loaded concentrated *Jv. Int. J. Eng. Sci., 34*, 943–950.
5. Eringen, A., & Wegner, J., (2003). *Nonlocal Continuum Field Theories, 56*. https://doi.org/10.1115/1.1553434.
6. Chen, P. J., & Gurtin, M. E., (1968). On a theory of heat conduction involving two temperatures. *Zeitschrift Für Angew Math Und Phys ZAMP., 19*, 614–627. https://doi.org/10.1007/BF01594969.
7. Youssef, H. M., (2006). Theory of two-temperature-generalized thermoelasticity. *IMA J. Appl. Math., 71*, 383–390. Institute Math Its Appl. https://doi.org/10.1093/imamat/hxh101.
8. Youssef, H. M., & Al-Lehaibi, E. A., (2007). State-space approach of two-temperature generalized thermoelasticity of one-dimensional problem. *Int. J. Solids Struct., 44*, 1550–1562. https://doi.org/10.1016/j.ijsolstr.2006.06.035.
9. Yu, Y. J., Hu, W., & Tian, X. G., (2014). A novel generalized thermoelasticity model based on memory-dependent derivative. *Int. J. Eng. Sci., 81*, 123–134. https://doi.org/10.1016/j.ijengsci.2014.04.014.
10. Ezzat, M. A., El-Karamany, A. S., & El-Bary, A. A., (2015). A novel magneto-thermoelasticity theory with memory-dependent derivative. *J. Electromagn. Waves Appl., 29*, 1018–10131. https://doi.org/10.1080/09205071.2015.1027795.
11. Ezzat, M. A., El-Karamany, A. S., & El-Bary, A. A., (2017). On dual-phase-lag thermoelasticity theory with memory-dependent derivative. *Mech. Adv. Mater. Struct., 24*, 908–916. https://doi.org/10.1080/15376494.2016.1196793.
12. Peddieson, J., Buchanan, G. R., & McNitt, R. P., (2003). Application of nonlocal continuum models to nanotechnology. *Int. J. Eng. Sci., 41*, 305–312. https://doi.org/10.1016/S0020-7225(02)00210-0.
13. Khurana, A., & Tomar, S. K., (2013). Reflection of plane longitudinal waves from the stress-free boundary of a nonlocal, micropolar solid half-space. *J. Mech. Mater. Struct., 8*, 95–107. https://doi.org/10.2140/jomms.2013.8.95.
14. Kumar, R., Sharma, N., & Lata, P., (2016). Thermomechanical interactions due to hall current in transversely isotropic thermoelastic with and without energy dissipation with two temperatures and rotation. *J. Solid Mech., 8*, 840–858.
15. Khurana, A., & Tomar, S. K., (2017). Rayleigh-type waves in nonlocal micropolar solid half-space. *Ultrasonics, 73*, 16216–16218. https://doi.org/10.1016/j.ultras.2016.09.005.
16. Kaur, G., Singh, D., & Tomar, S. K., (2018). Rayleigh-type wave in a nonlocal elastic solid with voids. *Eur. J. Mech. A/Solids., 71*, 134–150. https://doi.org/10.1016/j.euromechsol.2018.03.015.
17. Sarkar, N., Ghosh, D., & Lahiri, A., (2019). A two-dimensional magneto-thermoelastic problem based on a new two-temperature generalized thermoelasticity model with

memory-dependent derivative. *Mech. Adv. Mater. Struct., 26*, 957–966. https://doi.org/10.1080/15376494.2018.1432784.
18. Sarkar, N., De, S., & Sarkar, N., (2019). Memory response in plane wave reflection in generalized magneto-thermoelasticity. *J. Electromagn. Waves Appl., 33*, 1354–1374. https://doi.org/10.1080/09205071.2019.1608318.
19. Mondal, S., (2020). Memory response for thermal distributions moving over a magneto-thermoelastic rod under Eringen's nonlocal theory. *J. Therm. Stress, 43*, 72–89. https://doi.org/10.1080/01495739.2019.1676682.
20. Saeed, T., & Abbas, I., (2022). Effects of the nonlocal thermoelastic model in a thermoelastic nanoscale material. *Mathematics, 10*, 1–10. https://doi.org/10.3390/math10020284.
21. Singh, S., Lata, P., & Singh, S. B., (2021). Recent developments in the theory of nonlocality in elastic and thermoelastic mediums. *Appl. Mechatronics Mech.*, 111–122.
22. Lata, P., & Singh, S., (2021). Effects due to two temperature and hall current in a nonlocal isotropic magneto-thermoelastic solid with memory dependent derivatives. *Coupled Syst. Mech., 10*, 351–369. https://doi.org/10.12989/csm.2021.10.4.351.
23. Lata, P., & Singh, S., (2022). Effect of rotation and inclined load in a nonlocal magneto thermoelastic solid with two temperature. *Adv. Mater. Res. (South Korea), 11*, 23–39. https://doi.org/10.12989/amr.2022.11.1.023.
24. Lata, P., & Singh, S., (2022). Effects of hall current and nonlocality in a magneto-thermoelastic solid with fractional order heat transfer due to normal load. *J. Therm. Stress, 45*, 51–64. https://doi.org/10.1080/01495739.2021.2018954.
25. Lata, P., & Singh, S., (2022). Axisymmetric deformations in a nonlocal isotropic thermoelastic solid with two temperature. *Forces Mech., 6*, 100068. https://doi.org/10.1016/j.finmec.2021.100068.
26. Lata, P., & Singh, S., (2022). Rayleigh wave propagation in a nonlocal isotropic magneto-thermoelastic solid with multi-dual-phase lag heat transfer. *GEM – Int. J. Geomathematics, 13*, 1–17. https://doi.org/10.1007/s13137-022-00195-5.
27. Lata, P., & Singh, S., (2022). Disturbances due to normal heat in a nonlocal thermoelastic solid with two temperatures and without energy dissipation. *Mater. Model. Macro to Micro/Nano Scale Syst.*, 203–218.
28. Honig, G., & Hirdes, U., (1984). A method for the numerical inversion of Laplace transforms. *J. Comput. Appl. Math., 10*, 113–132. https://doi.org/10.1016/0377-0427(84)90075-X.
29. Press, W. H., Flannery, B. P., Teukolsky, S. A., & Vetterling, W. T., (1992). *Numerical Recipes in Fortran: The Art of Scientific Computing* (Vol. 2). Cambridge university press, New York; 1992.
30. Dhaliwal, R. S., & Singh, A., (1980). *Dynamic Coupled Thermoelasticity.* Hindustan Publ. Corp., New Delhi

CHAPTER 5

Mechanical and Tribological Characteristics of Two-Dimensional (2D) Nanomaterials

AVINASH V. BORGAONKAR[1], SHITAL B. POTDAR[2], and SONALI KALE[3]

[1]*Department of Mechanical Engineering, Pimpri Chinchwad College of Engineering (PCCOE), Pune, Maharashtra, India*

[2]*Department of Chemical Engineering, National Institute of Technology, Warangal, Telangana, India*

[3]*Department of Applied Sciences and Humanities, Pimpri Chinchwad College of Engineering (PCCOE), Pune, Maharashtra, India*

ABSTRACT

Currently, efforts have been made to pursue research on a special class of materials which possesses superior mechanical and tribological properties. The 2D nanomaterials have excellent features in this perspective since they exhibits higher young's modulus, high strength, and lower friction. The recent literatures have illustrated that the nano-additives are potential candidates for enrichment into the mechanical properties. The 2D nanomaterials can be incorporated into material matrix to prepare 2D nanomaterial-based composites (2DNBCs). In present context, the tribological and mechanical characteristics of such 2DNBCs have been reviewed. The various matrix materials like metals, ceramics, polymers, and the various 2D nanomaterials as reinforcement materials like Molybdenum Disulfide (MoS_2) graphene, hexagonal boron nitride (hBN) as well as transition metal carbides and

New Advances in Materials Technologies. Hossein Hariri Asli, Ali Pourhashemi, Ann Rose Abraham, & A. K. Haghi (Eds.)

nitrides have been reviewed in the present work. A brief summary and future scope for the further advances in the field of 2D nanocomposites are included.

5.1 INTRODUCTION

The tribological properties of the interacting surfaces significantly dominate the energy consumption and operational life of the components. The controlling of friction and wear is desirable in many circumstances impacts on a variety of material applications such as engine components, nano- to micro-scaled devices (MEMS, NEMS), and medical implants used in joint replacement [1]. As almost 23% of total world's energy usage is consumed by interacting surfaces, this perhaps motivates researchers to contribute to the area of development of nanolubricants and composite coatings [2, 3]. In the recent years, solid lubricants molybdenum disulfide (MoS_2), graphene, hexagonal boron nitride (hBN) have been notably used in several applications due to their significant tribological properties. Although for achieving superior tribological performance recently several researchers worked with combined 2D materials with nanomaterials as additives. The properties such as easy shearing ability, low surface energy, good corrosion resistance and superior thermal conductivity of nanomaterials make them appropriate reinforcement for both nano- and macro-scale tribological applications [4, 5]. Depending on the structural parameters, these nano-additives have been classified such as zero-dimensional (0D), one-dimensional (1D), two-dimensional (2D), and three-dimensional (3D) nanomaterials [6].

The discovery of 2D materials such as compounds of graphite, MoS_2, hBN, and other transition layered materials have been evolved and used the field of tribology. These 2D materials, with few atomic layers exhibited significant reduction in friction coefficient (COF) well below 0.01, which is often termed as super-lubricity [7]. Beside low frictional characteristics on surfaces of 2D materials; their crystallographic structure resulted into nearly zero friction (termed as super-lubricity) [8]. The achievement of super-lubricity against the applied load and travel distance was a challenging task in the past; in the recent researchers developed nanocomposites and observed super-lubricity at macro- and meso-scales [9]. The improvement in the frictional and wear properties of nano-additive liquid and solid lubricants majorly depends on their inherent characteristics like structural and chemical properties, dispersion rates, etc. [10, 11]. The weak Van-der-Waals forces between layered structures of 2D-nanomaterials possess low shear strength due to which they exhibit excellent lubrication performance [12, 13]. Figure 5.1 represents the layered crystal structure of the 2D nanomaterials like molybdenum disulfide, graphite, h-BN [14].

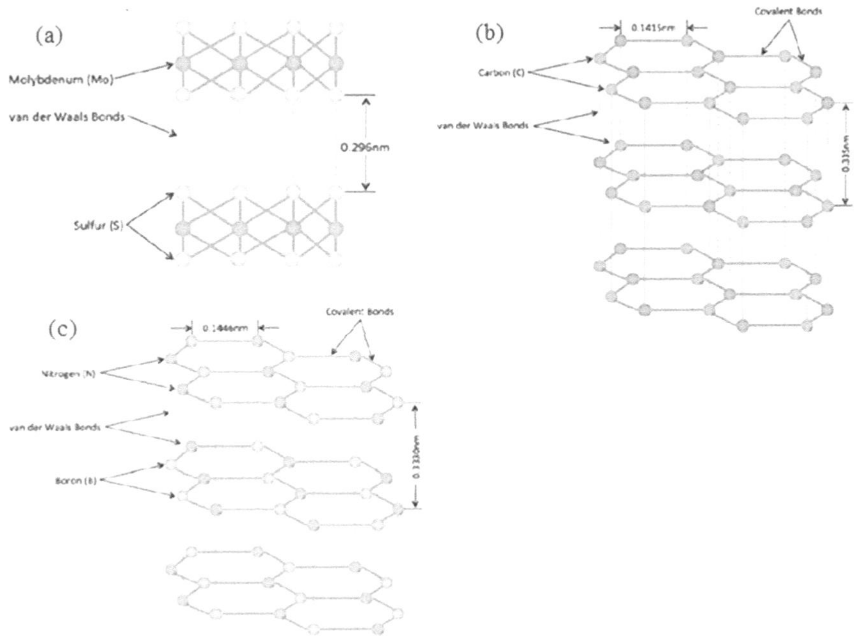

FIGURE 5.1 Schematic layered crystal structures of (a) molybdenum disulfide; (b) graphite; and (c) h-BN.

The 2D nanomaterials are classified as per the chemical elements present and atomic structure such as Xenes, transition metal carbides and nitrides (MXenes), transition metal dichalcogenides (TMDs), organic frameworks and nitrides. The Xenes are comprised of only single element like phosphorous, carbon, and silicon. An example of Xenes is graphene, which is a carbon-based material [15]. The MXenes, where M shows transition metals like Ti, V, Mo, and X stands for nitrogen or carbon, these are recently discovered novel 2D nanomaterials [16]. The TMDs are composed of hexagonal layered metal atoms (M) sandwiched within two layers of chalcogen atoms (X), the examples are WS_2, MoS_2, $MoSe_2$ and hBN is an example of 2D nitride material.

5.2 MECHANICAL PROPERTIES OF 2D NANOMATERIALS

Generally, in composites the interatomic bonds influences the mechanical properties. In recent years, 2D nanomaterials have been mixed into base metal matrices to enhance the mechanical properties. For efficient use of 2D nanomaterials their mechanical properties need to be considered fundamentally, but still these are not fully explored and understood. Figure 5.2 represents the mechanical properties and wear mechanism of 2D nanomaterials and their usage in various engineering applications.

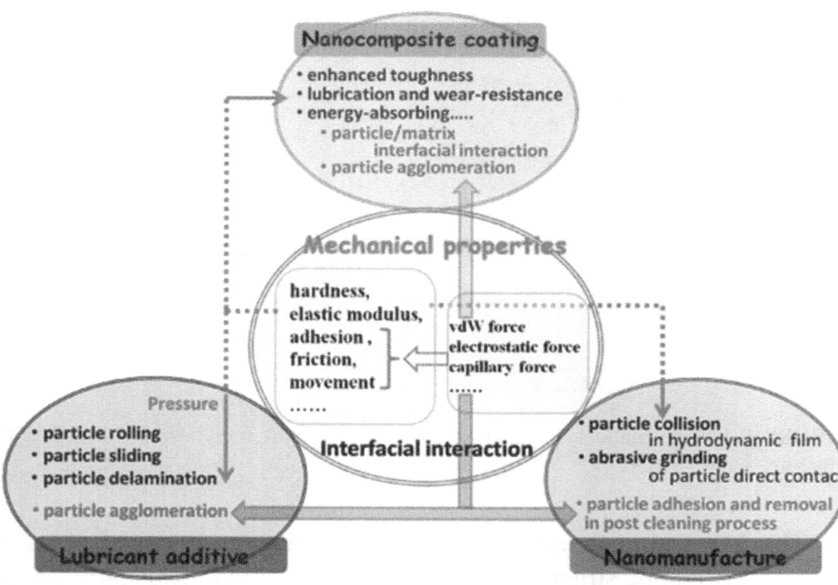

FIGURE 5.2 Mechanical properties and wear mechanisms of nanocomposite coatings.

Source: Reprinted with permission from Ref. [17]. © Copyright. IOP Publishing. https://creativecommons.org/licenses/by/3.0/

The 2D organic framework materials includes metal-organic and covalent-organic frameworks [18]. These materials having special characteristic features, i.e., they are having mesopores or micropores within the layers. The

2D nanomaterials possess high in-plane strength due to covalent bonding [19]. The mechanical characteristics of 2D nanomaterials are described in Table 5.1.

TABLE 5.1 Mechanical Characteristics of 2D Nanomaterials

2D Nanomaterials	Thickness	Tensile Strength (GPa)	Modulus of Elasticity (GPa)	References
MoS_2	Monolayer	23	270	[20]
Graphene	–	125	1,000	[21, 25]
h-BN	Monolayer	35	865	[22]
Silicone	–	–	82	[21]
Ti_2C	Monolayer	–	597	[23]
Phosphorene	Monolayer	–	41	[21]
MOF	–	–	5	[24]

5.3 TRIBOLOGICAL PROPERTIES OF 2D NANOMATERIALS

5.3.1 TRIBOFILMS OF COMPOUND SOLID LUBRICANTS

Solid or liquid lubricants individually and in combination of both assist to improve the tribological properties. However, the friction between the interacting surfaces not only depends on properties of the lubricants employed, but also on the chemical and physical processes occurs on the surfaces [22, 26]. The solid lubricants worn out due to tribochemical reactions and surface deformation at the interacting surfaces. Among the use of various solid lubricants, the composite solid lubricants films are in possession of superior tribological performance due to their unique enhanced properties. In this section, tribological properties of the solid lubricants reinforced with other nanomaterials are discussed. The tribological properties of various composite solid lubricants have been reported in Table 5.2.

5.3.2 TRIBOFILMS OF LIQUID LUBRICANTS MIXED WITH 2D NANOMATERIALS

It is observed that nanomaterials as reinforcement are suitable for various tribological applications as solid and in liquid lubrication [44, 45]. Similar

TABLE 5.2 Tribological Properties of Various Composite Solid Lubricants

Matrix	Reinforcement	Findings	References
Ni coating	0.3 g/L GO	An effective layer of lubricant was developed during operation, which resulted into reduced friction and wear.	[27]
Epoxy coating	20 wt.% $NbSe_2$ nano-sheet	At low speed with 0.5 wt.% addition of MWCNTs reduces the COF by 42.6%.	[28]
MoS_2 coating	5 to 25 wt.% of TiO_2	15 wt.% addition of TiO_2 into the MoS_2 matrix resulted in enhanced tribological characteristics of composite MoS_2–TiO_2 coating material.	[29, 30]
Epoxy	1.5 wt.% graphite 1.5 wt.% CNTs 1.5 wt.% (1/3 CNTs + 2/3 graphite)	The composite epoxy + CNT material exhibits excellent tribological properties.	[31]
316 L stainless steel	hBN	The addition of hBN exhibited improved tribological properties.	[32]
Copper	0 to 10 wt.% of graphite and 0 to 10 wt.% of h-BN	The graphite exhibits superior tribological properties compared to hBN. Graphite along with lower contents of hBN helped to stabilize friction and wear.	[32]
Cu-graphite-SiC wt.% (graphite with 0, 5, 10, and 15 wt.%	Graphite at 5 and 10 wt.%	10 wt.% gr. and 15 wt.% SiC exhibited excellent tribological properties.	[33]
Nickel	Ag = 12.5 wt.% Baf_2/CaF_2 = 5 wt.%, Graphite with 0, 0.5, 1, 2 wt.%	Graphite (0.5 wt.%) demonstrated lowest COF at all testing temperatures except at 800°C. The composite material with Graphite (2 wt.%) exhibited lowest COF at 800°C.	[34]

TABLE 5.2 *(Continued)*

Matrix	Reinforcement	Findings	References
Ni	hBN=1.25 wt.% nano-Cu = 5 wt.%	The composite material exhibits reduction in COF from 0.48 to 0.35 when temperature changes from 25°C to 500°C.	[35]
Ni Cr 80–20 wt.%	MoS_2 Graphite	Ni Cr with 10 wt.% MoS_2 exhibits lowest.	[36]
WC-Ni-Cr composite	WS_2 = 5 wt.%	The sintered composite material demonstrated lowest COF at contact pressure of 250 MPa.	[37]
Ni	CaF_2 = 20%	Reduction in COF was observed with increase in temperature.	[38]
Fe-0.3C-2Ni based composites	WS_2 with 3, 5, 7, and 9 wt.%)	The composite material with 9 wt.% of WS_2 exhibited lowest COF.	[39]
PEEK	Micro- and nano-MoS_2, WS_2	Reduction in COF was observed with addition of micro- and nano-MoS_2, WS_2.	[40]
M50 steel	TiO_2 10 wt.% TiO_2/Graphite powder (10 wt.% TiO_2 + 5 wt.% Graphite)	Reduction in COF was observed with TiO_2/Graphite with increase in temperature.	[41]
Tin-bronze	Graphite, MoS_2 or PTFE	With PTFE-20 wt.% and Graphite-40 wt.% the composite material exhibited excellent tribological properties.	[42]
Fe–Cu 5 wt.%–Sn_3 wt.%	MoS_2 (0–3 wt.%)	Reduction in COF was observed with increase of MoS_2 content.	[43]

to solid nanocomposites, the tribological properties of liquid lubricating systems need to enhance [46–48]. The multidimensional nanoadditives as a reinforcement found to be efficient for improving the tribological properties, adhesion strength and bearing capacity of the tribofilms [49–51]. Here, the role of multidimensional reinforcement material in the formation and behavior of tribofilms in liquid lubricating systems is described. The tribological properties of lubricant with nanomaterial reinforcement are depicted in Table 5.3.

The primary intention to use solid lubricants is that in some applications liquid lubricants are ineffective like space exploration and aircraft [40]. Under such applications, liquid lubricants are affected due to its stability and design constrains. However, currently very few solid lubricants have been employed in industries to enhance the tribological properties like coatings on engine-piston assembly and antifriction bearings. The tribofilm in composite lubricants (solid-liquid), the chemical and physical and phenomena related to deformation, damage, and reaction of the materials at the interacting surfaces significantly influences the formation mechanism [62, 63]. Among these, currently the multidimensional nanoadditives are getting growing interests [64, 65].

5.4 TRIBOFILM FORMATION

Whenever the two interacting surfaces have the relative motion the friction occurs due to material molecules collision in localized contact region. This leads to conversion of mechanical energy into thermal energy and this build-up of thermal energy plays an important role in controlling friction, specifically in dry conditions. Tribo-chemical reactions occurring due to high frictional heat and stresses significantly affect the tribological properties of the tribo-system [66, 67]. The worn-out material from the surfaces gets mixed up with the plastically deformed matrix and forms a superior tribo-chemical film. As compared to dry friction, the tribofilm developed under fluid lubrication is classified into two phenomenon physical adsorption film and tribochemical film. In physical adsorption phenomenon, the "stress assistance" and "thermal effect" are absent during the film formation [68, 69] (Figure 5.3).

It has been observed that compared to pure MoS_2 coating, the composite MoS_2–TiO_2 coating exhibits excellent tribological properties. The reinforcement percentage, i.e., TiO_2 addition into MoS_2 matrix also affects the

TABLE 5.3 Tribological Characteristics of Nanolubricant

Base Matrix	Reinforcement Additive	Findings	References
Glycerol	0.05 wt.% GO	With addition of GO improvement in wear resistance is observed.	[52]
Deionized water	0.005 wt.% PEI-RGO	With addition of RGO the improvement in tribological properties has been observed.	[53]
Water-based lubricant	4 wt.% TiO_2	The addition of TiO_2 helps to reduce friction and wear rate.	[54]
PEG	1 wt.% CDs-PF6	With addition of CD improvement in wear resistance is observed.	[55]
Liquid paraffin	1 wt.% OA/CuS nanorods	With addition of OA/CuS the improvement in tribological properties has been observed.	[56]
PAO	5 wt.% MWNTs	With addition of MWNTs improvement in wear resistance is observed.	[57]
SN 500 mineral oil	MoS_2 nanosheets (0.5 wt.%)	With addition of MoS_2 nanosheets improvement in wear resistance is observed.	[58]
Rapeseed oil	Cu/PDA/CNTs (0.2 wt.%)	With addition of Cu/PDA/CNTs improvement in wear resistance is observed.	[59]
Sunflower oil	Cu/PDA/MoS_2 (0.5 wt.%)	With addition of Cu/PDA/CNTs improvement in wear resistance is observed.	[60]
Hydroisomerization base oil	0.030 mg/mL GO-GPTS-GPTS-AlOOH	The addition of GO-GPTS-GPTS-AlOOH helps to reduce friction and wear rate.	[61]

tribological properties of the composite coating. The test results showed that composite MoS_2–TiO_2 coating with 15 wt.% (weight percentage), shown excellent tribological properties among all the samples [2, 70].

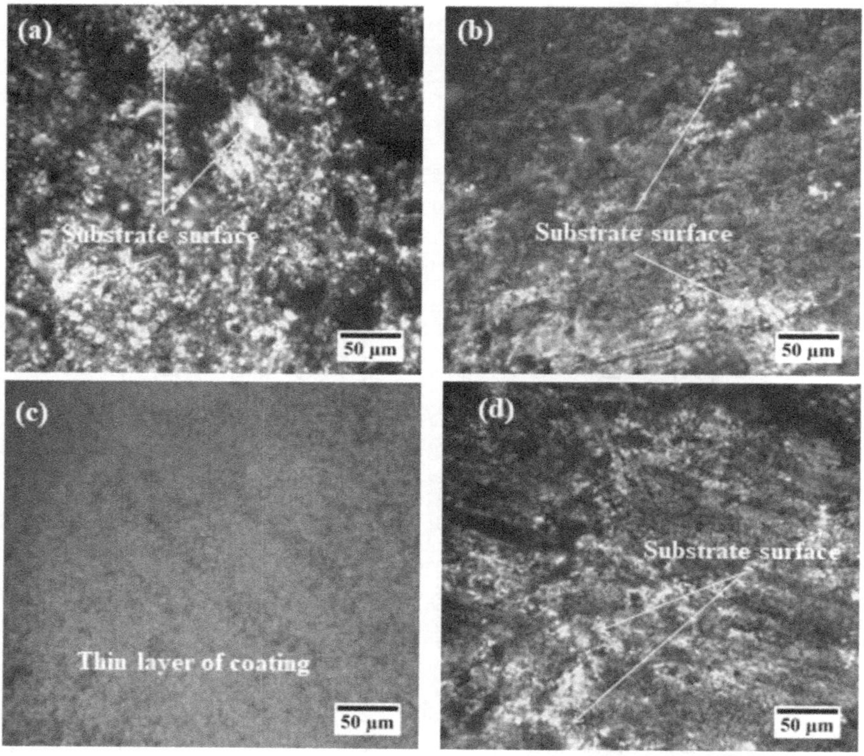

FIGURE 5.3 Specimen surface after wear test (a) pure MoS_2, and composite MoS_2–TiO_2 coating; (b) 5 wt.% TiO_2; (c) 15 wt.% TiO_2; and (d) 25 wt.% TiO_2 (with TiO_2 sample C).

5.5 FACTORS AFFECTING ON NANO-TRIBOFILMS

In the recent, researchers are attracted towards 2D nanomaterials as an additive material to improve the tribological characteristics of base lubricants. The tribological properties of 2D nanomaterials found to be affected by various parameters like substrate surface, chemical interactions between the interacting surfaces, and dispersibility of reinforcement material. The chemical bonding and van-der-Waal forces between the substrate and coatings of 2D materials affects mechanical characteristics of the coating material. The other environmental factor such as humidity reduces the adhesion strength resulting into increased frictional coefficient. The operational conditions like load, speed, and temperature also affects the performance of the tribofilm. Higher operating speed at low load initiates the tribochemical interaction between reinforcement and base matrix material results into production of a thin tribofilm. Whereas in vice-versa condition the film will rupture and fail, since the soft nanomaterials are unable to withstand the stresses induced due to friction. In such cases, the hard nanoadditives can be employed to reduce the contact areas of the interacting surfaces, optimizing the tribological properties of the film. In case of nanoadditives mixed in base lubricant the uniform dispersion of reinforcement material both in solid and liquid matrix results into improved tribological properties of the lubricant. Beside surface morphology also plays an important role while investigating the frictional and wear properties of the interacting surfaces [69].

The operating temperature also affects the tribological characteristics of the coating as shown in Figure 5.4. From Figure 5.4 it is observed that with increase in temperature the transfer of worn-out coating material from specimen surface to counter body. Due to which increase in wear rate was observed but at the same time coating material transferred onto the counter body it reduces the COF [2, 70].

5.6 CONCLUSION AND FUTURE SCOPE

There is growing interest in use of 2D nanomaterials as a reinforcement material with base solid and liquid lubricants but still there is scope for improvement in research. It is observed that the tribological behavior can be improved with the inclusion of nanoparticles as additives. Basically, to improve wear properties of material, the research need to be extended in all aspects based on economically as application point of view and also health issues. In addition, the 2D nanomaterials as additive in the solid lubricants assist to explore its use in thermoelectric coatings which possess high wear resistance, and are beneficial in fabrication of thermoelectric device. The polymer-based nanocomposites are gaining interests due to their important properties like non-toxic, higher resilience and availability. It is remarkably noted that the 2D nanoadditives plays vital role in the fabrication of high-performance tribofilms. Recent studies observed that

MXene-based tribo-systems and its derivatives are most suitable for fabrication of the ultra-wear resistant tribofilms. Still there is huge scope for research to investigate various wear mechanisms to produce highly efficient tribo-system. The novel findings of nanocomposite material will facilitate their use in various industrial applications.

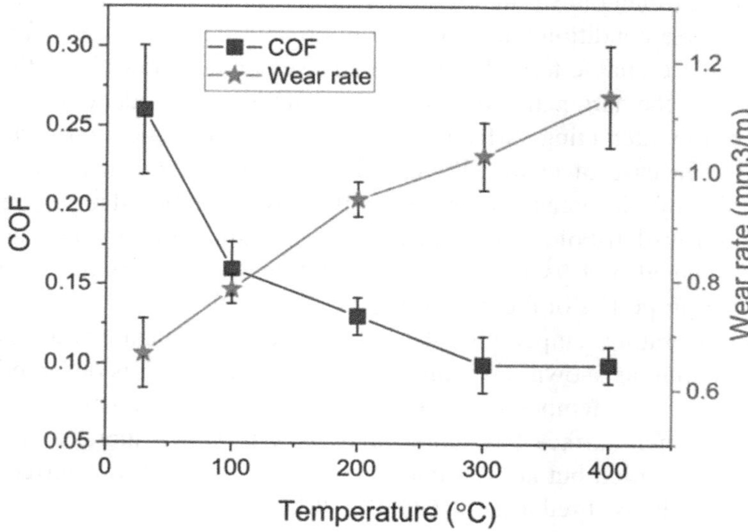

FIGURE 5.4 Influence of temperature on COF and wear rate of the composite coating.

KEYWORDS

- **graphene**
- **hexagonal boron nitride**
- **mechanical properties**
- **molybdenum disulfide**
- **nanocomposites**
- **tribological properties**
- **two-dimensional (2D)**

REFERENCES

1. Spear, J. C., Ewers, B. W., & Batteas, J. D., (2015). 2D-nanomaterials for controlling friction and wear at interfaces. *Nano Today, 10*(3), 301–314.
2. Borgaonkar, A., & Syed, I., (2020). Friction and wear behavior of composite MoS_2–TiO_2 coating material in dry sliding contact. *J. Braz. Soc. Mech. Sci. Eng., 43*, 1–13.
3. Borgaonkar, A. V., Syed, I., & Sonawane, S. H., (2020). Effects of lubrication on tribological properties coating material of composite MoS_2-TiO_2. In: Hameed, S. M. T., et al., (eds.), *Tribological Applications of Composite Materials* (pp. 267–281). Springer Singapore.
4. Li, N., Peng, J., Ong, W. J., Ma, T., Arramel, Zhang, P., et al., (2021). MXenes: An emerging platform for wearable electronics and looking beyond. *Matter, 4*(2), 377–407.
5. Guo, Y., Zhang, L., Zhao, F., Li, G., & Zhang, G., (2021). Tribological behaviors of novel epoxy nanocomposites filled with solvent-free ionic SiO_2 nanofluids. *Compos B Eng., 215*, 108751.
6. An, Y., Tian, Y., Wei, C., Tao, Y., Xi, B., Xiong, S., et al., (2021). Dealloying: An effective method for scalable fabrication of 0D, 1D, 2D, 3D materials and its application in energy storage. *Nano Today, 37*, 101094.
7. Berman, D., Erdemir, A., & Sumant, A. V., (2018). Approaches for achieving superlubricity in two-dimensional materials. *ACS Nano, 12*(3), 2122–2137.
8. Liu, S. W., Wang, H. P., Xu, Q., Ma, T. B., Yu, G., Zhang, C., & Luo, J., (2017). Robust microscale superlubricity under high contact pressure enabled by graphene-coated microsphere. *Nature Communications, 8*(1), 1–8.
9. Berman, D., Deshmukh, S. A., Sankaranarayanan, S. K., Erdemir, A., & Sumant, A. V., (2015). Macroscale superlubricity enabled by graphene nanoscroll formation. *Science, 348*(6239), 1118–1122.
10. Di Maro, M., Duraccio, D., Malucelli, G., & Faga, M. G., (2021). High density polyethylene composites containing alumina-toughened zirconia particles: Mechanical and tribological behavior. *Compos B Eng., 217*, 108892.
11. Avinash, V. B., Deepak, R., & Ismail, S., (2022). Tribological investigation of developed nano-composite MoS_2-TiO_2-ZrO_2 coating material. In: Sarfraj, A., & Vinayak, S. D., (eds.), *Tribology and Characterization of Surface Coatings* (pp. 33–47). Wiley-Scrivener, USA.
12. Fan, X., Yan, H., Cai, M., Song, S., Huang, Y., & Zhu, M., (2022). Achieving parallelly-arranged Ti_3C_2Tx in epoxy coating for anti-corrosive/wear high-efficiency protection. *Compos. B Eng., 231*, 109581.
13. Borgaonkar, A. V., & Syed, I., (2021). Computational analysis of composite MoS_2-TiO_2-ZrO_2 soft coating on tribological performance in dry sliding contact. In: Deepak, B. B. V. L., et al., (eds.), *Advanced Manufacturing Systems and Innovative Product Design: Lecture Notes in Mechanical Engineering* (pp. 978–981). Singapore Springer.
14. Borgaonkar, A. V., & Syed, I., (2020). Effect of coatings on rolling contact fatigue and tribological parameters of rolling/sliding contacts under dry/lubricated conditions: A review *Sādhanā., 45*(30), 1–16.

15. Wang, W., Xie, G. X., & Luo, J. B., (2018). Black phosphorus as a new lubricant. *Friction 6*(1), 116–142.
16. Halim, J., Kota, S., Lukatskaya, M. R., Naguib, M., Zhao, M. Q., Moon, E. J., Pitock, J., Nanda, J., May, S. J., Gogotsi, Y., et al., (2016). Synthesis and characterization of 2D molybdenum carbide (MXene). *Adv. Funct. Mater., 26*(18), 3118–3127.
17. Guo, D., Xie, G., & Luo, J., (2013). Mechanical properties of nanoparticles: Basics and applications. *Journal of Physics D: Applied Physics, 47*(1), 013001.
18. Dai, W. Y., Shao, F., Szczerbiński, J., McCaffrey, R., Zenobi, R., Jin, Y. H., Schlüter, A. D., & Zhang, W., (2016). Synthesis of a two dimensional covalent organic monolayer through dynamic imine chemistry at the air/water interface. *Angew. Chem. Int. Ed. Engl., 55*(1), 213–217.
19. Liu, L. C., Zhou, M., Jin, L., Li, L. C., Mo, Y. T., Su, G. S., Li, X., et al., (2019). Recent advances in friction and lubrication of graphene and other 2D materials: Mechanisms and applications. *Friction, 7*(3), 199–216.
20. Bertolazzi, S., Brivio, J., & Kis, A., (2011). Stretching and breaking of ultrathin MoS_2. *ACS Nano, 5*(12), 9703–9709.
21. Sha, Z. D., Pei, Q. X., Ding, Z. W., Jiang, J. W., & Zhang, Y. W., (2015). Mechanical properties and fracture behavior of single layer phosphorene at finite temperatures. *J. Phys. D: Appl. Phys. 48*(39), 395303.
22. Zhi, C. Y., Bando, Y., Tang, C. C., Kuwahara, H., & Golberg, D., (2009). Large-scale fabrication of boron nitride nanosheets and their utilization in polymeric composites with improved thermal and mechanical properties. *Adv. Mater., 21*(28), 2889–2893.
23. Pang, J. B., Mendes, R. G., Bachmatiuk, A., Zhao, L., Ta, H. Q., Gemming, T., Liu, H., et al., (2019). Applications of 2D MXenes in energy conversion and storage systems. *Chem. Soc. Rev., 48*(1), 72–133.
24. Hermosa, C., Horrocks, B. R., Martínez, J. I., Liscio, F., Gómez-Herrero, J., & Zamora, F., (2015). Mechanical and optical properties of ultralarge flakes of a metal-organic framework with molecular thickness. *Chem. Sci., 6*(4), 2553–2558.
25. Li, H., Xie, Y., Li, K., Huang, L., Huang, S., Zhao, B., et al., (2014). Microstructure and wear behavior of graphene nanosheets-reinforced zirconia coating. *Ceram Int., 40*(8), 12821–12829. Part B).
26. Borgaonkar, A. V., & Syed, I., (2020). Effect of coating material properties on the lubrication performance of rolling contacts under TEHL regime. *Australian Journal of Mechanical Engineering*, 1–8.
27. Zhang, H., Zhang, N., & Fang, F., (2020). Fabrication of high-performance nickel/graphene oxide composite coatings using ultrasonic-assisted electrodeposition. *Ultrason. Sonochem., 62*, 104858.
28. Chen, J., Yang, J., Chen, B., Shuai, L., Dong, J., & Li, C., (2016). Large-scale synthesis of NbSe2 nanosheets and their use as nanofillers for improving the tribological properties of epoxy coatings. *Surf. Coating Technol.*, 305.
29. Borgaonkar, A., Sonawane, S., & Syed, I., (2021). Fabrication and characterization of composite MoS_2–TiO_2 coating material. *Proc. IMechE, Part C: Journal of Mechanical Engineering Science*, 236(3), 1–12.
30. Borgaonkar, A., & Syed, I., (2021). Tribological behavior prediction of composite MoS2-TiO2 coating using Taguchi coupled artificial neural network approach. *Proc. IMechE, Part C: Journal of Mechanical Engineering Science*, 236(12), 1–16.

31. Sakka, M. M., Antar, Z., Elleuch, K., & Feller, J. F., (2017). Tribological response of an epoxy matrix filled with graphite and/or carbon nanotubes. *Friction, 5*(2), 171–182.
32. Chen, B., Bi, Q., Yang, J., Xia, Y., & Hao, J., (2008). Tribological properties of solid lubricants (graphite, h-BN) for Cu-based P/M friction composites. *Tribology International, 41*(12), 1145–1152.
33. Sundararajan, S., Rajadurai, J. S., & Antony, V. C., (2017). Impact of solid self-lubricant and hard ceramic reinforcements on tribological characteristics of microwave-sintered Cu hybrid composites processed by solid-state mixing technique. *Proc. Inst. Mech. Eng. Part J. J. Eng. Tribol., 231*, 523–542.
34. Zhen, J., Zhu, S., Cheng, J., Li, M., Lu, Y., Qia, Z., & Yang, J., (2017). Influence of graphite content on the dry sliding behavior of nickel alloy matrix solid lubricant composites. *Tribol. Int., 114*, 322–328.
35. Zhao, Y., Feng, K., Yao, C., Nie, P., Huang, J., & Li, Z., (2019). Microstructure and tribological properties of laser cladded self-lubricating nickel-base composite coatings containing nano-Cu and h-BN solid lubricants. *Surf. Coat. Technol., 359*, 485–494.
36. Chen, B., Zhang, J., Jia, J., Hu, W., & Gong, C., (2018). Tribological properties of solid lubricants (graphite, MoS_2) for Ni based materials. *IOP Conf. Ser. Earth Environ. Sci., 186*, 012030.
37. Yang, H. X., Wang, J., Zhang, H., Ren, J., Zan, Q., Gong, Q., & Wu, B., (2017). WC-Ni-Cr-based self-lubricating composites fabricated by pulsed electric current sintering with addition of WS2 solid lubricant. *Int. J. Refract. Met. Hard Mater., 66*, 158–162.
38. Kotkowiak, M., Piasecki, A., & Kulka, M., (2019). The influence of solid lubricant on tribological properties of sintered Ni–20%CaF_2 composite material. *Ceram. Int., 45*, 17103–17113.
39. Gupta, A., Mohan, S., Anand, A., Haq, M. I. U., Raina, A., Kumar, R., Singh, R. A., et al., (2019). Tribological behavior of Fe-C-Ni self-lubricating composites with WS2 solid lubricant. *Mater. Res. Express, 6*, 126507.
40. Zalaznik, M., Kalin, M., Novak, S., & Jakša, G., (2016). Effect of the type, size, and concentration of solid lubricants on the tribological properties of the polymer PEEK. *Wear, 364*, 31–39.
41. Ali, M. K. A., & Xianjun, H., (2019). M50 matrix sintered with nanoscale solid lubricants shows enhanced self-lubricating properties under dry sliding at different temperatures. *Tribol. Lett., 67*, 1–16.
42. Li, T., Tao, J., Cong, P., & Liu, X., (2001). Study on the tribological characteristics of solid lubricants embedded tin-bronze bearings. *J. Appl. Polym. Sci., 80*, 2394–2399.
43. Mushtaq, S., Wani, M. F., Nadeem, M., Najar, K. A., & Mursaleen, M., (2019). A study on friction and wear characteristics of Fe–Cu–Sn alloy with MoS_2 as solid lubricant under dry conditions. *Sadhana, 44*, 1–7.
44. Liu, Y., Ge, X., & Li, J., (2020). Graphene lubrication. *Appl. Mater. Today, 20*, 100662.
45. Chen, Z., Guo, L., Yan, H., Yao, H., Li, L., & Liu, Q., (2019). Amino functionalization of graphene/ graphene-like MoSe2 hybrids as lubricant additives for bismaleimide composites: Preparation, mechanical and tribological properties. *Compos. B Eng., 161*, 263–271.
46. Karthikeyan, K. M. B., Vijayanand, J., Arun, K., & Seshagiri, R. V., (2020). Thermophysical and wear properties of eco-friendly nano lubricants. *Mater Today Proc., 39*, 285–291.

47. Dai, W., Kheireddin, B., Gao, H., & Liang, H., (2016). Roles of nanoparticles in oil lubrication. *Tribol. Int., 102*, 88–98.
48. Senthilkumar, A., Abhishek, P. V., Adithyan, M., & Arjun, A., (2020). Experimental investigation of CuO/SiO_2 hybrid nano-lubricant in R600a vapor compression refrigeration system. *Mater. Today Proc., 45*, 6083–6086.
49. Mai, Y. J., Li, Y. G., Li, S. L., Zhang, L. Y., Liu, C. S., & Jie X. H., (2019). Self-lubricating Ti_3C_2 nanosheets/ copper composite coatings. *J. Alloys Compd., 770*, 1–5.
50. Li, G., Lü, W., Liu, S., Li, C., Zhou, Y., & Wang, Q., (2021). Multilayer-growth of TiAlN/WS self-lubricating composite coatings with high adhesion and their cutting performance on titanium alloy. *Compos. B Eng., 211*, 108620.
51. Tabandeh-Khorshid, M., Ajay, K., Omrani, E., Kim, C., & Rohatgi, P., (2020). Synthesis, characterization, and properties of graphene reinforced metal-matrix nano-composites. *Compos. B Eng., 183*, 107664.
52. Novikova, A. A., Burlakova, V. E., Varavka, V. N., Uflyand, I. E., Drogan, E. G., & Irkha, V. A., (2019). Influence of glycerol dispersions of graphene oxide on the friction of rough steel surfaces. *J. Mol. Liq., 284*, 1–11.
53. Liu, C., Guo, Y., & Wang, D., (2019). PEI-RGO nanosheets as a nanoadditive for enhancing the tribological properties of water-based lubricants. *Tribol. Int., 140*, 105851.
54. Wu, H., Zhao, J., Cheng, X., Xia, W., He, A., Yun, J. H., et al., (2018). Friction and wear characteristics of TiO_2 nano-additive water-based lubricant on ferritic stainless steel. *Tribol. Int., 117*, 24–38.
55. Zhao, L., Cai, T., Ye, M., Liu, D., & Liu, S., (2019). The regulation of the microstructure, luminescence and lubricity of multi-element doped carbon nanodots with alkylated diquaternary 1, 4-diazabicyclo[2.2.2]octane derived dicationic ionic liquids inserted in carbon skeleton. *Carbon, 150*, 319–333.
56. Chen, L., & Zhu, D., (2017). Preparation and tribological properties of unmodified and oleic acid-modified CuS nanorods as lubricating oil additives. *Ceram Int., 43*(5), 4246–4251.
57. Kalin, M., Kogovˇsek, J., & Remˇskar, M., (2012). Mechanisms and improvements in the friction and wear behavior using MoS_2 nanotubes as potential oil additives. *Wear, 280, 281*, 36–45.
58. Rajendhran, N., Palanisamy, S., Periyasamy, P., & Venkatachalam, R., (2018). Enhancing of the tribological characteristics of the lubricant oils using Ni-promoted MoS_2 nanosheets as nano-additives. *Tribol. Int., 118*, 314–328.
59. Wang, Z., Ren, R., Song, H., & Jia, X., (2018). Improved tribological properties of the synthesized copper/carbon nanotube nanocomposites for rapeseed oil-based additives. *Appl. Surf. Sci., 428*, 630–639.
60. Jia, X., Huang, J., Li, Y., Yang, J., & Song, H., (2019). Monodisperse Cu nanoparticles @ MoS_2 nanosheets as a lubricant additive for improved tribological properties. *Appl. Surf. Sci., 494*, 430–439.
61. Zhang, L., He, Y., Feng, S., Zhang, L., Zhang, L., Jiao, Z., et al., (2016). Preparation and tribological properties of novel boehmite/graphene oxide nano-hybrid. *Ceram Int., 42*(5), 6178–6186.
62. Borgaonkar, A., & Syed, I., (2021). Tribological investigation of composite MoS_2-TiO_2-ZrO_2 coating material by response surface methodology approach. *Journal of Tribology, 144*(3), 1–9.

63. Busch, C., (2017). Solid lubrication. *Lubricants and Lubrication* (pp. 843–880).
64. Xu, J., Chen, X., Grützmacher, P., Rosenkranz, A., Li, J., Jin, J., et al., (2019). Tribochemical behaviors of onion-like carbon films as high-performance solid lubricants with variable interfacial nanostructures. *ACS Appl. Mater Interfaces, 11*(28), 25535–25546.
65. Xu, J., Luo, T., Chen, X., Grützmacher, P., Rosenkranz, A., & Luo, J., (2021). Influence of structural evolution on sliding interface for enhancing tribological performance of onion-like carbon films via thermal annealing. *Appl. Surf. Sci., 541*, 148441.
66. Borgaonkar, A. V., & Potdar, S. B., (2022). Frictional and wear behavior of epoxy resin based nano-composite in dry sliding contact. In: Jena, H., Katiyar, J. K., & Pattnaik, A., (eds.), *Tribology of Polymer and Polymer Composite for Industry 4.0* (pp. 1–13). Springer Singapore.
67. Borgaonkar, A. V., & Potdar, S. B., (2021). CNTs as new emerging lubricant additives for enhancing energy efficiency. In: Haghi, A. K., et al., (eds.), *Carbon Nanotubes for Energy and Environmental Applications.* Apple Academic Press, CRC Press, Taylor and Francis group, Canada. ISBN: 9781774637173. Proof correction made on 16th September 2021, Status-in Press.
68. Li, Z., Dolocan, A., Morales-Collazo, O., Sadowski, J. T., Celio, H., Chrostowski, R., Brennecke, J. F., & Mangolini, F., (2020). Lubrication mechanism of phosphonium phosphate ionic liquid in nanoscale single-asperity sliding contacts. *Advanced Materials Interfaces, 7*(17), 2000426.
69. Tang, H., Sun, J., He, J., & Wu, P., (2021). Research progress of interface conditions and tribological reactions: A review. *J. Ind. Eng. Chem., 94*, 105–121.
70. Borgaonkar, A. V., & Syed, I., (2020). Effect of temperature on the tribological performance of MoS_2-TiO_2 coating material. In: Voruganti, H., Kumar, K., Krishna, P., & Jin, X., (eds.), *Advances in Applied Mechanical Engineering: Lecture Notes in Mechanical Engineering* (pp. 611–618). Singapore Springer.

PART II

Energy Materials and Structures

CHAPTER 6

Conjugated Polymers as Active Layers in Organic Solar Cells

LAURA CROCIANI

Institute of Condensed Matter Chemistry and Technologies for Energy (ICMATE), National Council of Research (CNR), Padova, Italy

ABSTRACT

Organic solar cells are systems producing electricity from sunlight by the photovoltaic effect where light absorption and charge transport occur through small organic molecules or conductive organic polymers. In particular, their performance is highly affected by the design of conjugated polymers (CPs) containing alternated π electron abundant and π electron deficient units (donor-acceptor-donor D-A-D) acting as active layers in organic photovoltaics (OPVs) which thanks to low energy payback time (EPBT), result to be a feasible option for energy market. The synthesis and application of suitable D-A-D fashion conjugated polymers is therefore fundamental to increase the power conversion efficiencies (PCE) of active layers, boosting the EPBT and therefore, increasing OPVs participation in energy market.

6.1 INTRODUCTION

The photovoltaic technology is based on the conversion of sunlight energy to electrical energy by the photovoltaic effect, first published by Becquerel in 1839, in which free electron generated by the absorption of photon by

New Advances in Materials Technologies. Hossein Hariri Asli, Ali Pourhashemi, Ann Rose Abraham, & A. K. Haghi (Eds.)

the semiconductor can be channeled into an electrical current [1]. The potential and importance of the photovoltaic technology is revealed by a robust and continuous growth offering promises of renewable energy. Beyond the first-generation technology based on crystalline silicon and the second one based on thin layers of semiconductors materials (cadmium telluride, copper, indium diselenide and copper, indium, gallium selenide) new photovoltaic technologies such as concentrator photovoltaic, organics, quantum dot cells, organic-inorganic hybrid and other technologies are emerging as a third generation being however not yet commercialized at large scale [2].

In particular, the organic solar cells based on organic semiconductors consisting of polymers or small molecules present various advantages such as low cost materials and processes, efficiency under dim light or indoor light, shorter EPBT (less than 1 year) [3]; moreover new applications can be envisioned (flexible, low weight, colorful…) as well as large area panels can be produced by printing technologies on both rigid and flexible substrates [4].

Most organic photovoltaic cells are polymer solar cells whose interest greatly increased when the bulk heterojunction (BHJ) solar cell was discovered in the early 1990s (Figure 6.1) [5].

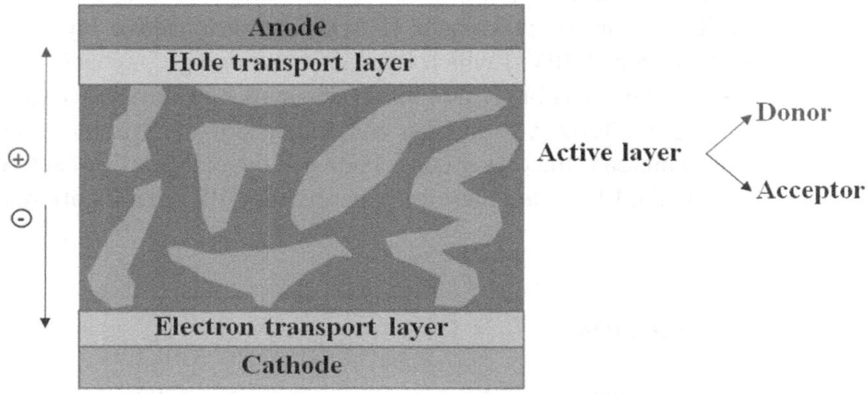

FIGURE 6.1 Bulk heterojunction organic solar cells scheme.

Typically, in the BHJ architecture a conjugated polymer (CP) serving as a donor is blended with an electron acceptor to form bicontinuous interpenetrating networks at the nanoscale, enabling a sufficient photoinduced charge separation process [6, 7]. The principle of a BHJ solar cell is shown in Figure 6.2.

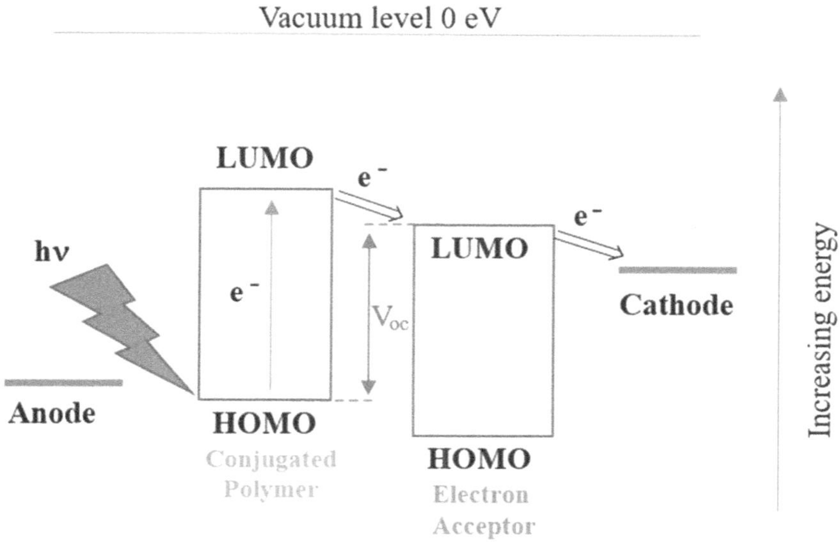

FIGURE 6.2 Energy diagram of a polymer solar cell photoenergy conversion [V_{oc}: open-circuit voltage].

The incident photons pass through the transparent anode and are absorbed by the active layer. Upon the absorption of light in the electron donor material, an exciton is formed [8]: actually, if a photon has an energy larger than the bandgap of the polymer, an electron will be excited from the highest occupied molecular orbital (HOMO) to the lowest unoccupied molecular orbital (LUMO), leaving a positive charge on the HOMO. Once the exciton reaches the interface between donor and acceptor, it is thermodynamically more favorable for the electron to be situated in the LUMO of the electron acceptor material lying below the LUMO of the donor and for the hole to remain in the HOMO of the electron donor material. Such an intermolecular charge transfer (CT) state must evolve to free charge carriers upon dissociation of the exciton minimizing charge recombination in the organic cells. As charge carriers are generated, electrons are transported to the cathode and the holes to the anode. Hence, the active layer has to provide large interfacial area and continuous pathways for the separated charges to reach their respective electrodes.

Besides the morphological control of the BHJ layer, and interface engineering for efficient charge collection by the electrodes [9–15] a big deal of the research concerns the development of low-band gap CPs with strong absorptions in the visible and near-infrared (NIR) region-in order to cover an additional part of the solar emission spectrum and thus harvest more incident photons [16].

Polymers with band gaps above 2 eV only absorb in the ultraviolet (UV) and the green part of the visible range. Radiation at long wavelengths (>600 nm) passes through the solar cell and makes little contribution to the photocurrent. Since a critical factor for improving the efficiency of solar cells is to match the photon flux spectrum from the sun with the absorption of the donor/acceptor blend [17], the design of new materials that absorb at higher wavelengths is an ongoing issue for synthetic chemists [18].

In the "band gap engineering" field [19] besides attaching electron-donating or electron-accepting substituents as side groups of the polymers, band-gap control can be achieved by alternating sequence of electron-donating and electron-accepting aromatic units within a chromophore on the basis of the D-A-D concept.

In recent years, special attention has been paid to chromophores with a donor–acceptor–donor (D–A–D) structure because of the electronic coupling between the donor and the acceptor through a π-conjugated bridge. The charge resonance is a characteristic feature of these systems due to a push–pull mechanism [20].

It is interesting to observe that the D-A-D concept is at the base of many copolymers containing not a proper structure of repeating donor-acceptor-donor units but more precisely a D-A-D segment, like in the polyfluorene species [21; such system, which can be regarded more properly as D'-A-D copolymers, have been largely studied and reviewed [22–24]: on the contrary in this chapter, we intend to present those species consisting of strictly precise D-A-D repeating units reported in the literature and their performance in OPVs.

6.2 D-A-D CONJUGATED POLYMERS

D-A-D conjugated polymers have been synthesized and studied mostly in the first decade of 2000's using thiophene derivatives and pyrrole as donor group and different acceptor groups.

Pyrrole was used by Baran et al. [25] who prepared a novel polypyrrole derivative with 2-dodecyl benzotriazole in the main chain which was shown to be both p- and n-dopable by cyclic voltammetry and spectroelectrochemical experiments.

In order to obtain large scale processable polymer, chemical polymerization was performed in the presence of $FeCl_3$ (Scheme 6.1).

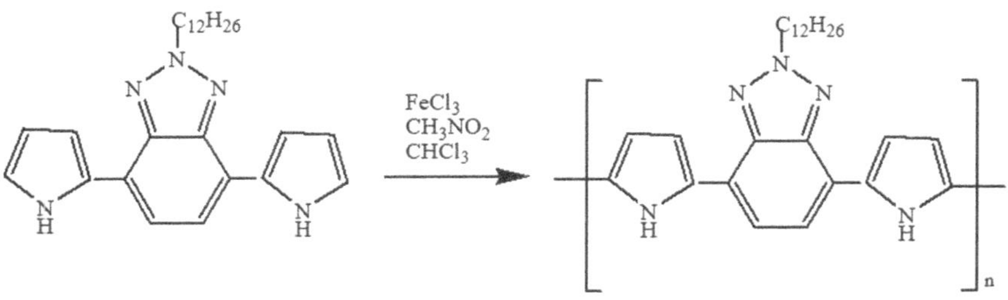

SCHEME 6.1 Synthesis of polymer PPyBT.

The dodecyl group on the benzotriazole unit enables solubility in common organic solvents and the polymer PPyBT (M_n = 11,900 and M_w = 30,600) was dissolved in chloroform and mixed with 1-(3-methoxycarbonyl)propyl-1,1-phenyl-(6,6)C61 (PCBM) as acceptor to form blends as active layers of photovoltaic devices: there is enough energy difference (ca 0.3–0.4 eV) between the LUMO of the donor and the acceptor for a CT taking place from PCBM.

Even if PPyBT:PCBM blend can generate electrons almost over the whole range of the absorption spectrum of pristine polymer as evidenced by photocurrent measurements, the BHJ devices under white light illumination (100 mW/cm^2) showed short circuit current densities (J_{sc}) of 0.3 and 0.43 mA/cm^2, an open-circuit voltage (V_{oc}) of 0.15 V and fill factors (FFs) of 0.26 and 0.27, respectively. The low rectification ratios can be attributed to the rather poor film formation of the active layer.

Hence, Baran et al. moved to a thiophene derivative as donor group [26] and by combining it with benzotriazole they produced the poly(2-dodecyl-4,7-bis(4-hexylthiophen-2-yl)-2H-benzo[d][1,2,3]triazole) (PHTBT) (Scheme 6.2), a polymer soluble in common organic solvents which revealed reversible both p and n doping properties.

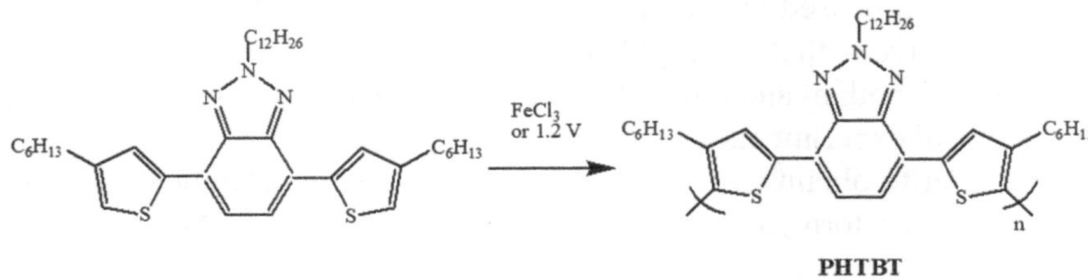

SCHEME 6.2 Synthesis of polymer PHTBT.

The polymer was used as donor and mixed with PCBM to form blends as active layers for BHJ photovoltaic cells with ITO/PEDOT:PSS/PHTBT: PCBM/Al (100 nm) configuration [ITO, indium tin oxide; PEDOT, poly(3,4-ethylenedioxythiophene); PSS, poly-(styrenesulfonate)], whose energy diagram constructed from the EVS-estimated energy levels (EVS= electrochemical voltage spectroscopy) of PHTBT was calculated (*see* Figure 11) [26].

Analyzing the photovoltaic performance of a series of devices fabricated with different PCBM loadings it results that the J_{sc} increases with increasing PCBM content in the devices and reaches a maximum value of 2.35 mA/cm^2 when the blend ratio is 1:3; moreover the incident photon to current efficiency spectrum, used to get information on the number of photons that contributes to charge generation in a solar cell, spans from 350 to 900 nm and has a maximum at 450 nm with a peak value of 30%. The lower J_{sc} values compared to poly(3-hexylthiophene) (P3HT):PCBM devices [27] may stem from a fast recombination of the separated charges, which is a common problem in polymer:fullerene solar cells [28].

Nevertheless, thiophene derivatives represent a relevant donor group because also of the opportunity of introducing alkyl chains which impart better solubility and strengthen electron donor properties.

Other thiophene derivatives have been copolymerized with different acceptor monomer such as thieno(3,4-b)pyrazine [17, 29] and diketopyrrolo-pyrrole [30].

Campos et al. [17] attached octyl groups to the thiophene ring and produced the polymer poly[5,7-bis-(3-octylthiophen-2-yl)thieno{3,4-bpyrazine] (PB3OTP) (Scheme 6.3).

The photovoltaic devices were prepared by spin coating the blend from a solution in chlorobenzene onto a PEDOT/PSS covered glass-ITO substrate to achieve a layered structure of glass-ITO/PEDOT-PSS/active layer/LiF/Al. The metal electrodes were deposited under vacuum. The active layer was composed of a varied weight-to-weight (w/w, 1:1, 1:2, 1:3, 1:4) mixture of PB3OTP and the soluble fullerene derivative PCBM. Increasing the concentration of PCBM in the blend has the effect of lowering J_{sc} and the FF, while the V_{oc} drops to a constant value after the addition of more than 50% PCBM. The J_{sc} and the FF may be affected by the increased resistance in the blend or an increased charge recombination due to the excessive presence of PCBM in close proximity with the polymer. The device that gave the best results contained a ratio of 1:1 PB3OTP/PCBM in the active layer with a J_{sc} of 1.0 mA/cm^2, a V_{oc} of 0.220 V, and a FF of 0.394. The active layer in such device absorbs photons to generate current from 300 to 900 nm, with a peak in the red region at 660 nm but modifying further the thiophene and thienopyrazine ring a photoresponse up to 1 μm could be obtained by Wienk et al. [29] who synthesized PBEHTT and PTBEHT (Figure 6.3) from the corresponding dibrominated monomers via a condensation polymerization using bis(1,5-cyclooctadiene)nickel(0).

SCHEME 6.3 Synthesis of polymer PB3OTP.

Despite their high molecular weights [M_n = 23,650 and M_w = 97,000 (polydispersion index, PDI=4.1) for PBEHTT, and M_n = 52,000 and M_w = 1,60,000 (PDI=3.1) for PTBEHT], both polymers are soluble in most common organic solvents (e.g., tetrahydrofuran, chloroform, toluene, chlorobenzene) and can be easily processed into thin films.

It is interesting to observe that the optical band gap of PTBEHT (1.20 eV) is slightly lower than that of PBEHTT (1.28 eV) because of an improved conjugation, mostly caused by a more coplanar structure of the main chain in PTBEHT due to the absence of bulky substituents compared to PBEHTT. For PBEHTT and PTBEHT the energy of the charge separated state with PCBM, estimated (in a first approximation) from the difference in oxidation and reduction potentials, is 0.83 and 1.01 eV, respectively, and therefore below the optical band gaps of the pure polymers.

Spin coating mixtures of either polymer with PCBM onto glass substrates covered with ITO and a 60 nm film of PEDOT:PSS and vacuum deposition of LiF (1 nm) and Al (150 nm) as back contacts produced photovoltaic cells with cell areas of 0.10 and 0.15 cm^2. For both polymers, a 1:4 weight ratio of polymer:PCBM gave the best performance in terms of power conversion efficiency.

In particular, the PTBEHT:PCBM cells exhibit V_{oc} = 0.56 V, J_{sc} = 3.1 mA/cm^2 and FF = 0.58 under white light illumination (75 mW/cm^2), resulting in significantly improved energy conversion efficiency of 1.1% compared to PBEHTT:PCBM. The increased V_{oc} arises from to the higher oxidation potential of PTBEHT and V_{oc} is close to the value that can be derived from the redox levels after correction for contact losses (0.61 V). Interestingly, the V_{oc} value of PTBEHT:PCBM cells (0.56 V) is close to the value observed in state-of-the-art P3HT:PCBM solar cells (V_{oc} = 0.61–0.63 V) [31, 32]. For PTBEHT:PCBM blends the spectral response extents to 1 μm, with external quantum efficiency (EQE) exceeding 15% in the 700–900 nm region.

The improved J_{sc} of PTBEHT:PCBM compared to PBEHTT:PCBM is somewhat surprising because even if the driving force for photoinduced electron transfer to PCBM is small [the energy of the charge separated state (~1.01 eV) is close to the band gap energy (1.20 eV)] PTBEHT:PCBM has an improved *J*sc compared to PBEHTT:PCBM. Tapping mode atomic force microscopy (AFM) on both blends revealed a rough surface and coarse phase separation for PBEHTT:PCBM layers compared to a rather smooth surface and more intimate mixing for PTBEHT:PCBM. The improved miscibility of PTBEHT and PCBM compared to PBEHTT and PCBM explains, at least in part, the increased photocurrent seen in Figure 6.3(a) of Ref. [29] taking into account the nanometer range exciton diffusion lengths in conjugated polymers.

In order to improve the cell performance Wienk et al. developed a new narrow-bandgap polymer: poly[3,6-bis-(40-dodecyl-[2,20]bithiophenyl-5-yl)-2,5-bis-(2-ethyl-hexyl)-2,5-dihydropyrrolo[3,4-]pyrrole-1,4-dione] (pBBTDPP2) (Figure 6.4) [30], evidencing that the morphology of the mixed film, and particularly the crystallinity of the materials, are key parameters for optimization of BHJ solar cells as well as the bandgap of the materials.

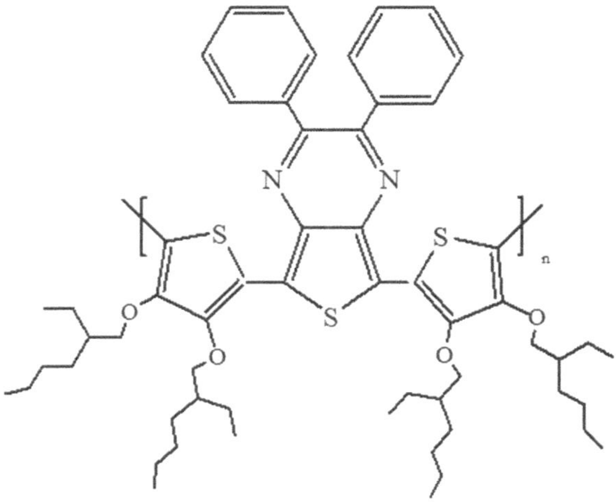

PBEHTT

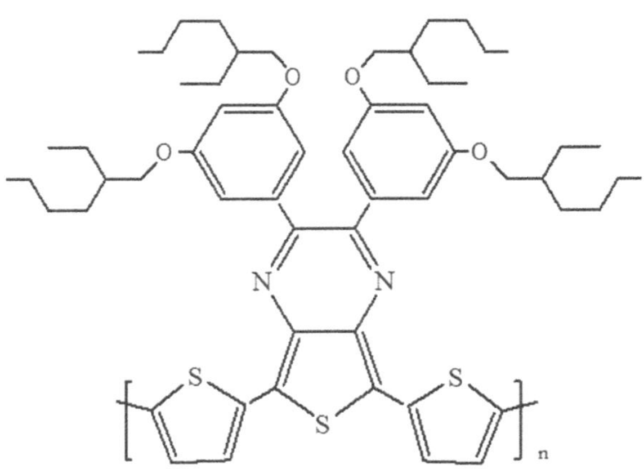

PTBEHT

FIGURE 6.3 Polymer PBEHTT and PTBEHT.

The new polymer combines electron-rich quaterthiophene (BBT) segments with electron-poor diketo pyrrolo-pyrrole (DPP) units to lower the optical bandgap to 1.4 eV in thin films. The polymer has significant molecular weights [M_n = 20,000 and M_w = 67,000 (DPI = 3.35)].

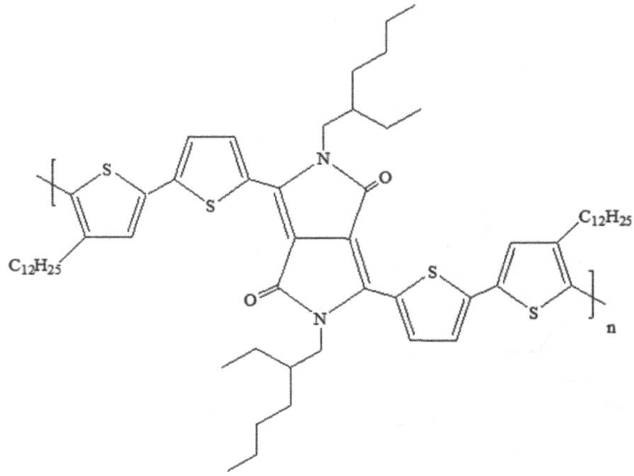

FIGURE 6.4 Polymer pBBTDPP2.

The polymer is soluble in chloroform: the fast evaporation of chloroform does not allow the pBBTDPP2 to aggregate or crystallize during spin-casting and an amorphous chain arrangement is maintained even if a certain degree of crystallization can be achieved by heating the film to 130°C, as inferred from a red-shifted absorption onset. On the contrary, the polymer is poorly soluble in ortho-dichlorobenzene (ODCB) and to solve this problem and make semi-crystalline thin films, a polymer dispersion rather than a solution can be used. The best solution to deposit mixed films with a large degree of semi-crystalline polymer was by using a chloroform:ODCB mixture (4:1) where the polymer and fullerene completely dissolve at room temperature, allowing easy and reproducible processing. Therefore, photovoltaic devices were made by coating the appropriate pBBTDPP2: fullerene film on patterned indium-oxide-coated glass substrates covered with 60 nm of PEDOT:PSS, using lithium fluoride (1 nm) and aluminum (100 nm) as metal electrode. The best results were achieved using a 1:2 pBBTDPP2:PCBM ratio, a photoactive-layer thickness in the range between 100 and 120 nm and processing film from chloroform:ODCB solvent mixtures; in this way an efficiency of 3.2% was obtained, with the highest FF of 0.54, compared to 0.47 and 0.41 for the dispersion- and chloroform-processed cells, respectively and the highest current density of 9.6 ma/cm^2 under AM1.5G illumination. The higher performance of cells processed from chloroform:ODCB solvent mixture is related to the amorphous nature of the as-deposited layer from chloroform and the unfavorable morphology, evidenced by AFM measurements, reducing the generation of charges and thereby the effectiveness of these layers [33].

Once set the optimal conditions, it was used as acceptor of the active layer [6,6]-phenyl C71 butyric acid methyl ester [70]PCBM}: in devices, this leads to a significantly higher quantum efficiency at wavelengths below 600 nm, with a higher short-circuit current density of 11.3 mA/cm^2 under AM1.5G light. Combined with a FF of 0.58 and a V_{oc} of 0.61 V, this yields a power conversion efficiency of 4.0%.

It is interesting to observe that the polymers with a delocalized π−electron system that comprises alternating electron-rich and electron-deficient repeating units can be valid as well as n-type material and be used in alternative to the fullerene based common acceptor materials in OPV. Actually, C61-PCBM has a number of drawbacks including costly synthetic procedure, poor photochemical stability and poor absorption properties in the visible spectrum [34]. Therefore, the development of alternative polymer-based acceptor materials and the construction of an all-polymer OPV represent a more and more relevant scientific challenge. In the case oh D-A-D polymers it is known the preparation of cyclopentadithiophene-naphthalenediimide polymers (Scheme 6.4) with high molecular weight (M_n = 22.000–66.000) by Li et al. [35] which were blended with P3HT as a p-type material to fabricate OPV devices.

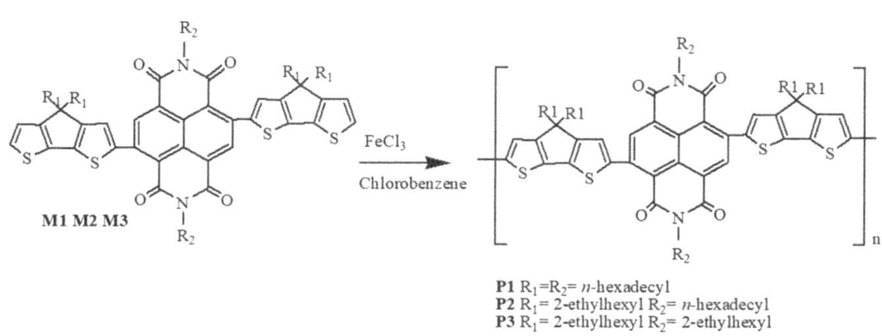

SCHEME 6.4 Synthesis of polymer P1, P2, and P3.

Photovoltaic devices of P1–P3 were fabricated on glass substrates with an (ITO)/PEDOT:PSS anode while the active layers were spinning cast from chlorobenzene solutions of the polymer mixed with P3HT (30 mg mL^{-1}) using different amounts of polymers.

The best OPV performance was obtained for P2 containing a 1:1 blend of P2:P3HT where a PCE of 0.25% was measured (0.15% PCE for P3 and 0.04 for P1): not satisfactory results have been explained in terms of undesirable morphology of the device. Large domains were observed for all film prepared from a solution with P3HT with diameters in the size ranging from 300 nm for P3 based device to 80 nm for P2 based device. It appeared that increasing the branched alkyl chain ratios decreased the domain size, reducing the film surface became and thus enhancing OPV performance.

6.3 CONCLUSIONS

In this chapter, we have focused the attention to the narrow field of pure D-A-D polymers that have been used as effective part of the active layer of BHJ solar cell either as donor or acceptor species. The intense synthetic efforts have produced low band-gap polymers but they have not been unfortunately awarded by exciting OPV performance: further work needs to be done to match suitable donor and acceptor group in the polymers and at the same time construct suitable device to optimize the photovoltaic performance.

KEYWORDS

- **chemical synthesis**
- **conjugated polymers**
- **electron**
- **organic solar cell**
- **photovoltaic technology**
- **polymers**
- **thiophene**

REFERENCES

1. Parida, B., Iniyan, S., & Goic, R., (2011). A review of solar photovoltaic technologies. *Renewable & Sustainable Energy Reviews, 15*(3), 1625–1636.

2. Bagnall, D. M., & Boreland, M., (2008). Photovoltaic technologies. *Energy Policy, 36*(12), 4390–4396.
3. Hoppe, H., & Sariciftci, N. S., (2004). Organic solar cells: An overview. *Journal of Materials Research, 19*(7), 1924–1945.
4. Hou, W. J., Xiao, Y. M., Han, G. Y., & Lin, J. Y., (2019). The applications of polymers in solar cells: A review. *Polymers, 11*(1).
5. Yu, G., Gao, J., Hummelen, J. C., Wudl, F., & Heeger, A. J., (1995). Polymer photovoltaic cells - enhanced efficiencies via a network of internal donor-acceptor heterojunctions. *Science, 270*(5243), 1789–1791.
6. Loi, M. A., Toffanin, S., Muccini, M., Forster, M., Scherf, U., & Scharber, M., (2007). Charge transfer excitons in bulk heterojunctions of a polyfluorene copolymer and a fullerene derivative. *Adv. Funct. Mater., 17*(13), 2111–2116.
7. Henson, Z. B., Mullen, K., & Bazan, G. C., (2012). Design strategies for organic semiconductors beyond the molecular formula. *Nat. Chem., 4*(9), 699–704.
8. Wadsworth, A., Hamid, Z., Kosco, J., Gasparini, N., & McCulloch, I., (2020). The bulk heterojunction in organic photovoltaic, photodetector, and photocatalytic applications. *Advanced Materials, 32*(38).
9. You, J. B., Li, X. H., Xie, F. X., Sha, W. E. I., Kwong, J. H. W., Li, G., Choy, W. C. H., & Yang, Y., (2012). Surface plasmon and scattering-enhanced low-bandgap polymer solar cell by a metal grating back electrode. *Advanced Energy Materials, 2*(10), 1203–1207.
10. Dou, L. T., Chen, C. C., Yoshimura, K., Ohya, K., Chang, W. H., Gao, J., Liu, Y. S., et al., (2013). Synthesis of 5H-dithieno 3,2-b:2,'3 '-d pyran as an electron-rich building block for donor-acceptor type low-bandgap polymers. *Macromolecules, 46*(9), 3384–3390.
11. Li, X. H., Choy, W. C. H., Huo, L. J., Xie, F. X., Sha, W. E. I., Ding, B. F., Guo, X., et al., (2012). Dual plasmonic nanostructures for high performance inverted organic solar cells. *Advanced Materials, 24*(22), 3046–3052.
12. He, Z. C., Zhong, C. M., Su, S. J., Xu, M., Wu, H. B., & Cao, Y., (2012). Enhanced power-conversion efficiency in polymer solar cells using an inverted device structure. *Nature Photonics, 6*(9), 591–595.
13. He, Z. C., Zhong, C. M., Huang, X., Wong, W. Y., Wu, H. B., Chen, L. W., Su, S. J., & Cao, Y., (2011). Simultaneous enhancement of open-circuit voltage, short-circuit current density, and fill factor in polymer solar cells. *Advanced Materials, 23*(40), 4636-+.
14. Amb, C. M., Chen, S., Graham, K. R., Subbiah, J., Small, C. E., So, F., & Reynolds, J. R., (2011). Dithienogermole as a fused electron donor in bulk heterojunction solar cells. *Journal of the American Chemical Society, 133*(26), 10062–10065.
15. Price, S. C., Stuart, A. C., Yang, L. Q., Zhou, H. X., & You, W., (2011). Fluorine substituted conjugated polymer of medium band gap yields 7% efficiency in polymer-fullerene solar cells. *Journal of the American Chemical Society, 133*(12), 4625–4631.
16. Steinberger, S., Mishra, A., Reinold, E., Mena-Osteritz, E., Muller, H., Uhrich, C., Pfeiffer, M., & Bauerle, P., (2012). Synthesis and characterizations of red/near-IR absorbing A-D-A-D-A-type oligothiophenes containing thienothiadiazole and thieno-pyrazine central units. *Journal of Materials Chemistry, 22*(6), 2701–2712.

17. Campos, L. M., Tontcheva, A., Gunes, S., Sonmez, G., Neugebauer, H., Sariciftci, N. S., & Wudl, F., (2005). Extended photocurrent spectrum of a low band gap polymer in a bulk heterojunction solar cell. *Chemistry of Materials*, *17*(16), 4031–4033.
18. Colladet, K., Fourier, S., Cleij, T. J., Lutsen, L., Gelan, J., Vanderzande, D., Nguyen, L. H., et al., (2007). Low band gap donor-acceptor conjugated polymers toward organic solar cells applications. *Macromolecules*, *40*(1), 65–72.
19. Beaujuge, P. M., Amb, C. M., & Reynolds, J. R., (2010). Spectral engineering in pi-conjugated polymers with intramolecular donor-acceptor interactions. *Accounts of Chemical Research*, *43*(11), 1396–1407.
20. Painelli, A., & Terenziani, F., (2000). Optical spectra of push-pull chromophores in solution: A simple model. *J. Phys. Chem. A*, *104*(47), 11041–11048.
21. Inganas, O., Zhang, F. L., & Andersson, M. R., (2009). Alternating polyfluorenes collect solar light in polymer photovoltaics. *Accounts of Chemical Research*, *42*(11), 1731–1739.
22. Zhao, Y., Zhang, S., Bi, F. Z., Tian, Y., Gu, C. T., Li, Y. H., & Bao, X. C., (2022). Progress review of asymmetric polymers for organic solar cells. *J. Mater. Chem. C*, *10*(25), 9419–9440.
23. Gopalakrishnan, V., Balaji, V. D., & Dangate, M. S., (2022). Review-conjugated polymer photovoltaic materials: Performance and applications of organic semiconductors in photovoltaics. *ECS Journal of Solid State Science and Technology*, *11* (3).
24. Pati, P. B., (2016). Benzazole (B, N, O, S, Se and Te) based D-A-D type oligomers: Switch from electropolymerization to structural aspect. *Organic Electronics*, *38*, 97–106.
25. Baran, D., Balan, A., Esteban, B. M., Neugebauer, H., Sariciftci, N. S., & Toppare, L., (2010). Spectroelectrochemical and photovoltaic characterization of a solution-processable n-and-p type dopable pyrrole-bearing conjugated polymer. *Macromolecular Chemistry and Physics*, *211*(24), 2602–2610.
26. Baran, D., Balan, A., Celebi, S., Esteban, B. M., Neugebauer, H., Sariciftci, N. S., & Toppare, L., (2010). Processable multipurpose conjugated polymer for electrochromic and photovoltaic applications. *Chemistry of Materials*, *22*(9), 2978–2987.
27. Park, S. H., Roy, A., Beaupre, S., Cho, S., Coates, N., Moon, J. S., Moses, D., et al., (2009). Bulk heterojunction solar cells with internal quantum efficiency approaching 100%. *Nature Photonics*, *3*(5), 297-U5.
28. Gunes, S., Neugebauer, H., & Sariciftci, N. S., (2007). Conjugated polymer-based organic solar cells. *Chemical Reviews*, *107*(4), 1324–1338.
29. Wienk, M. M., Turbiez, M. G. R., Struijk, M. P., Fonrodona, M., & Janssen, R. A. J., (2006). Low-band gap poly(di-2-thienylthienopyrazine): Fullerene solar cells. *Applied Physics Letters*, *88*(15).
30. Wienk, M. M., Turbiez, M., Gilot, J., & Janssen, R. A. J., (2008). Narrow-bandgap diketo-pyrrolo-pyrrole polymer solar cells: The effect of processing on the performance. *Advanced Materials*, *20* (13), 2556-+.
31. Shrotriya, V., Li, G., Yao, Y., Moriarty, T., Emery, K., & Yang, Y., (2006). Accurate measurement and characterization of organic solar cells. *Adv. Funct. Mater.*, *16*(15), 2016–2023.

32. Ma, W. L., Yang, C. Y., Gong, X., Lee, K., & Heeger, A. J., (2005). Thermally stable, efficient polymer solar cells with nanoscale control of the interpenetrating network morphology. *Adv. Funct. Mater.*, *15*(10), 1617–1622.
33. Hoppe, H., & Sariciftci, N. S., (2006). Morphology of polymer/fullerene bulk heterojunction solar cells. *Journal of Materials Chemistry*, *16*(1), 45–61.
34. Han, S. H., Kim, G. M., & Oh, S. Y., (2015). Synthesis and characteristics of fullerene derivatives with hexyl perylene moieties as N-type materials in organic solar cells. *Journal of Nanoscience and Nanotechnology*, *15*(7), 5446–5449.
35. Li, C. H., Kettle, J., & Horie, M., (2014). Cyclopentadithiophene-naphthalenediimide polymers; synthesis, characterization, and n-type semiconducting properties in field-effect transistors and photovoltaic devices. *Materials Chemistry and Physics*, *144*(3), 519–528.

CHAPTER 7

Carbon Nanotubes for Hydrogen Storage Applications

MAMATHA SUSAN PUNNOOSE[1] and BEENA MATHEW[2]

[1]*Bishop Chulaparambil Memorial College, Kottayam, Kerala, India*

[2]*School of Chemical Sciences, Mahatma Gandhi University, Kottayam, Kerala, India*

ABSTRACT

Recently, nanostructured carbon materials, especially carbon nanotubes (CNTs), is of stupendous interest due to its high potential for storing large volumes of hydrogen. The high surface area to volume ratio along with exceptional optical, mechanical, electronic, magnetic, and surface properties are the key attributes of this new class of nanomaterials for energy storage. This chapter represents the structural description of CNTs and its application as convenient adsorbent material for atomic or molecular hydrogen storage systems. The synergistic effects of nanotubes doped with metal or complex hydrides for improving the kinetics and thermodynamics of hydrogen reaction are also discussed.

7.1 INTRODUCTION

The severe rise in threats of pollution and global warming associated with combustion of fossil fuels is a major concern in the present scenario [1]. The scarcity of petroleum-related fuels have directed attention towards the exploitation of inexhaustible sources of energy [2]. Hydrogen is of

New Advances in Materials Technologies. Hossein Hariri Asli, Ali Pourhashemi, Ann Rose Abraham, & A. K. Haghi (Eds.)

tremendous interest because of its clean, sustainable, and renewable energy option [3]. Hydrogen energy systems are envisaged to progressively replace the existing exhaustible sources of fuels with a significant environmental impact such as reduction in emission of greenhouse gases and global dependence on fossil fuels. It also increases the efficacy of energy conversion process for both internal combustion engines and proton exchange membrane fuel cells [4, 5]. It acts as a clean fuel in power transport vehicles and portable electronic devices like mobile phones, chargeable or dischargeable electrolytic cells of high energy density [6, 7]. Hydrogen is an ideal alternative fuel for various energy converters due to its clean-burning nature, light weight, zero pollutant emission, high conversion efficiency and abundant production from renewable resources [8, 9]. In spite of the various remarkable advantages of hydrogen, it also exhibits much limitation in its large-scale utilization because of several substantial constrains in the development of hydrogen economy including its high cost of production and storage characteristics. Its explosive nature and large volume furnish hindrances for the transportation and storage purposes. These troubles acted as great obstacles for the scientific world for the past few decades [10, 11].

Carbon nanotubes (CNTs) can be used as convenient adsorbent materials that could form the basis of viable hydrogen storage systems because of its high mechanical strength, thermal stability and electronic properties. The catalytically active metal nanoparticles can be executed in the cavities or on the external walls of the CNTs [12]. This chapter involves major accomplishments in the research of CNTs in connection with their possible applications of hydrogen storage. Large-scale production and purification of CNTs with remarkable improvement of hydrogen storage capacity represent significant technological challenges in the years to come [8, 10].

7.2 CARON NANOTUBES (CNTS)

CNTs are one among the allotropes of carbon, that possess a unique nanotubular like structure of inner hollow rolled graphene planes with a high length to diameter ratio. CNTs have various exciting properties such as high aspect ratio, very light weight, high thermal conductivity, enormous strength and remarkable electronic properties ranging from metallic to semiconducting [13]. The CNTs were first discovered by Iijima in 1991 using high resolution electron microscopy (HREM) has prompted intense experimental and theoretical studies on CNTs [14].

Depending on the number of graphene layers that are rolled CNTs are classified as single-walled carbon nanotubes (SWCNTs) and multiwalled carbon nanotubes (MWCNTs). SWCNTs are graphene sheets seamlessly rolled up to form hollow cylinders. While MWCTs can be viewed as a concentric arrangement of SWCNTs, i.e., consisting of multiple layers of graphene rolled up seamlessly into a tube shape [13, 15] (Figure 7.1).

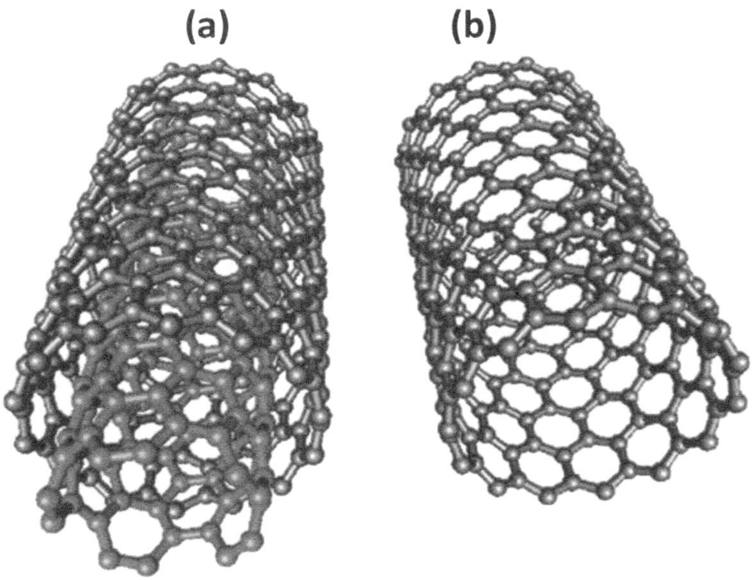

FIGURE 7.1 Structure representation of (a) MWCNT; and (b) SWCNT.
Source: Reproduced from: Ref. [13], ©2009, American Chemical Society.

The SWCNTs contains a single cylindrical carbon layer with a diameter in the range of 0.4–2 nm, based on the temperature of synthesis. It was observed that the higher the growth temperature larger is the diameter of synthesized CNTs. The distance between the multilayers that constitute the MWCNTs is generally about 3.3 Å. This distance is almost equal to that of the graphene layers that make up graphite [16]. The outer diameter of MWCNTs are in the 2–100 nm range, while the inner diameter is of the 1–3 nm range, and their length extends up to one to several micrometers.

Depending on the length, diameter, and number of multiple layers, properties of these structures varies from those of SWCNTs [17]. The delocalized electron cloud along the nanotube wall and sp^2 hybridization of MWCNTs are responsible for the interactions between adjacent cylindrical layers in MWCNTs resulting in more structural defects with less flexibility.

7.3 STRUCTURE OF CNTS

The structure of SWCNTs and MWCNTs are usually determined by the number of graphene layers rolled which varies from one to many. The rolling can be done in various ways and directions along the graphene sheet. The geometry of nanotubes is described by a vector called, the chiral vector of the original hexagonal lattice. The chiral vector is determined by two integers (*n, m*). Two carbon atoms in the planar graphene sheet are selected and one represents the origin. The chiral vector C_h is pointed from the first atom towards the second one and is defined by the relation $C_h = na_1 + ma_2$, *n* and *m* are integers, a_1 and a_2 are the unit cell vectors of the two-dimensional lattice formed by the graphene sheets [18] (Figure 7.2).

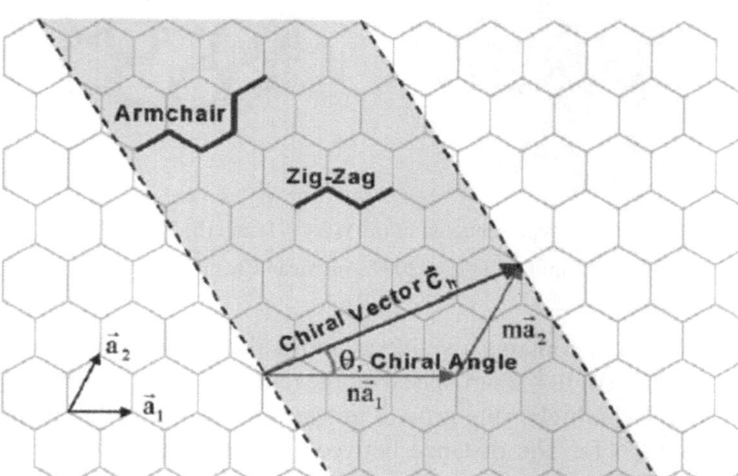

FIGURE 7.2 Diagram of 2D graphene sheet representing the vector structure classification used for defining CNTs structure.

Source: Reproduced from: Ref. [18], ©2001 Elsevier.

The CNT axis obtained is in a direction perpendicular to this chiral vector. The angle between C_h (chiral vector) and ma_2 (i.e., the zig zag tube axis) is named the chiral angle (θ). Based on this, nanotube structures are classified into three:

- Zig-zag tubes where $m = 0$ and $\theta = 30°$;
- Armchair tubes where $n = m$ and $\theta = 0°$; and
- Chiral tubes where $n \neq m$ and $0° < \theta < 30°$.

The values of *n* and *m* determines the chirality of CNT and acutely manipulate its optical, electronic, and mechanical properties. CNTs with $|n - m| = 3i$ are metallic like as in (10,10) tube, and those with $|n - m| = 3i \pm 1$ are semiconducting like as in (10,0) tube (*i* is an integer) (Figure 7.3).

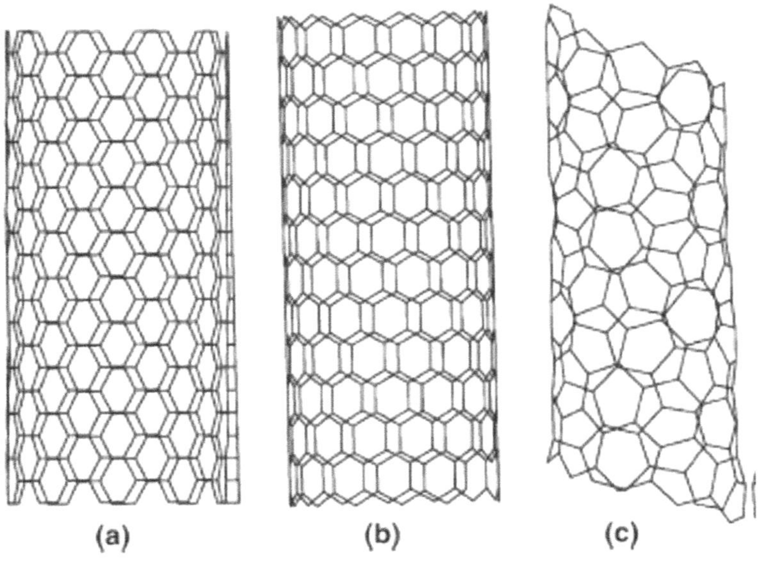

FIGURE 7.3 Schematic representation of various SWCNTs armchair (10,10) (a); zigzag (14,0) (b); and chiral (7,3) (c).

Source: Reproduced from: Ref. [19], ©2022 Elsevier.

Arc discharge, laser ablation, hydrothermal, sonochemical, and chemical vapor deposition are the main methods of CNTs preparation [20]. Each method has its own merits and demerits, resulting in various final properties.

7.4 CNTS FOR HYDROGEN STORAGE

Carbon is an esteemed adsorbent for gases due to its ability to exist as very fine powder with a highly porous structure which relays the unique interactions between carbon atoms and gas molecules. Various carbonaceous materials with optimized structures were investigated for hydrogen adsorption. The recent research shows immense interests in CNTs due to its high hydrogen storage capacity. CNTs can be used for onboard hydrogen storage with large capacity, high energy density and transition flexibility at room temperature [21, 22]. These carbon nanomaterials are very promising because of their chemical stability, bulk adsorption, large surface area, high pore volume, regular nanometric structure, low mass density and wetting properties. They are able to siphon off liquid or gas into its porous surface by capillary action [23].

The remarkable characteristics for a solid material to act as good hydrogen storing involves light weight, reduction in cost, high availability, fast adsorption-desorption process, easy activation, relevant thermodynamic properties, low decomposition temperature, high degree of reversibility, long term cycling stability, high volumetric and gravimetric density of hydrogen [24]. Moreover, the nanoscale materials offer a considerable designing of nanomaterials by controlling the tailoring parameters independently from their own bulk counterparts. And they lead to the development of novel carbon nanomaterials as light weight hydrogen storage devices with better hydrogen storage features (Figure 7.4).

Despite the immense studies on hydrogen storage by carbonaceous materials, the definite mechanism of the storage process is not yet lucid. It may be due to the weak Van der Waals forces of interaction which form the basis of physisorption. A linear correlation is obtained for the maximum gravimetric excess adsorption density with the pore volume and the specific surface of the adsorbent [26]. At a given temperature, the amount of gas adsorbed is a function of pressure only and is easily desorbed on decrease of pressure, as the phenomenon is highly reversible with pressure [27]. Or maybe because of the overlap between the HOMO of carbon with occupied electronic wave function of the hydrogen electron. The interaction of molecular orbitals of carbon and electronic wave function of the hydrogen molecule, overcoming the activation energy barrier for hydrogen dissociation forms the foundation of chemisorption. In the mechanism of physisorption (molecular adsorption) less than one hydrogen atom per two carbon atoms is realized. While a ratio of two hydrogen atoms per one carbon atom is maintained in the process of chemisorption (atomic adsorption). Physisorbed hydrogen normally shows a binding energy in the order of 0.1 eV (9.6 kJ/mol)). While a binding energy of more than 2–3 eV (192.5–288.7 kJ/mol)) is obtained for hydrogen on chemisorption [28–30]. Hydrogen trapping on CNTs originate from C–H covalent bonding and determination of its exclusive physisorption or chemisorption mechanism is highly difficult.

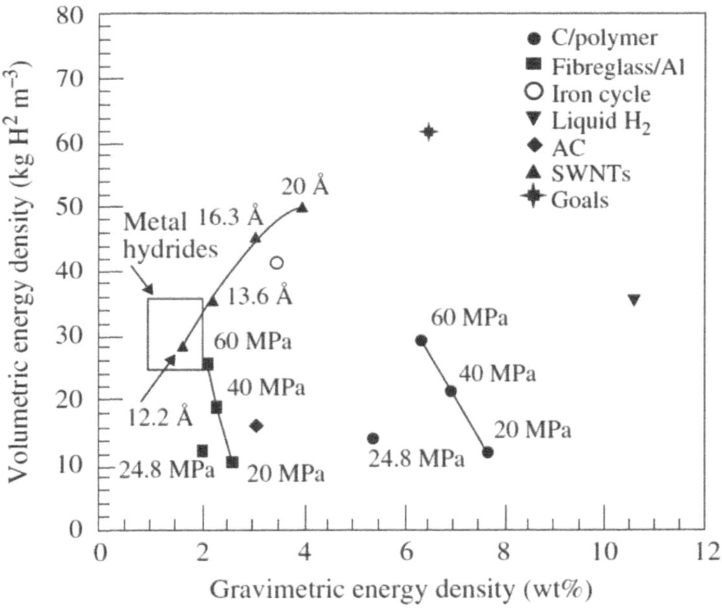

FIGURE 7.4 Energy densities for several vehicular hydrogen storage technologies.
Source: Reproduced from: Ref. [25], ©2001 Elsevier.

Different studies proved that hydrogen can be stored in the interior surface of CNTs, shaping a cylindrical monolayer form, or on the exterior surface, or between the nanotubes in the case of bundles of CNTs [7, 8]. From the surface topological studies, hydrogen adsorption on CNTs are favored only on some particular sites [31]. Gayathri et al. proposed three different places of bonding where hydrogen molecular axis is positioned at: (a) parallel and above C–C bond; (b) perpendicular and above carbon ring;

and (c) parallel and above carbon ring, respectively [10]. Two chemisorption sites at the ends of the tube and one inclusion site within the hollow space of nanotubes were predicted from density function calculations [32]. Active sites, effective accessible surface area, surface topology, molecular configurations, nanopore size, defects, dopant, chemical composition of the surface, applied pressure and temperature are the important factors that greatly affect the hydrogen adsorption by CNTs.

7.4.1 HYDROGEN STORAGE BY SWCNTS

The first report on hydrogen storage in SWCNTs was presented by Dillon et al. could reach a hydrogen gravimetric storage density of around 5–10% [25]. During the study, a mass spectrometer was used to analyze the amount of hydrogen gas released by the carbon sample. The temperature programmed desorption (TPD) analysis proposed that mechanism of adsorption is physisorption. And desorption peaks were observed at 150 K and 300 K. Careful work at National Renewable Energy Laboratory (NREL) indicates a maximum capacity of hydrogen adsorption on SWNTs is about 10% nanotube weight [33]. A hydrogen adsorption rate of 8 wt.% at 40 atm and 80 K was obtained for laser generated SWCNTs. Adsorption measurements on these SWNTs were done by volumetric method, using a Sievert's apparatus [34]. Liu et al. observed the effects of pretreatments of SWNT on their hydrogen adsorption capacity [35]. Large diameter SWNTs with high purity were prepared using charge-arc method. The SWNTs were pickled in an aqueous solution of HCl and heated under vacuum conditions can adsorb hydrogen up to 4.2 wt.% at room temperature. The high hydrogen uptake was attributed to the presence of defects or cavities which originated from the acid treatment. Nishimiya et al. reported a novel torch arc method for the preparation of SWNTs with averaged diameter of 1.32 nm [36]. They measured isotherms of hydrogen sorption by volumetric method and the maximum hydrogen concentrations reached up to 0.932 wt.% at 295 K under 106.7 kPa and 2.37% weight at 77 K under 107.9 kPa. The studies suggested the presence of an undefined mechanism other than physisorption on carbon hexagons.

Anson et al. estimated the hydrogen storage values by a volumetric procedure for as-grown and heat-treated arc-generated SWNTs [37]. At atmospheric pressure, the amounts of hydrogen adsorbed are approximately

0.01 wt.% at 298 K and 1 wt.% at 77 K. The oxidative processes of heat treatment in air resulted in the generation of micropores of SWCNTs which much contributed to gas adsorption at 77 K. However, majority among them have an adsorption energy that is probably too low to facilitate adsorption at ambient temperature. If the CNTs alone are not effective, different additives can be used to increase the kinetics and capacity of hydrogen storage. CNTs admixed with $NaAlH_4$ shows increase in hydrogen desorption rate [38]. Hydrogen storage properties of mechano-chemically prepared MgH_2/SWNT composites were highly dependent on the amount of SWCNT and milling time [39]. The composite MgH_2/5 wt.% SWNTs milled for 10 h is optimum, absorb 6.7 wt.% hydrogen within two minutes at 573 K, and desorb 6 wt.% hydrogen in 5 min at 623 K. On extending the time to 10 h leads to a major degradation on hydrogen storage property of the $MgH_{2/}$SWNT composite. This property degradation arises from the structure destruction of the SWNTs.

On doping CNTs with alkalis stimulated non-dissociative binding of hydrogen occurs which is more efficient than the Van der Waals forces of attraction. The best hydrogen storage capacity of 2.61% was reported for potassium doped SWNTs at 0.1 MPa pressure and 33 K [40]. Metal such as palladium is used to catalyze the dissociation of the hydrogen molecule and the resultant gas is stored as hydrogen atom [41].

The hydrogen adsorption/desorption studies of nanocrystalline Pt dispersed SWCNTs was reported by Reddy & Ramaprabhu [7]. Hydrogen storage data of as-grown, purified SWNT and nanocrystalline Pt dispersed SWNT were measured using high-pressure Seivert's apparatus in the pressure range 0.1–10 MPa and at 125 and 298 K. The effects of purification and dispersion of metal nanoparticles on CNTs for gas storage was based on the process of chemisorption. The results suggest that the hydrogen adsorption ability of the purified CNTs was substantially enhanced than that of the as-grown counterpart in both Pt dispersed and non-dispersed SWNT. The exclusion of impurities increased the surface area exposure of SWCNTs which successively caused the increment of storage capacity of the pure CNTs. The Pt dispersed SWNT showed a hydrogen uptake up to 3.03 wt.% at 125 K and at 7.8 MPa. The high catalytic activity of the Pt particles brought augmentation in hydrogen storage by dissociation of dihydrogen into atomic hydrogen which get to strongly adsorbed on the defect sites of each nanotube.

7.4.2 HYDROGEN STORAGE BY MWCNTS

MWCNTs are excellent adsorbent material because the adsorption occurs at various sites including external surface, interstitial space and inside the channel [42, 43]. An improvement in hydrogen adsorption by metal or metal oxide doped MWCNTs can be described by the generation of new adsorption sites, surface defects, stronger substrate–hydrogen interactions and spillover mechanism [44–46].

A comparison of adsorption performances of Ti/MWCNT composite prepared using two ball milling and sputtering techniques was reported. Sievert's volumetric measurement apparatus showed a 0.43–2.0 wt.% adsorption of hydrogen at 298 K and 16 atm. These results were 5 and 25 larger than those achieved by pristine MWCNTs [47]. Co and Cu decorated MWCNTs, respectively showed hydrogen uptake up to 0.8 wt.% and 0.9 wt.% at a pressure of 23 atm. and temperature of 25°C. The hydrogen retention enhanced about 10 times than MWCNTs alone [48]. A dissociative chemisorption was operated by dopant Ni nanoparticles on MWCNTs. The mass percentage of 0.87 wt.% hydrogen storage was observed at 298 K and 100 bar [49]. Silambarasan et al. demonstrated γ-rays irradiated MWCNTs at various dosages at 25, 50, 100, 150, and 200 kGy [50]. The irradiation caused improvement in hydrogen storage due to the physical effects and several defects produced on the CNTs. The hydrogen uptake of 1.2 wt.% was estimated for MWCNTs irradiated with a 150 kGy dose.

Acid treated h-BN/MWCNTs composite shows a hydrogen uptake of 1.4, 1.6, and 2.3 wt.% was measured, respectively for the samples prepared with 1, 3, and 5 wt.% of boron hydrides. After three hours of heating MWCNTs was acidified with sulfuric and nitric acids in the ratio three to one. The acid treated MWCNTs were mixed with boron nitride (h-BN) in specific weight ratios. The nanocomposite was dispersed in a dimethylformamide (DMF) and ultrasonicated for three hours. DMF was evaporated in various consecutive steps to obtain acid treated h-BN/MWCNTs. The desorption process shows a loss of 2.3 wt.% in the temperature range of 120–410°C [51] (Figure 7.5).

MgH_2 and 5 wt.% of MWCNTs mixture on milling showed enhanced hydrogen sorption on comparison with pure MgH_2, at 350°C and 0.1 bar [52]. Veron et al. showed the effect of co-addition of Co on MgH_2/MWCNTs also produced by high energy conditions of long milling time up to 50 hours [53]. Inspite of the destruction to the nanotubes, MgH_2

became nanocrystalline and remarkably upgraded the absorption capacity of hydrogen to 6.5%. The collateral addition of Co metal to MWCNT provided an additional refinement to the mixture. Mg-Ni alloy/MWCNT supported titania synthesized by the sol-gel method showed an uptake of hydrogen to be 5.60 wt.% H_2 at 373 K. The obtained results can be explained by the synergy between magnesium and MWCNTs [54]. Chemisorptive mode of hydrogen storage by lithium or potassium doped MWNTs were reported by Chen et al. The chemical dissociation was suggested to arise inside the nanotubes. Li and K doped CNTs showed capacities of 20 and 14 wt.% respectively at 653 K and room temperature. The great storage was attributed to the moisture coming out from wet hydrogen [55]. A comparison of hydrogen uptake by various composite materials based on MWCNTs are tabulated in Table 7.1.

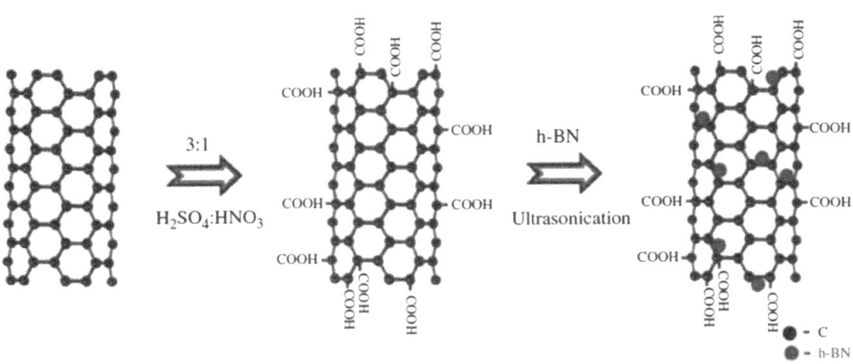

FIGURE 7.5 Schematic diagram for the preparation of acid treated h-BN/MWCNTs composite.

Source: Reproduced from: Ref. [51], ©2015 Elsevier.

7.5 CONCLUSIONS

CNTs show high hydrogen storage capacities due to several factors including surface interactions, fast kinetics, adsorption in addition to bulk absorption, low-temperature sorption, hydrogen atom dissociation, and molecular diffusion through the surface catalyst. The large surface area and exceptional adsorbing properties of nanomaterials assist the gaseous hydrogen dissociation and facilitates short diffusion paths into the nanosurface interiors. Nanosized dopants provides high dispersion of the catalytically active species which provides a provision for better mass transfer reactions. This chapter briefly discussed about the structure and properties of CNTs and its eminent application for hydrogen storage.

TABLE 7.1 Experimentally Reported Hydrogen Storage Capacities in Different MWCNTs

Composite	H_2 (% Weight)	Temperature	Pressure	References
Ni/MWCNT	0.4	Room temperature	Ambient pressure	[45]
Fe/MWCNT	0.75	Room temperature	Ambient pressure	[45]
Ca/MWCNT	1.05	Room temperature	Ambient pressure	[45]
Co/MWCNT	1.51	Room temperature	Ambient pressure	[45]
Mg/MWCNT	1.5	300°C	–	[56]
MWCNT/h-BN	2.3	100°C	–	[51]
PdNi/MWCNT	2.3	Room temperature	1.5 MPa	[57]
MWCNT/TiO_2	2.5	77 K	25 bar	[58]
MWCNT/Ni	2.8	330–520 K	–	[59]
MWCNT/$DyNi_2$	3.5	143 K	75.4 bar	[22]
Pt/MWCNT	3.1	298 K	90 bar	[7]
Pd/Ni/MWCNT	6.6	573 K	–	[60]
Mg/Pd/MWCNT	6.67	473 K	–	[61]

Source: Reproduced from: Ref. [51], ©2015 Elsevier.

KEYWORDS

- **caron nanotubes**
- **chemisorption**
- **diffusion**
- **hydrogen gas storage**
- **multiwalled carbon nanotubes**
- **nanosurface**
- **physisorption**
- **single walled carbon nanotubes**

REFERENCES

1. Satyapal, S., Petrovic, J., & Thomas, G., (2007). Gassing up with hydrogen. *Scientific American, 296*(4), 80–87.
2. Dresselhaus, M. S., & Thomas, I. L., (2001). Alternative energy technologies. *Nature, 414*(6861), 332–337.
3. Sakintuna, B., Lamari-Darkrim, F., & Hirscher, M., (2007). Metal hydride materials for solid hydrogen storage: A review. *International Journal of Hydrogen Energy, 32*(9), 1121–1140.
4. Kutz, M., (2007). *Environmentally Conscious Alternative Energy Production* (Vol. 3). John Wiley & Sons.
5. Sherif, S. A., Goswami, D. Y., Stefanakos, E. L., & Steinfeld, A., (2014). *Handbook of Hydrogen Energy*. CRC Press.
6. Zhang, H., Fu, X., Chen, Y., Yi, S., Li, S., Zhu, Y., & Wang, L., (2004). The electrochemical hydrogen storage of multi-walled carbon nanotubes synthesized by chemical vapor deposition using a lanthanum nickel hydrogen storage alloy as catalyst. *Physica B: Condensed Matter., 352*(1–4), 6–72.
7. Reddy, A. L. M., & Ramaprabhu, S., (2008). Hydrogen adsorption properties of single-walled carbon nanotube-nanocrystalline platinum composites. *International Journal of Hydrogen Energy, 33*(3), 1028–1034.
8. Ioannatos, G. E., & Verykios, X. E., (2010). H_2 storage on single-and multi-walled carbon nanotubes. *International Journal of Hydrogen Energy, 35*(2), 622–628.
9. Li, W., Wang, H., Ren, Z., Wang, G., & Bai, J., (2008). Co-production of hydrogen and multi-wall carbon nanotubes from ethanol decomposition over Fe/Al_2O_3 catalysts. *Applied Catalysis B: Environmental, 84*(3, 4), 433–439.
10. Gayathri, V., Devi, N. R., & Geetha, R., (2010). Hydrogen storage in coiled carbon nanotubes. *International Journal of Hydrogen Energy, 35*(3), 1313–1320.
11. Viswanathan, B., & Sankaran, M., (2009). Hetero-atoms as activation centers for hydrogen absorption in carbon nanotubes. *Diamond and Related Materials, 18*(2, 3), 429–432.
12. Eswaramoorthi, I., Sundaramurthy, V., & Dalai, A. K., (2006). Partial oxidation of methanol for hydrogen production over carbon nanotubes supported Cu-Zn catalysts. *Applied Catalysis A: General, 313*(1), 22–34.
13. Zhao, Y. L., & Stoddart, J. F., (2009). Noncovalent functionalization of single-walled carbon nanotubes. *Accounts of Chemical Research, 42*(8), 1161–1171.
14. Iijima, S., (1991). Helical microtubules of graphitic carbon. *Nature, 354*(6348), 56–58.
15. Jacobsen, N. R., Møller, P., Clausen, P. A., Saber, A. T., Micheletti, C., Jensen, K. A., Wallin, H., & Vogel, U., (2017). Biodistribution of carbon nanotubes in animal models. *Basic & Clinical Pharmacology & Toxicology, 121*, 30–43.
16. Aqel, A., Abou El-Nour, K. M., Ammar, R. A., & Al-Warthan, A., (2012). Carbon nanotubes, science and technology part (I) structure, synthesis and characterization. *Arabian Journal of Chemistry, 5*(1), 1–23.
17. Iijima, S., & Ichihashi, T., (1993). Single-shell carbon nanotubes of 1-nm diameter. *Nature, 363*(6430), 603–605.

18. Thostenson, E. T., Ren, Z., & Chou, T. W., (2001). Advances in the science and technology of carbon nanotubes and their composites: A review. *Composites Science and Technology, 61*(13), 1899–1912.
19. Dodziuk, H., & Dolgonos, G., (2002). Molecular modeling study of hydrogen storage in carbon nanotubes. *Chemical Physics Letters, 356*(1, 2), 79–83.
20. Prasek, J., Drbohlavova, J., Chomoucka, J., Hubalek, J., Jasek, O., Adam, V., & Kizek, R., (2011). Methods for carbon nanotubes synthesis. *Journal of Materials Chemistry, 21*(40), 15872–15884.
21. Suttisawat, Y., Rangsunvigit, P., Kitiyanan, B., Williams, M., Ndungu, P., Lototskyy, M. V., Nechaev, A., et al., (2009). Investigation of hydrogen storage capacity of multi-walled carbon nanotubes deposited with Pd or V. *International Journal of Hydrogen Energy, 34*(16), 6669–6675.
22. Rakhi, R. B., Sethupathi, K., & Ramaprabhu, S., (2008). Synthesis and hydrogen storage properties of carbon nanotubes. *International Journal of Hydrogen Energy, 33*(1), 381–386.
23. Shaijumon, M. M., & Ramaprabhu, S., (2005). Studies of yield and nature of carbon nanostructures synthesized by pyrolysis of ferrocene and hydrogen adsorption studies of carbon nanotubes. *International Journal of Hydrogen Energy, 30*(3), 311–317.
24. Niemann, M. U., Srinivasan, S. S., Phani, A. R., Kumar, A., Goswami, D. Y., & Stefanakos, E. K., (2008). Nanomaterials for hydrogen storage applications: A review. *Journal of Nanomaterials, 2008*, 1–9.
25. Dillon, A., Jones, K. M., Bekkedahl, T. A., Kiang, C. H., Bethune, D. S., & Heben, M. J., (1997). Storage of hydrogen in single-walled carbon nanotubes. *Nature, 386*(6623), 377–379.
26. Bénard, P., & Chahine, R., (2007). Storage of hydrogen by physisorption on carbon and nanostructured materials. *Scripta Materialia, 56*(10), 803–808.
27. Darkrim, F. L., Malbrunot, P., & Tartaglia, G. P., (2002). Review of hydrogen storage by adsorption in carbon nanotubes. *International Journal of Hydrogen Energy, 27*(2), 193–202.
28. Nijkamp, M. G., Raaymakers, J. E. M. J., Van, D. A. J., & De Jong, K. P., (2001). Hydrogen storage using physisorption-materials demands. *Applied Physics A., 72*(5), 619–623.
29. Sudan, P., Züttel, A., Mauron, P., Emmenegger, C., Wenger, P., & Schlapbach, L., (2003). Physisorption of hydrogen in single-walled carbon nanotubes. *Carbon, 41*(12), 2377–2383.
30. Viswanathan, B., Sankaran, M., & Schibioh, M. A., (2003). Carbon nanomaterials: Are they appropriate candidates for hydrogen storage. *Bull. Catal. Soc. India, 2*, 12–32.
31. Kojima, Y., & Kawai, Y., (2005). IR characterizations of lithium imide and amide. *Journal of Alloys and Compounds, 395*(1, 2), 236–239.
32. Yürüm, Y., Taralp, A., & Veziroglu, T. N., (2009). Storage of hydrogen in nanostructured carbon materials. *International Journal of Hydrogen Energy, 34*(9), 3784–3798.
33. Costa, P. M., Coleman, K. S., & Green, M. L., (2005). Influence of catalyst metal particles on the hydrogen sorption of single-walled carbon nanotube materials. *Nanotechnology, 16*(4), 512.

34. Ye, Y., Ahn, C. C., Witham, C., Fultz, B., Liu, J., Rinzler, A. G., Colbert, D., et al., (1999). Hydrogen adsorption and cohesive energy of single-walled carbon nanotubes. *Applied Physics Letters*, *74*(16), 2307–2309.
35. Liu, C., Fan, Y. Y., Liu, M., Cong, H. T., Cheng, H. M., & Dresselhaus, M. S., (1999). Hydrogen storage in single-walled carbon nanotubes at room temperature. *Science*, *286*(5442), 1127–1129.
36. Nishimiya, N., Ishigaki, K., Takikawa, H., Ikeda, M., Hibi, Y., Sakakibara, T., Matsumoto, A., & Tsutsumi, K., (2002). Hydrogen sorption by single-walled carbon nanotubes prepared by a torch arc method. *Journal of Alloys and Compounds*, *339*(1, 2), 275–282.
37. Anson, A., Callejas, M. A., Benito, A. M., Maser, W. K., Izquierdo, M. T., Rubio, B., Jagiello, J., et al., (2004). Hydrogen adsorption studies on single wall carbon nanotubes. *Carbon*, *42*(7), 1243–1248.
38. Chen, Y., Wang, P., Liu, C., & Cheng, H. M., (2007). Improved hydrogen storage performance of Li–Mg–N–H materials by optimizing composition and adding single-walled carbon nanotubes. *International Journal of Hydrogen Energy*, *32*(9), 1262–1268.
39. Wu, C. Z., Wang, P., Yao, X., Liu, C., Chen, D. M., Lu, G. Q., & Cheng, H. M., (2006). Hydrogen storage properties of MgH_2/SWNT composite prepared by ball milling. *Journal of Alloys and Compounds*, *420*(1, 2), 278–282.
40. Challet, S., Azaıs, P., Pellenq, R. M., Isnard, O., Soubeyroux, J. L., & Duclaux, L., (2004). Hydrogen adsorption in microporous alkali-doped carbons (activated carbon and single wall nanotubes). *Journal of Physics and Chemistry of Solids*, *65*(2, 3), 541–544.
41. Lachawiec, A. J., Qi, G., & Yang, R. T., (2005). Hydrogen storage in nanostructured carbons by spillover: Bridge-building enhancement. *Langmuir*, *21*(24), 11418–11424.
42. Park, S. J., & Lee, S. Y., (2010). Hydrogen storage behaviors of platinum-supported multi-walled carbon nanotubes. *International Journal of Hydrogen Energy*, *35*(23), 13048–13054.
43. Reyhani, A., Golikand, A. N., Mortazavi, S. Z., Irannejad, L., & Moshfegh, A. Z., (2010). The effects of multi-walled carbon nanotubes graphitization treated with different atmospheres and electrolyte temperatures on electrochemical hydrogen storage. *Electrochimica Acta*, *55*(16), 4700–4705.
44. Chang, J. K., Chen, C. Y., & Tsai, W. T., (2009). Decorating carbon nanotubes with nanoparticles using a facile redox displacement reaction and an evaluation of synergistic hydrogen storage performance. *Nanotechnology*, *20*(49), 495603.
45. Reyhani, A., Mortazavi, S. Z., Mirershadi, S., Moshfegh, A. Z., Parvin, P., & Golikand, A. N., (2011). Hydrogen storage in decorated multiwalled carbon nanotubes by Ca, Co, Fe, Ni, and Pd nanoparticles under ambient conditions. *The Journal of Physical Chemistry C.*, *115*(14), 6994–7001.
46. Chen, C. Y., Chang, J. K., & Tsai, W. T., (2012). Improved hydrogen storage performance of defected carbon nanotubes with Pd spillover catalysts dispersed using supercritical CO_2 fluid. *International Journal of Hydrogen Energy*, *37*(4), 3305–3312.
47. Rather, S. U., & Hwang, S. W., (2016). Comparative hydrogen uptake study on titanium–MWCNTs composite prepared by two different methods. *International Journal of Hydrogen Energy*, *41*(40), 18114–18120.

48. Rather, S. U., (2017). Hydrogen uptake of cobalt and copper oxide-multiwalled carbon nanotube composites. *International Journal of Hydrogen Energy, 42*(16), 11553–11559.
49. Han, Y. J., & Park, S. J., (2017). Influence of nickel nanoparticles on hydrogen storage behaviors of MWCNTs. *Applied Surface Science, 415*, 85–89.
50. Silambarasan, D., Surya, V. J., Iyakutti, K., Asokan, K., Vasu, V., & Kawazoe, Y., (2017). Gamma (γ)-ray irradiated multi-walled carbon nanotubes (MWCNTs) for hydrogen storage. *Applied Surface Science, 418*, 49–55.
51. Muthu, R. N., Rajashabala, S., & Kannan, R., (2016). Hexagonal boron nitride (h-BN) nanoparticles decorated multi-walled carbon nanotubes (MWCNT) for hydrogen storage. *Renewable Energy, 85*, 387–394.
52. Campos, R. B. V., Camargo, S. A. D. S., Brum, M. C., & Santos, D. S. D., (2017). Hydrogen uptake enhancement by the use of a magnesium hydride and carbon nanotubes mixture. *Materials Research, 20,* 85–88.
53. Verón, M. G., Troiani, H., & Gennari, F. C., (2011). Synergetic effect of Co and carbon nanotubes on MgH_2 sorption properties. *Carbon, 49*(7), 2413–2423.
54. Tan, Y., Zhu, Y., & Li, L., (2015). Excellent catalytic effects of multi-walled carbon nanotube supported titania on hydrogen storage of a Mg–Ni alloy. *Chemical Communications, 51*(12), 2368–2371.
55. Chen, P., Wu, X., Lin, J., & Tan, K. L., (1999). High H_2 uptake by alkali-doped carbon nanotubes under ambient pressure and moderate temperatures. *Science, 285*(5424), 91–93.
56. Reyhani, A., Mortazavi, S. Z., Mirershadi, S., Golikand, A. N., & Moshfegh, A. Z., (2012). H_2 adsorption mechanism in Mg modified multi-walled carbon nanotubes for hydrogen storage. *International Journal of Hydrogen Energy, 37*(2), 1919–1926.
57. Ren, J., Liao, S., & Liu, J., (2006). Hydrogen storage of multiwalled carbon nanotubes coated with Pd-Ni nanoparticles under moderate conditions. *Chinese Science Bulletin, 51*(24), 2959–2963.
58. Mishra, A., Banerjee, S., Mohapatra, S. K., Graeve, O. A., & Misra, M., (2008). Synthesis of carbon nanotube–TiO_2 nanotubular material for reversible hydrogen storage. *Nanotechnology, 19*(44), 445607.
59. Kim, H. S., Lee, H., Han, K. S., Kim, J. H., Song, M. S., Park, M. S., Lee, J. Y., & Kang, J. K., (2005). Hydrogen storage in Ni nanoparticle-dispersed multiwalled carbon nanotubes. *The Journal of Physical Chemistry B, 109*(18), 8983–8986.
60. Gao, L., Yoo, E., Nakamura, J., Zhang, W., & Chua, H. T., (2010). Hydrogen storage in Pd-Ni doped defective carbon nanotubes through the formation of CH_x (x= 1, 2). *Carbon, 48*(11), 3250–3255.
61. Yuan, J., Zhu, Y., Li, Y., Zhang, L., & Li, L., (2014). Effect of multi-wall carbon nanotubes supported palladium addition on hydrogen storage properties of magnesium hydride. *International Journal of Hydrogen Energy, 39*(19), 10184–10194.

CHAPTER 8

Inorganic Nanomaterials in Organic Solar Cells: A Renewable Energy Application

R. GEETHU, ANJU NAIR, and M. V. SANTHOSH

Department of Basic Science, SCMS School of Engineering and Technology, Karukutty, Ernakulam, Kerala, India

ABSTRACT

Concept and view of energy usage and energy sources are undergoing a transition in this scenario from the conventional to Green Energy sources. Among green energy sources, most prominent and reliable energy source – THE SUN. Solar energy is the perennial and inexhaustible source of energy which can cater the mounting energy requirement of mankind. Conversion of this regenerative source directly to electricity, harnessing the photovoltaic technologies has always been of paramount importance [1]. Silicon solar cell is the most commercialized and well known to common man. But the price of those devices are still not affordable. Huge fabrication cost of silicon solar cell is the reason for the cost of the product. Now immense number of solar cells materials were developed that make use of very simple and cost-effective deposition technique leading to low-cost final product (solar cells). Thin film solar cell, Organic photovoltaics (OPVs), perovskite solar cells, quantum dot solar cells, etc., are a few among them which comes under second and third generation solar cells. Here OPVs is taken into consideration. Unlike inorganic solar cells, device configuration as well as working of OPV are different. Various

New Advances in Materials Technologies. Hossein Hariri Asli, Ali Pourhashemi, Ann Rose Abraham, & A. K. Haghi (Eds.)

device structures, brief explanation on the working of organic solar cells are explained in this chapter. Comparing to Si solar cells, organic solar cell fabrication cost as well as production time is low, owing to the availability of low cost and simple technique. But the stability and life time of these devices were not reached the requirement level yet. To enhance the stability of organic devices, one of the best methods is to incorporated inorganic materials into organic solar cells as thin film, nanoparticles, etc. Among these current chapter focused on the research works on incorporation of nanoparticles like metal or compounds of oxides, sulfide, etc., into OPV. Nanoparticles are incorporated in the device structure as hole transport layer (HTL) or electron transport layer (ETL). Light trapping, scattering, absorption, etc., owing to surface plasmon resonance (SPR), localized surface plasmon resonance (LSPR) are the root cause for the enhancement in the performance parameter of the solar cell devices.

8.1 INTRODUCTION

In current scenario the world is under a transformation from conventional to green energy (renewable) and more reliable energy sources. Among various energy convertors photovoltaics (PV) possess the prominent place. Even though silicon solar cell is most common and highly commercialized now, this picture will change within a few decades. Researcher introduced numerous semiconducting materials suitable for solar cell application. These materials include both inorganic and organic. Highly suitable inorganic materials for PV application comes under II-VI and I-III-VI chalcogenide like indium sulfide, indium selenide, zinc oxide, titanium oxide, copper indium selenide, copper indium sulfide, copper gallium indium selenide, copper indium gallium sulfide, copper indium zinc sulfide, copper indium zinc selenide, etc., were suitable for buffer as well as absorber layer in solar cell. Organic materials are another group that proves their potential in this field. Organic materials has the power to overcome disadvantages of silicon as well as inorganic material based solar cells. Even though, efficiency and life span of OPVs devices were lower compared to other, its cost effectiveness, easily manufacturability, flexible nature, etc., overcome those said disadvantages mainly for small-scale applications. Many more materials are also developed seeking application in this field like perovskite, nanomaterials (nanoparticles, nanofilms, nanoclusters, etc.), dyes, etc.

Hybrid type device are another version in which a combination of organic and inorganic materials are used for developing solar cells. These types of solar cells overcome the disadvantage of both type of materials and make use of their advantages to enhance the device parameters and hence its performance. Hybrid type can be developed mainly by two ways one by depositing as two separate layers of film, i.e., transparent electrode/inorganic material/organic material/metal electrode or by blending inorganic material in organic one, i.e., transparent electrode/buffer layer/ organic: inorganic blend/metal electrode. Active region of later one is fabricated by blending inorganic and organic material in the solution followed but thin film deposition.

In this chapter, organic solar cells were taken into consideration and a review on the works based on nanomaterial incorporation (as explained in the last part of previous paragraph) and its effect on those devices were analyzed. Since the device structure, fabrication techniques, working and theory behind organic solar cell differ from the conversional silicon solar cell, a short description on the device structure and theory of OPV is explained here.

8.2 DEVICE CONFIGURATIONS FOR ORGANIC PHOTOVOLTAICS (OPVS)

A basic structure of organic solar cell is given below. Even though the organic solar cell were found to fabricate using various structures like single layer, bilayer, and blend, later two gives the better device. Hence here considering the only two types of structures. Schematic diagram of bilayer and BHJ device is shown in Figures 8.1(a) and (b). Bilayer and blend devices are further modified by incorporating inorganic layer resulting in hybrid organic solar cells. Schematic structure of hybrid (combination of inorganic and organic) type device is given in Figures 8.1(c) and (d). Along with this devices structure can be conventional or inverted (sample device structure as given in Figures 8.1(e) and (f).

8.3 DEPOSITIONS TECHNIQUE AVAILABLE FOR ORGANIC SOLAR CELL FABRICATION

Highlight of OPV is the availability of simple deposition technique. A large number of simple, vacuum free, cost-effective techniques are developed for depositing materials as thin film. A few commonly available techniques are given below:

- Spin coating;
- Doctors blading;
- Roll-to-roll;
- Screen-printing;
- Spray coating;
- Dip coating;
- Drop-casting.

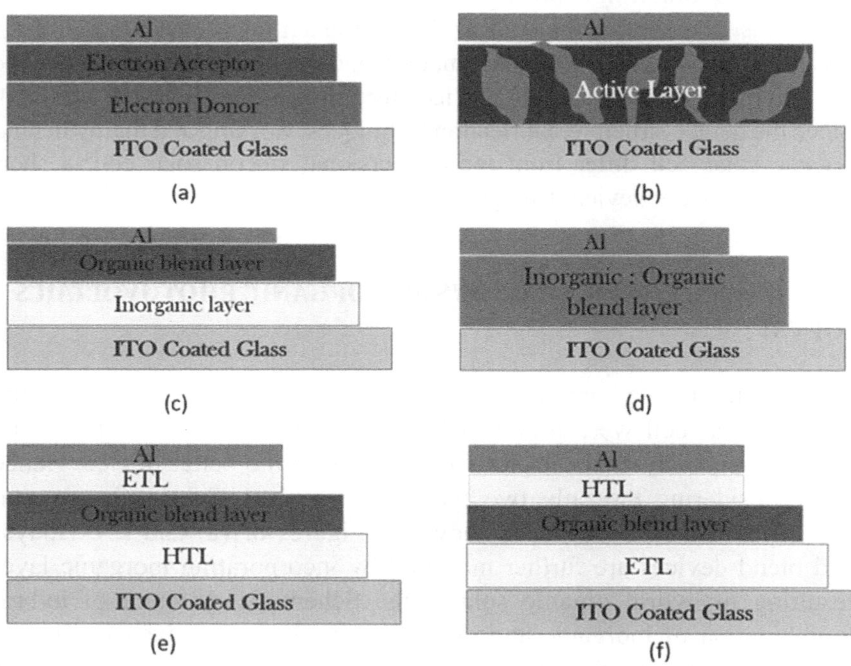

FIGURE 8.1 (a) Bilayer device; (b) bulk device; (c) and (d) hybrid devices; (e) conventional organic solar cell; and (f) inverted organic solar cell.

Among these techniques, spin coating is most common in lab scale device fabrication whereas roll-to roll or roll coating, screen printing are usually meant for the fabrication of large area device [2]. One among the new comer in the above said technique is spray coating. Spray coating or chemical spray pyrolysis is well-know and established technique for the inorganic solar cell fabrication [3–10]. There are reports in the literature regarding organic material deposition and device fabrication using spray technique [11–14]. Even though spray coating has a large potential for fabricating devices on flexible area size, but it is not yet absolutely recognized for the OPV fabrication. More possibilities of spray coating technique for organic device fabrication are yet to be unveiled.

8.4 WORKING OF ORGANIC SOLAR CELLS

Brief explanation on working in general to all devices is given below:

- Light incident on the device will pass through glass (substrate) and is collected by the transparent electrode (ITO).
- Glass and ITO have higher band gap, hence the enter light will pass through those layers and reaches buffer layer/carrier (either hole or electron) transport layer.
- Electron donor like material possess wider band gap and hence the white light without much loss will reach "active layer."
- Electron acceptor/active layer much have a band gap less than 2 eV (most preferable band gap ~ 1.5 eV).
- In the active region the light was absorbed and exciton (electron-hole pair) was generated, this electron hole pair will travel for a short distance without recombination.
- Exciton reaching the interface before recombination will split into electron and hole and collected by corresponding carrier transport layer/electrodes:
    - For example: Say the exciton reaches the PEDOT:PSS/Active layer interface, it splits into electrons and holes and the holes will be collected by PEDOT:PSS. Then the electron travel back through the active region and reach the other electrode near active region to get collected.
    - Another example: Say-ZnO/active region, here charge carrier separated at interface and electron will be collected by the ZnO and hole has to travel back as mentioned in the above case.
- These collected charge carriers will generate photocurrent in the outer circuit.

The mobility of these charge carriers are largely dependent on the type of the junction used [15]. A significant potential drop is provided when a

BHJ bilayer bearing a stacked donor (D) and acceptor (A) materials [16], is bracketed between the cathode and anode. Various materials structures and properties are designed as interlayers with suitable work functions equal to the energy levels of donor and acceptor materials that are inserted between the active layer and electrodes, which improve the collection of charge carriers. This reduces the diffusing distance for the exciton and makes it more feasible for the electron to reach the electrode which means the separated charge carriers are transported in the active interface layers till they are taken by the opposite electrodes [17].

The most popularly discussed anode interface layer (AIL) is the poly(3,4-ethylenedioxythiophene):poly(styrene sulfonate) (PEDOT:PSS) modified the ITO electrode, due to its high conductivity and so extensive research on enhancing its properties have been undertaken. However, to reduce the poor lifetime and quick degradation of PEDOT: PSS, inverted OSCs with reversed polarity is used. Owing to less oxidative decomposition, Au and Ag are used as high work function top electrodes in inverted OSC, whereas a low work function metal oxide become transparent cathode, lead to highly stable device [18].

To characterize a fabricated solar cell, voltage–current characterization has to be done. The said characterization provide four core parameters of the device. Open circuit voltage (V_{oc}), Short circuit current (I_{sc}) or short circuit current density (J_{sc}), FF and Power Conversion Efficiency (PCE).

8.5 RELEVANCE AND ADVANTAGE OF NANOMATERIALS INCORPORATION

Major issues with PEDOT:PSS:

- Unwanted series resistance offered owing to thick PEDOT:PSS layer lower FF and current density of solar cell [19].
- Causing corrosion to ITO-PEDOT:PSS as a result of acidic nature.
- Hygroscopic nature–make device unstable [20–23].

In spite of various advantages of organic solar cell, major drawn back is its short life span and low efficiency. Enhancing lifespan and efficiency can be achieved by incorporating highly stable inorganic materials into the device either as separate layer to form bilayer hybrid device or blending both organic and inorganic material and depositing it as single film so as to form bulk hybrid device. Our group (G.R.) has developed hybrid solar cells with ZnO as the highly stable inorganic material and thiophene based polymer blended with carbon fullerene as the active layer. The device and device structure was studied thoroughly for a number of years and enhancement in the device performance on the incorporation of ZnO was staggering. Unlike normal case of using highly smooth thin films of electron transport layer (ETL) device fabrication done then was by incorporated a highly rough surfaced ZnO and the named given was surface modified ZnO. The following were the highlights of those type of structures.

- Efficient photon trapping and making photons more available to active region;
- Enhancing surface area of the active region with in the same substrate area;
- Blocking the diffusion of indium from ITO to active region.

Efficient photon trapping and enhancing surface area can be proved or confirmed by checking the morphology ZnO thin film and by XPS analysis of the sample for different cycles of etching. SEM (scanning electron microscopy) of the surface modified ZnO and XPS data showing blocking of ITO by ZnO-SM is depicted in Figures 8.2 and 8.3.

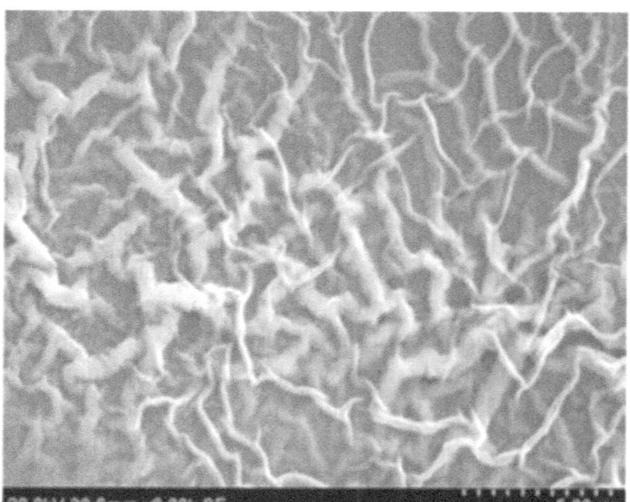

FIGURE 8.2 SEM images of ZnO-SM. The prepared sample appeared like a tangled root structure.

Source: Reprinted with permission from Ref. [5]. Copyright 2015. Elsevier Ltd.

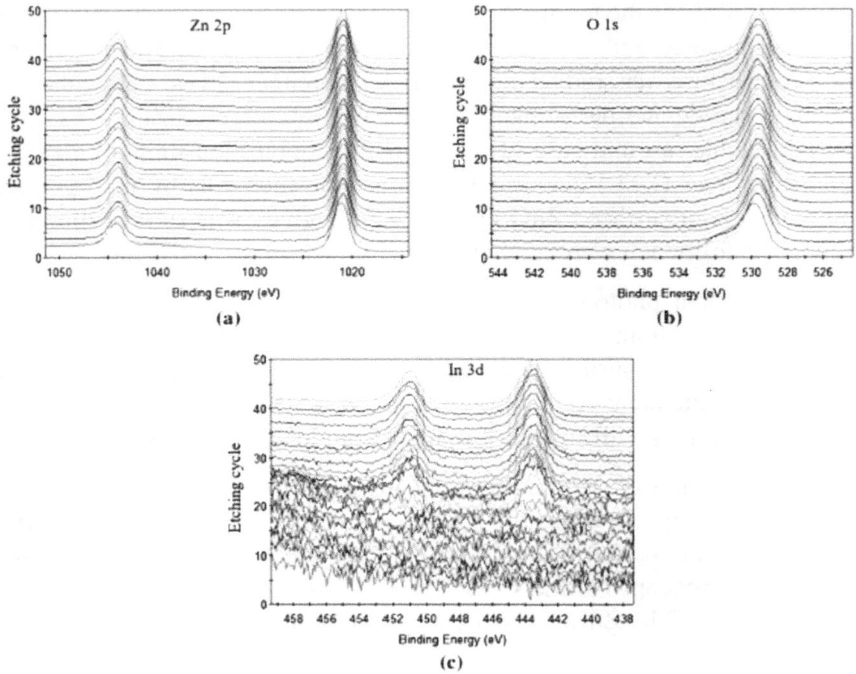

FIGURE 8.3 XPS analysis of the device showing blocking of indium diffusion from ITO to polymer layer.

Source: Reprinted with permission from Ref. [25]. Copyright 2019. Springer Nature.

Further enhancement in the device performance might be achieved by doing doping in organic solar cell. Photon trapping, as well as surface area enhancement, can be achieved by doping. Injection of charge carriers or collection of charge carriers leading to the enhancement in device efficiency can be achieved by incorporating materials or molecules. This doping in the active region of organic solar cells will commendably reduce the probability of charge carrier recombination. This enhanced separation of charge carriers owing to the upsurge in phase separation of polymer and fullerene derivatives [26]. Say, if we are incorporating inorganic materials, into the active region, this can act as center for dissociation of exciton and function as a channel for CT. Moreover, energy level get more aligned so that it promote transfer of separated charge carrier resulting in reduce recombination, enhance charge carrier collection finally leading to betterment in short circuit current density and efficiency [27–32].

While dealing with nanoparticle incorporation of it is significant to mention and explain SPR and local SPR. "Surface plasmon resonance (SPR) is the collective oscillation of conduction band electrons that are in resonance with the oscillating electric field of incident light, which will produce energetic plasmonic electrons through non-radiative excitation" [33]. If the electrons are localized or with in a prescribed geometry or bounded in a region which is small than the wavelength of light, the electrons excited by the electromagnetic radiation will under oscillation further led to polarization charge build-up on the surface of the particle. This is known as LSPR. SPR and LSPR occurring in metal nanoparticles leads to light scattering and light absorption. As size of the nanoparticles decreases the light absorption dominate over light scattering, i.e., light scattering is high if nanoparticle has large size [34].

Nanomaterials can be integrated into the organic devices mainly by two ways: (i) as separate thin film layers between conducting electrode and donor or active layer; (ii) by blending with electron donor or active region of the device as shown in Figure 8.1(d). Here latter one is taken into consideration, in which incorporation can be using different materials. It can be as nano-metals, nanoparticle of some compounds like ZnO, TiO_2, MoO_3, CuS, CdS, etc. Based on nanoparticles here the sections are classified and discussed. Relevant publication in the field with in a decade is also discussed to understand the applicability of these materials in solar cell fabrication.

8.6 METAL NANOPARTICLES INCORPORATED ORGANIC PHOTOVOLTAICS (OPVS)

Excitons are short-lived and the organic conducting polymers (CPs) have a reduced carrier mobility which limits the active layer thickness ~ 100 nm. Increasing the thickness escalates the recombination probabilities which reduces PCE. Hence, increasing the optical path length is decisive to enhance the PCE. Better photo-absorption can be achieved by inserting noble metal nanoparticles (metal nanoparticles (MNP)) that limit resonant photons to induce surface plasmon oscillations (Figure 8.4) [35, 36]. Such oscillations are successively brought by noble metal nanoparticles in their conduction band electrons, through confining resonant photons [37]. Better photo-absorption is achieved when noble metal nanoparticles confine the resonant photons and induce coherent surface plasmon oscillation of their conduction band electrons [35, 37, 39, 41–43]. Incorporation of Au and Ag nanoparticles are attributed to the LSPR effect as shown in Figure 8.5 [44, 45]. The text summarizes recent advances in this regard.

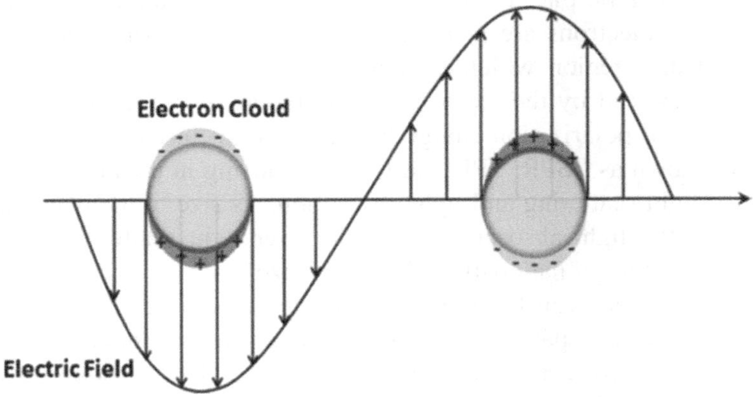

FIGURE 8.4 Schematic representation of SPR, exhibiting the oscillation of metal nanoparticles.

Source: Reprinted with permission from Ref. [46]. Copyright 2014. Elsevier Ltd.

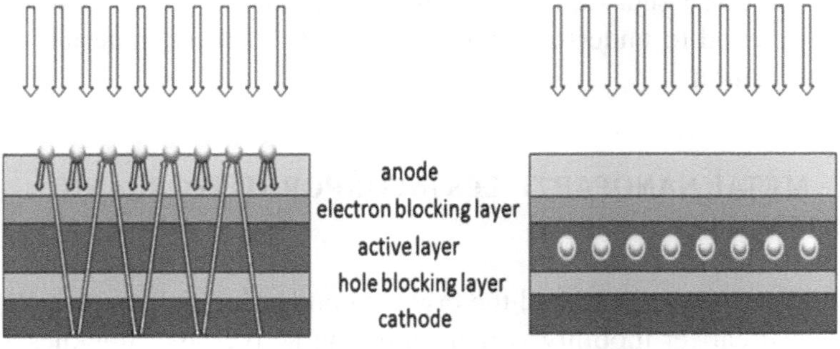

FIGURE 8.5 (a) OPV devices with nanoparticles (metallic) on its surface is trapping light into the device. (b) LSPR of nanoparticle incorporated in the active region.

Source: Reprinted with permission from Ref. [46]. Copyright 2014. Elsevier Ltd.

8.6.1 INCORPORATION OF AU NANOPARTICLES

Au nanoparticle exhibits strong near-field electromagnetic fields and far-field propagating waves leading to enhancement of photon absorption ability of active region as well as possibility of exciton generation and dissociation. Umpteen reviews and findings demonstrate the doping of Au in various nanoparticle morphologies and the corresponding improved PCEs.

Incorporating Au nanospheres of various dimensions and the subsequent enhancement in PCEs has also been studied. Au nanosphere ~45 mm diameter was embedded to buffer layer of a P3HT:PC60BM OPV device, and an increase in PCE of the device from 1.3 to 2.2% [47] reported by Morfa et al: further backed by a PCE of 3.57% to 4.24% [48] when introduced in the anodic layer, studied by Wu et al. Au nanoparticles (~15 nm diameter) introduced into the PEDOT:PSS layer using poly(2-methoxy-5(20-ethylhexyloxy)-1,4-phenylenevinylene (MEH-PPV) as the active layer enhanced the PCE from 1.99 to 2.36% was seen in study by Qiao et al. [49] More recently, an inverted tandem polymer solar cell with Au nanoparticles (~72 nm diameter) were deposited in the interconnecting layer showed a 20% increase in PCE (from 5.22 to 6.24%).

Blending of Au/SiO_2 nanospheres and nanorods into the active layers (P3HT:PC60BM and PBDTT-DPP:PC60BM) yielded device with enhanced short circuit current density and efficiency of solar cells. In addition, the dependence of concentration on the amount of Au/SiO_2 was studied. This displayed a negative correlation on increasing the concentration was observed after an initial hike, which is attributed to the distortion that would affect the OPV morphology, which was confirmed by AFM measurements.

Baek et al. attempted a hybrid plasmon structure by integrating AuNPs with the broad spectral response and AgNPs combined with the sharp scattering power as an Au@Ag core-shell nanocube (NC) [50]. The thickness Ag shell diameter was varied (0 to 30 nm) after fixing Au core at 45 nm. It was noted that with increase in shell thickness, the scattering power enhanced significantly while slightly blue shift was happened for LSPR peak (Figure 8.6). Beyond 20 nm thickness, the scattering power and LSPR peak approaches to those AgNPS which was an indicative of the diminishing core-shell properties. In addition to a PCE of 9.2%, the meticulously fabricated Au@Ag NCs showed exceptional scattering efficiency at the long wavelength region compared to the AuNPs, while minimizing the blue shift. This confirmed a positive correlation between Au@AgNC LSPR spectral peaks and low absorption region of control devices towards maximizing the EQE and absorption enhancement.

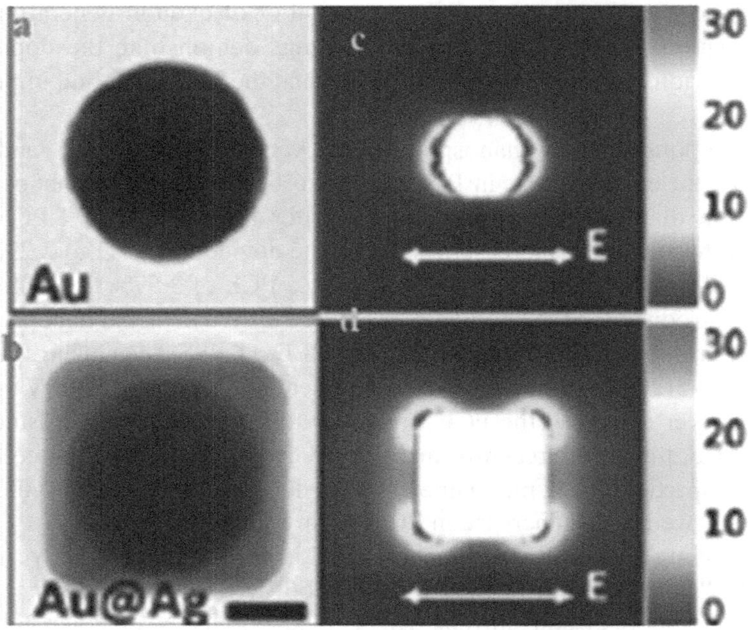

FIGURE 8.6 (a) TEM image of gold nanoparticles AuNPs; (b) TEM image of Au@Ag coreshell nanocube at scale bar of 20 nm. The scattered electric field $|E|^2$ distribution of (c) AuNPs of ~45 nm; and (d) 45 nm@10 nm Au@Ag NCs.

Source: Reprinted with permission from Ref. [50]. Copyright 2014. American Chemical Society.

In the work reported Wenxin Niu et al. Au nanostars seems to be another remarkably explored AuNPs due to their fascinating properties. A gold nanostar has a branched structure with projecting sharp tips and a central core. It could be defined as a branched nanostructure with a central core and several protruding arms with sharp tips. Field strengths of these sharp tips could be improved by several orders of magnitude for providing a plasmon-enhanced spectroscopy [51]. Devices including Au NSs showed better charge separation/transfer, decreased charge recombination rate, and efficient charge transport. Embedding Au NSs improved the PCE by 6% due to improved light absorption. In addition, AuNS renders strong field for effective charge separation within the active layer of OSCs [52].

Phetsang et al. investigated the symbiotic effect of AuQDs blended with AuNPS in the HTLs of the OSCs. LSPR effect of the AuNPs combined with the emission of fluorescence of AuQDs enhances the phototrapping capacity in OSCs. In the study, an elevation in the photovoltaic efficiency was achieved and best synergistic effect was observed on combining green-emitting AuQD layer and a HTL containing AuNPs (Figure 8.7), which showed a PCE of 13.0% [53, 54].

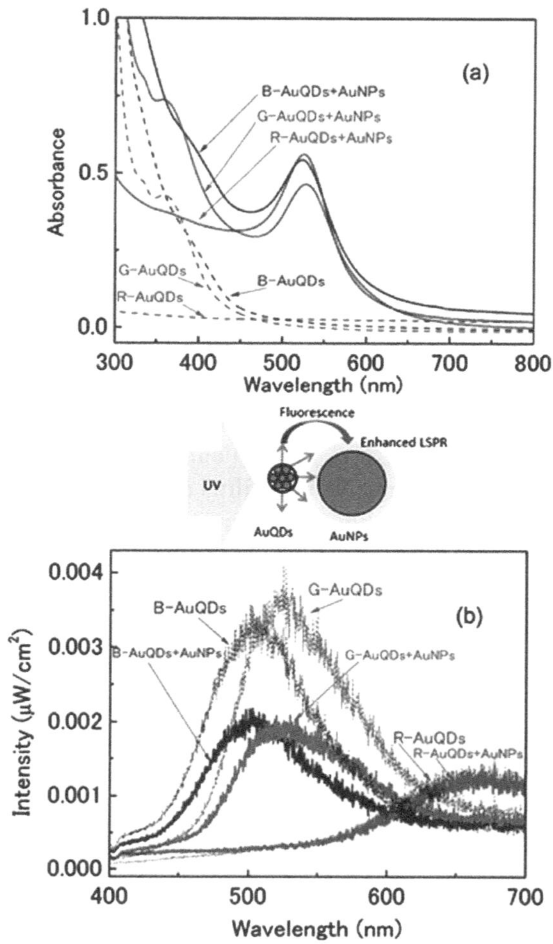

FIGURE 8.7 Schematic showing enhanced LSPR (top) (a) absorption spectra of samples; (b) florescence spectra of AuQDs and AuQD: AuNP solution showing an enhanced fluorescence for green-AuQDs.

Source: Reprinted with permission from Ref. [53]. Copyright 2019 © The Royal Society of Chemistry. https://creativecommons.org/licenses/by/3.0/

Recent report by Virginia Cuesta et al. shows an impressive innovation of AuNPs embedded in OSCs as Gold porphyrins [55]. Porphyrin rings are the planar, conjugated macrocyclic compounds with a metallic center that are well-known for their light harvesting capacity in both blue- and red-regions of the Visible spectra and a vital point in photosynthetic process. Low absorption coefficients in the visible region, massive cost production and difficulty in tailoring the Frontier MOs have limited the PCEs of fullerene based OSCs [56]. Au(III) porphyrin-based materials owing to its remarkable stability and an appreciable redox property, can be used as a non-fullerene acceptor with a conjugated polymer as a donor that shows a PCE of 9.25% [57]. The finding has steered a new path towards scrupulous molecular design by modulating the FMOs.

8.6.2 INCORPORATION OF AG NANOPARTICLES

MNP incorporation to OPVs harvests the light inside the photoactive material layer manifold due to localized SPRs. Low cost and enhancement of light absorption in UV region has rendered silver nanoparticles (AgNPs) incorporation in OPVs a promising area to explore [58]. Gupta et al. proposed discontinuous Ag thin films as an easy to design, optically tunable surface plasmon active system in the visible spectrum [59]. Size monitored plasmonic effect on PCE and their spectral responses were studied by Baek et al. which illustrated an enhancement of the EQE at wavelengths between 400 and 500 nm at all sizes. AgNP-67 showed a rise in EQE to 35% while larger AgNP-94 decreased the highly efficient plasmonic OPVs embedded to anodic PEDOT: PSS layer. The dependence of size on scattering efficiency is further confirmed by the Near-field Optical Scanning Microscopy [60, 61].

A new analytical parameter, Generation-to-recombination rate ratio (G/R) to gage the performance on MNP incorporation in OSCs. Improvement in light absorption depends on the size of particles and the concentration of the embedded MNPs. Sakshi Koul et al. formulated generation rate for AgNPs of various sizes. First, G/R is calculated for a single Ag-NP placed at the center of a unit cell, in the active layer and further extended to study the performance to randomly dispersed NPs in the active layer with a precise concentration. The model can be further extended for a study of varying concentrations of the MNPs in the active layer [62]. A comparison is then carried out between modified G/R ratio and the G/R ratio found for the cell without any NPs in the active layer, through simulations [63].

Yao et al. attempted to embed silver nanoprisms at the top and bottom interface layers of the OSCs. Gold nanospheres possessed high fraction of absorption loss due to scattering and hence a prismatic geometry displays stronger antenna effects in addition to higher strong plasmon resonances. The flat shapes of these prism-like particles maximize optical interactions and reduces the chance of shorting [64].

A notable improvement of BHJ performance from 7.7% PCE up to a maximum 9.0% PCE, by merging silver nanoprisms at both top and bottom interfacial layers. The study has been decisive for designing new absorber materials of tuned optical property [65].

An investigation by Tam et al. shows a printable silver nanoparticle (AgNP) film top electrode that shows a parallel performance as that of evaporated ones (EvapAg) with ~ 90% PCEs. Such cells perform better than EvapAg due to their lower leakage currents. Poor PCEs and reflectance of AgNPs were circumvented by Han et al. by depositing UV sinterable silver paste as top electrode. The UV sensitive photoactive layer is secured by a Hole Extraction Layer to prevent photo-oxidation during the photo sintering process [66, 67].

Shabani et al. studied embedding Ag plasmonic nanocrystal hexagonal array in the P3HT:PCBM active layer at the bottom of the OPV solar cell was examined. Lumerical FDTD simulations were conducted to further the investigation in which the amount of scattered light and the size of Ag plasmonic nanocrystals could be studied. I–V curve shows a sharp enhancement achieved with the Ag plasmonic nanocrystals implanted in the active layer of the OPV solar cells [68].

8.6.3 INCORPORATION OF AL NANOPARTICLES

Aluminum nanoparticles exhibit high plasma frequency, which results in better absorption at a broad range of wavelengths. Though these are prone to oxidation; the stabilization could be achieved by proper surface functionalization. Overlap of plasmon resonance and the natural absorption bands of the photovoltaics gives better percolation thresholds. This utilization has paved for the pursuit of cost advantageous Aluminum nanoparticles in place of gold and silver [69]. Sygletou et al. has further backed this notion which states the incorporation of these nanoparticles blocks the polymer donor photo-oxidation by the formation of a triplet excitons assisted by oxygen. This could probably be due to the overlap of such excitonic levels with the plasmon resonance of the incorporated Al NPs. This has opened the way to a more durable polymer devices including light emitting diodes, detectors, and sensors [70]. A remarkable PCE of 30% is demonstrated by AlNP incorporated OPVs by George Kakavelakisa and coworkers, mainly attributed to multiple scattering effects and improved structural stability of the photoactive blend. AlNP incorporated devices were seen die after ~150 hrs. of operation under constant illumination which is considered as a better lifetime than the pristine devices which die after ~30 hrs., This is due to improved absorption caused by scattering effects by the embedded AlNPs, which shows a lifetime hike by 5 times. Hence AlNPs incorporation to OPVs are effective technological solution to alleviate photo-degradation effects [71].

8.7 SEMICONDUCTING NANOPARTICLES INCORPORATED ORGANIC PHOTOVOLTAICS (OPVS)

Above said metal nanoparticles are capable of enhancing photon absorption by LSPR or by photon scattering. But using semiconducting nanoparticles instead of metal nanoparticle has an added advantage. Semiconducting nanoparticle like CdSe, CdS, ZnO, TiO_2, CuS, NiS, PbS, etc., has absorption in spectral region [72–76]. Moreover, they can become an efficient electron cascade layer, provided the work functions are suitable. But usage of selenium like elements for solar cell fabrication is banned over many countries. Here safer, less harmful, environmentally friendly inorganic semiconducting nanomaterials (oxide and sulfide based) that have the potential in improving the performance parameter of photovoltaic devices are taken into consideration.

8.7.1 OXIDE-BASED NANOPARTICLE INCORPORATED ORGANIC SOLAR CELLS

Common, well studied and most exploited inorganic nano-semiconductor for OPVs is ZnO. TiO_2 is equally important and well known in the field of PVs. Both materials can be synthesized easily via different simple deposition techniques. They possess a wider band gap and is easily tunable by doping. While dealing with OPVs, utilizing TiO_2 as ETL is rarely reported. Another oxide semiconducting nanomaterials widely used for OPV is CuO_x. Numerous reports are in the literature that utilize copper oxide as thin film/nanoparticle to enhance device efficiency. Along with this tungsten oxide, nickel oxide, vanadium pentoxide, molybdenum oxide are another set of material that seek application in OPVs. A few among the oxide nanomaterials, along with a few relevant and recent report within a decade are discussed below.

8.7.1.1 ZINC OXIDE (ZNO)

ZnO is one among the most common and highly potential semiconductor with wide variety of application. ZnO thin films are highly transparent, resistivity of the material can easily be tunable by doping and they possess less hazard. Along with these the give below nature of ZnO make it usable of numerous application [77]:

- Energy band gap 3.37 eV;
- Conductivity type is 'n;'
- Environment friendly, low-cost material;
- Availability of numerous simple deposition techniques.

Many reports are in the literature explain the devices fabricated using ZnO nanoparticle. Among which works explained in a few recently published attractive reports are discussed below.

ZnO nanoparticle applying as nanodiffusers to reduce photon reflection and to improve availability of light for active region. Here research used computational simulation to analyze the contribution of these nanodiffusers in organic solar cells. Nanospheres of ZnO was dispersed on the top of organic solar cells. These particles reduces the reflection and improve absorption. Another specialty of the work is that they utilized a green synthesis technique to develop ZnO nanosphere. Optimized nanospheres possess a diameter oh 160 nm enhance the device performance via, trapping of light and anti-reflection [78].

Even though hybrid solar cell are more efficient, the interface between the inorganic and organic material is still ruining the performance of device. Tao et al. concentrated their work to develop self-assembled monolayer of conjugated molecules into ZnO nanoparticles. Thus, the passivized metal oxide (ZnO) nanoparticles incorporated organic solar cells yielded an efficiency ~ 17% (Figure 8.8) [79].

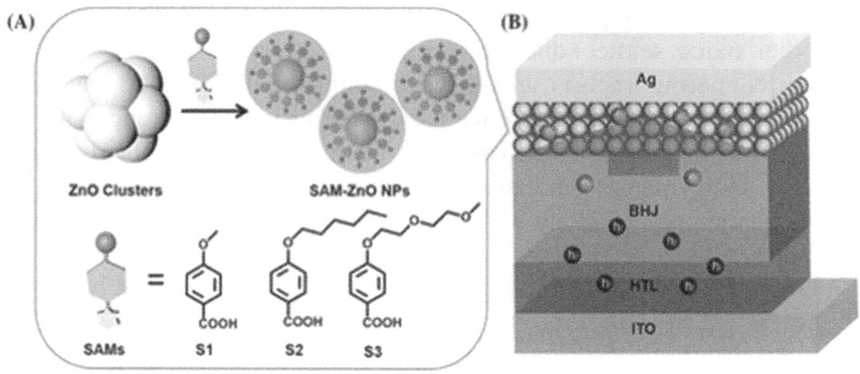

FIGURE 8.8 (A) Passivized metal oxide; (B) solar cell configuration used of the study by Tao et al. in 2022.

Source: Reprinted with permission from Ref. [79]. Copyright. © 2021 The Authors. InfoMat published by UESTC and John Wiley & Sons Australia, Ltd. https://creativecommons.org/licenses/by/3.0/

Olayinka et al. developed ZnO nanoparticles for green route from the leaf of Carica Papaya. Different stages of preparation technique is shown in figure. Developed ZnO nanoparticles possess all the properties suitable for photovoltaic application and suggest this material for the same [80].

8.7.1.2 COPPER OXIDE (CUO)

Major attraction of CuO nanoparticles are their cost effectiveness, low toxicity, high absorption, etc., CuO has an energy band gap ~ 1.5 eV (best suitable for solar cell application). This possess p-type conductivity [81–83]. A few recently published works are given below.

Wet chemical synthesis is chosen by Siddiqui et al. for developing nanoparticles of copper oxide. As-prepared NP is inserted as a thin film between P3HT and PCBM (Figure 8.9). A change in physics behind the working of the device was argued by the authors. As a result of morphological change in photoactive layer owing to the insertion of CuO nanoparticle device exhibit enhanced efficiency of 3.82% from 2.85% [84].

Another work with CuO nanoparticles by Wanninayake et al. make use of the device structure ITO/PEDOT:PSS/P3HT:PCBM:CuO-Np/Al. Incorporation of nanoparticle enhance the efficiency by 24% comparing to that without [85]. Glass/ITO/ZnO/ZnO-CuO/P3HT:PCBM/MoO_3/Al was fabricated by

Mahajan et al. Here the researchers make use of a mixed form of ZnO and CuO NPs as a thin film interface with P3HT:PCBM. Addition of inorganic layer was found to be effective to enhance device performance [86]. Oh et al. fabricated organic solar cell with P3HT:PCBM and further modified by incorporating CuO nanoparticles into PEDOT:PSS layer at different concentration. At optimized concentration, the device efficiency was better owing to suitable properties like electrical resistivity, absorption, etc. [87].

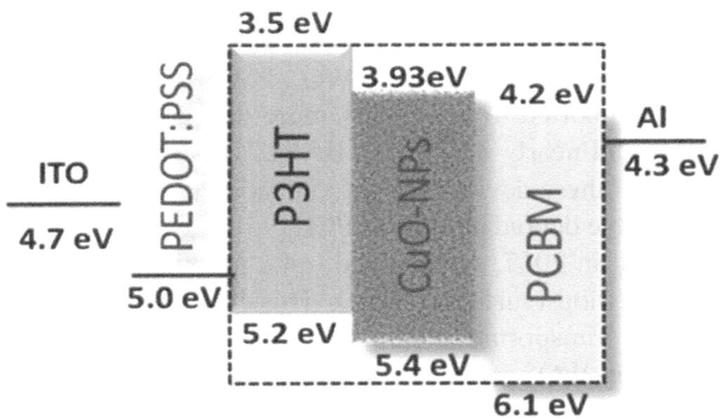

FIGURE 8.9 Device configuration and work function of solar cell fabricated by Siddiqui et al. showing the insertion of CuO-nanoparticles in between P3HT and PCBM [84].

Source: Reprinted with permission from Ref. [84]. Copyright © 1969, Elsevier. https://creativecommons.org/licenses/by/4.0/

8.7.1.3 TUNGSTEN OXIDE

Tungsten Oxide (WO_x) is capable to function as a hole collector as well as light absorber. Along with this the following properties enable the material to become virtuous:

- Conductivity type – n;
- Work function range – 4.7 to 6.4 eV [88–91];
- Even surface morphologies;
- Enhance carrier mobility;
- Increase the charge collection [92].

Only countable research works were reported on utilizing tungsten nanoparticles in OPV, perovskite, dye sensitized solar cells, etc. [93, 94].

A few research works utilizing tungsten oxide in OPV in past decade is discussed below.

Recently in 2020, Remya et al. synthesized nanosheets of WO_3 and was incorporated in OPVs. Work function of tungsten oxide was in such a way that it promote and facilitates transportation of hole in OPVs. They developed nanosheets of both dehydrated tungsten oxide and di-hydrated layered tungsten oxide. Solar cells fabricated with the conventional PEDOT:PSS as HTL and the above said forms of tungsten oxide as HTL layers. Even though both the device with tungsten oxide as HTL layer exhibit better performance than the other, noticeable enhancement in device performance was exhibited by di-hydrated layered nanosheets of WO_3. Power conversion efficiency of the device was reported to be slightly more than 5% (for active region – P3HT:PCBM) and nearly 8%, for PTB7:PC71BM as active region. Their report shows that these devices possess operational stability for around 240 h, with a negligible degradation in its PCE [95].

Brütsch et al. in 2017, synthesized ~ 4.6 nm sized tungsten oxide via hot-injection method (surfactant-free). This is applied in OSC as anode buffer. Efficient transportation of electron and hole was a benchmark of sub-stoichiometry WO3-x, due to the lack of surfactants. PCE obtained for the device using tungsten oxide layer was 6.3% whereas that without has only 3.7%. Along with this the report obviously shows that these devices does not require thermal annealing making it suitable for large area device fabrication [96].

8.7.1.4 TITANIUM DIOXIDE (TIO_2)

Report in the literature shows that polymer solar cells make use of TiO_2 as optical spacers, as shield to prevent moisture, hole blocking/ETL. The thermal instability due to post-annealing limit the application thin film titanium dioxide [97]. But utilizing TiO_2 nanoparticles reduce this risk. But still stability of device with pure titanium dioxide nanoparticle is low. Hence researcher seek different way to improve the material stability and hence the device's.

Xiong et al. make use of solution-based method to prepare titanium dioxide sol and were utilized for solar cell with configuration ITO/PEDOT:PSS/P3HT:PCBM/TiO_2NPs/Al. Enhancement in device efficiency above 30% is claimed by the authors, compared to the device without TiO_2 nanoparticles [98].

Another piece of exciting research work on titanium dioxide in OPV was reported by Ikram in 2015. They incorporate nanoparticles of titanium into the active layer, P3HT:PCBM. Along with the incorporation by utilizing chlorobenzene as the solvent, amount of PCBM was reduced and they successfully replace PCBM with TiO_2. Significant enhancement in device efficiency and EQE were reported [99].

Another group of semiconducting nanomaterials that has the potential for becoming a part of organic solar cell are molybdenum oxide (MoOx), vanadium pentoxide V_2O_5), nickel oxide (NiOx). These materials have already proved their competency to be a better candidate for organic solar cells. MoO_3 owing to its stable and transparent nature is used as HTL instead of PEDOT:PSS. High temperature stability of these materials were reported by Lee et al. [100, 101]. Inserting spin coated NiO as HTL layer in OSC and improved device performance was reported by Parthiban et al. [102] Efficiency of 6.1% while using NiOx as HTL [103]. Cong et al. applied a green method to prepare vanadium oxide hydrate layers (VOx_nH_2O) to enhance the PCE in organic PTB7-Th:PC71BM- and P3HT:PC61BM-based polymer solar cells up to 8.11% and 3.24% [104].

8.7.2 SULFUR-BASED SEMICONDUCTING NANOPARTICLE INCORPORATED ORGANIC SOLAR CELLS

Along with oxide semiconductors, sulfur-based semiconductors also find wide range of application as carrier transport layer as well as efficient supporter for enhancing the performance of an organic solar cell. Here we are discussing a few research works reported in the literature in part decade on OPVs exploiting nanoparticles of sulfur-based semiconductors.

8.7.2.1 COPPER SULFIDE (CUS) – HOLE TRANSPORT MATERIAL

Most popular material used as HTL for OPV is PEDOT:PSS poly (3,4-ethylene dioxythiophene)-poly(styrenesulfonate) due to its easiness in deposition and high work function. But owing to the disadvantages like high R_s (series resistance) and hygroscopic nature of the above said, the combination of ITO and PEDOT:PSS become unstable and corrosive too. All these finally led to the poor device performance. As another option for HTM, CuS was preferred by many researchers. Copper Sulfide (CuS) has wide variety of application and place its signature in gas sensors, lithium-ion batteries, non-linear optical material and solar radiation, thin films and composite materials, in dye-sensitized solar cell. Plasmonic resonance effect of CuS make it good candidate for incorporating in active region of organic solar cells. Along with that CuS has:

- High transparency;
- High stability;
- Availability of easy deposition technique;
- Inexpensive;
- Solution processibility.

Adedeji and his coworkers in 2020 develop CuS nanoparticles and utilize it as HTL in OSC for photon collection/harvesting. Incorporation of CuS nanoparticles was done by dispersing the same in PEDOT:PSS at different concentrations. Device structure fabricated in this research work was ITO/PEDOT:PSS:CuS/P3HT:PC61BM/LiF/Al. CuS was incorporated into PEDOT:PSS solution and is deposited as single layer. Different concentration of CuS was dispersed and the better percentage by weight was optimized as 0.10%. Efficiency obtained for the device was 4.51%. This was higher compared to that obtained without HTL layer, owing to SPR absorption by CuS [105].

Bhargava and his co-workers in 2018 fabricated solar cells using CuS as the HTL. Basic device structure was Glass/ITO/CuS/Active layer/Al. PTB7:PC71BM and PCDTBT:PC17BM, were used as the blend for active region. On I-V characterization, polymer blend with PTB7:PC71BM as the active layer yielded and an efficiency of 4.32%, best efficiency from their studies. With same device structure except CuS, which was replaced by PEDOT:PSS yielded an efficiency half that of the above said. These obviously shows the potential of CuS to be a better choice for functioning as HTL in organic solar cell. Here, enhancement in device performance while using CuS film was claimed due to the better surface morphology of CuS sample. Highly uniform CuS film provide better contact further result in charge carrier collection and separation [106].

Aiming at the industrial application, mass production of CuS using simple and facile technique was introduced by Li et al. [107]. CuS colloidal nanocrystals, were developed using low-cost precursors at cost effective atmosphere. The prepared CuS nanocrystals were found to have suitable energy levels for the device application along hydrophilic, and non-corrosive nature. Prepared samples structure was hexagonal covellite CuS. Device was fabricated with the structure ITO/CuS/P3HT:PC61BM/Ca/Al. Yielded PV parameter of the devices were, V_{oc} = 0.57 V, J_{sc} – 10 mA/cm^2, FF of 0.54 and Power conversion efficiency (PCE) = 3.1% [107].

Nanoplates of Copper sulfide was developed by Kim et al. [108], by the method of hot injection. CuS was incorporated in the OPVs with different ratio with the active layer precursor. The group developed devices with different configuration as follows:

- ITO/PEDOT:PSS/P3HT:PCBM/Ca/Ag;
- ITO/PEDOT:PSS/P3HT:PCBM:CuS/Ca/Ag;
- ITO/ZnO/P3HT:PCBM/Ag;
- ITO/ZnO/P3HT:PCBM:CuS/Ag.

For the last said device, fabricated in conventional structure, the researchers go for various ratio of P3HT:PCBM:CuS. Among those P3HT:PCBM:CuS ratio – 1:0.8:0.2, possess better device efficiency (2.64%) due to enhancement in short circuit current density to 8.38 mA/cm^2 from 6.91 mA/cm^2 (device without CuS). Improvement in efficiency was only due to the betterment in J_{sc}. This might be due to more methodical charge carrier collection. CuS as HTL may collect holes more effectively from the excitons before getting recombined than the device with CuS nanoplates [108].

Lei et al. [109], tried and succeed in developing an inorganic complementary for transparent and conducting PEDOT:PSS. The material chosen was CuS and it was proven to be better hole collecting layer than PEDOT:PSS. In this work, nanosheet of CuS was fabricated using *in-situ* hydrothermal method was developed by Lei et al. and is applied for OPV resulting in device structure – FTO/CuS/P3HT: PCBM/LiF/Al. The properties exhibited by the CuS sample are:

- Transparent;
- Hexagonal structured 2D nanosheets with intercalated structure;
- p-type with high hole mobility.

Thickness of CuS film was varied from 0 to 100 nm and the devices were fabricated with each samples performance was measured. Better device performance was exhibited by the device with the CuS sheet with thickness in 60 nm. Authors highlighted the enhancement in J_{sc} as the reason for the betterment in efficiency. A good reduction in series resistance resulted in the increase in J_{sc}. But an increase in shut resistance was also increased and as a result V_{oc} shows slight enhancement. PV parameter for the best device (CuS – 60 nm thickness) exhibit V_{oc}= 0.55 V, J_{sc}=10.96 mA/cm^2, FF= 56%, η = 3.4% [109].

8.7.2.2 SILVER SULFIDE/SILVER GALLIUM SULFIDE (AGS)

Another sulfide compound that seeks application in OPV is silver sulfide and silver gallium sulfide (AGS). They support photon harvesting as well as carrier separation. Properties and advantages of these materials to utilize in OPV are:

- Optical band gap – 0–9 to 1.05 eV for AgS and 1.4 eV for AgGaS;
- Absorption co-efficient is high;
- Chemical stability good;
- Excellent electronic and photoconductivity [110–113].

Hamed et al. fabricated an inverted structure solar cell with device configuration: ITO/ZnO/P3HT:PC61BM-Ag_2S/MoO_3/Al. Nanoparticles of silver sulfide developed using wet chemical process was incorporated into the precursor solution for active layer P3HT:PC61BM at various ratios. Devices were fabricated with various ratios (1%, 3%, 5%, and 7% by weight) and a detail investigation was undergone to optimize the accurate ratio of AgS incorporation that provide the best performing device. Doping at 1% weight yielded best device of efficiency 5.15%. As mentioned in other cases here also the nanoparticles are designed to enhance photons harvesting, charge transport and reduced charge recombination. All these finally result in enhance current density of ~ 17 mA/cm^2. Authors also claimed about the better device stability due to the inverted structure and also mentioned about the necessity of device encapsulation [114].

AGS is another compound that place its signature as good supporter of light absorber as well as helps in enhancing charge carrier separation. The device structure choice for the study was: Glass/ITO/TiO_2/PCDTBT:PC71BM: AGS NCs/MoO_3/Ag. Here AGS with stoichiometry Ag_8GeS_6 in nanocrystal form with a band gap of 1.4 eV was developed for incorporating in the above said device structure. To study the effect of different concentration on solar cell performance, devices were fabricated with different weight ratio from 0 to 1 (as 0, 0.1, 0.4, 0.7, 1). Best device under this study yielded power conversion efficiency of 6.975% with 0.848 V as the open circuit voltage, 14.637 mA/cm^2 as short circuit current density with ~57% of FF [115].

8.7.2.3 CADMIUM SUFIDE

Cadmium sulfide (CdS) possess a band gap in the region 2.25 eV. CdS act as a better passage for electron form LUMO of donor to acceptor because of the conduction band at 3.9 eV. This position of conduction comes between the LUMO of PCBM and donor polymers, like PTB, PEHT, and PCPDTBT, etc. [116–118]. Other advantages of CdS material are:

- Strong absorption in visible region;
- Stability again temperature and photochemical reaction;
- Excellent charge carrier mobility [119].

Microwave synthesis of CdS nanoparticles were done by Alonso and co-workers in 2014 and utilize it for fabricating solar cell with device structure ITO/CdS-f/active layer/CP/Au. Thermal annealing was done to make proper alignment of layers and better device performance mainly short circuit current density [120]. Figure 8.10 shows the SEM image of spherical shaped CdS nanoparticles with similar size and containing carbonyl groups at their surface.

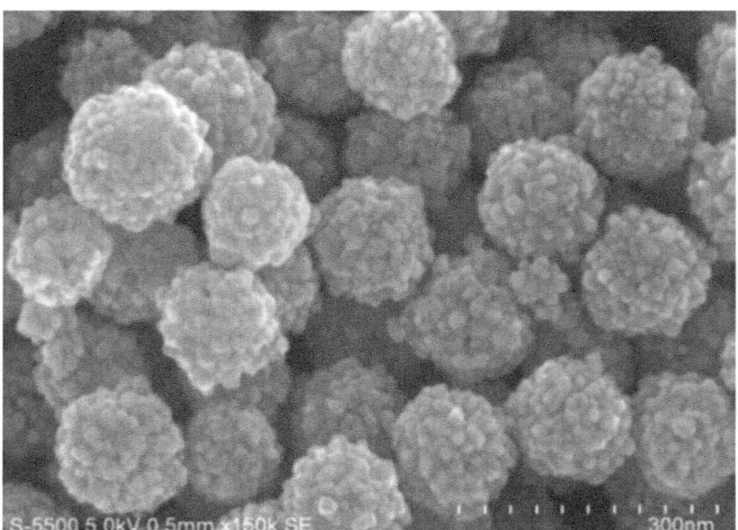

FIGURE 8.10 SEM image of CdS nanoparticles developed by Alonso and his co-workers in 2014 [120].

Wet chemical deposition technique was employed for developing CdS nanowires. Nanowires grown was ultrathin and was working as ETL in OSCs. CdS nanowires are as shown in Figure 8.11. Here device fabrication was done with the configuration: Glass/ITO/ZnO/CdS nanowire/P3HT:PCBM/PEDOT:PSS/Ag. It is obvious from the device structure that the energy band will better aligned owing to the presence of CdS in between ZnO and active region which smoothened the electron transportation. This in turn enhance the PCE by 22%, compared that without CdS nanowire layer. Besides, this U-CdS nanowires provide longer lifetime for the device [121].

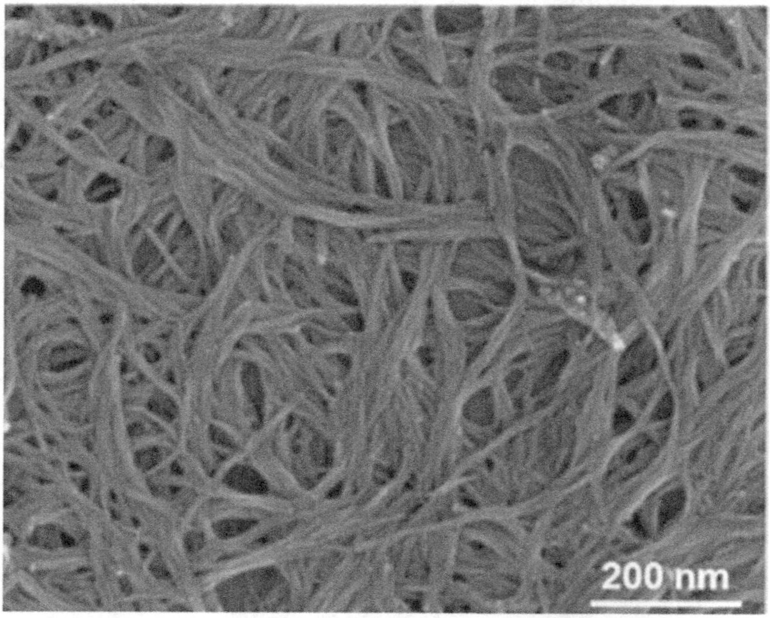

FIGURE 8.11 SEM image of ultrathin CdS nanowires. Developed by Seo et al. in 2020

Source: Reprinted with permission from Ref. [121]. Copyright 2020. Elsevier Ltd.

Sharma et al. reported the synthesis of CdS nanoparticles for integrating with PTB7:PCBM [poly[[4,8-bis[(2-ethylhexyl)oxy]benzo[1,2-b:4,5-b0] dithiophene-2,6-diyl][3-fluoro-2-[(2-ethylhexyl)carbonyl] thieno [3,4-b] thiophenediyl]] (PTB7): [6,6]-phenyl C71-butyric acid methyl ester (PCBM)], so as to develop a device with configuration ITO/ZnO/PTB7:CdS:

PCBM/MoO_3/Ag. This inverted structured solar cells exhibit an enhanced short circuit current density as well as efficiency by 10% [122].

8.7.2.4 NICKEL SULFIDE (NIS)

Another nickel sulfide (NiS) is another material suitable to use as HTL layer or as dopant in HTL because of it following properties and advantages:

- Low charge transfer resistance;
- High stability;
- Superior corrosion resistance;
- Low resistivity;
- Good optical transparency;
- Wide bandgap energy [123–126], excellent electrochemical activities;
- Irreversible transformation in several electrolytes [40, 127].

Hilal et al. [38] developed NiS through solvothermal technique having flower like morphology. Detailed study on the morphology of sample with respect to varying sulfur content was analyzed. These samples were utilized for developing a HTL for polymer solar cell application. This work aims at developing Nikel sulfide based HTL providing a low resistive path for charge carrier flow via offering an appropriate work function.

In this work researchers developed NiS having work function suitable to have a good alignment with HOMO level of P3HT. This allows the easy transport for hole to NiS from HOMO level of P3HT. Possibility of charge carrier recombination get reduced and better collection of charge carriers towards the electrode will increases owing to the above said better energy level alignment. In turn this reduces the series resistance also. All the above said will finally contribute to the enhanced power conversion efficiency of a solar cell.

Device structure used by Hilal and his co-worker was: Glass/ITO/HTL/P3HT:PCBM/PCBM/Al. NiS function as Anode interfacial layer in the device. Incorporation of NiS HTL layer was employed in OSC at different ratio of sulfur content. The highly stable nature of this device structure is more attractive, according to the authors [38].

Hilal et al. make use of NiS as HTL layer whereas Hamed et al. synthesized NiS nanocomposite and utilize this as a dopant in HTL of thin films OPV. HTL of this said device was PEDOT:PSS. The device structure was glass/ITO/PEDOT:PSS-NiS/P3HT:PCBM/LiF/AL. Characterization studies on the devices indicates an enhancement in the photo-generated current owing to the reduction in recombination of charge carriers leading to betterment in transportation of charge carriers at donor acceptor interface. A boost up in the efficiency for NiS incorporated HTL device compared that device without NiS doping 95% was observed by Hamed and co-workers. Local SPR absorption was considered to be the reason for the charge in device performance. An investigation on the doping quantity of NiS in PEDOT:PSS layer was also done to optimizing the doping level required for the best device (efficiency yielded – 6.03%) and was found to be 0.1% by weight [24].

Apart from the above said nanomaterials, numerous nanomaterials were developed by researchers round the word. Possibilities and potential of nanomaterials in photovoltaics as well as in all sorts of renewable energy sector is mammoth. Above mentioned studies shows that our current scarcity of energy sources will once be surely replace by these renewable energy devices. Let the coming future be rich with energy!

8.8 CONCLUSION

Role of nanoparticle in harvesting energy from Green Energy Source is inevitable. Here applying nanomaterials for harvesting solar energy is taken into consideration. The topic discussed above specifically pinpointed to OPVs. Metal, metal oxide, metal sulfide nanoparticle incorporation into organic solar cell (resulting in hybrid solar cell) to enhance the device efficiency was the major concern in this chapter. Numerous works done by many researchers in this field obviously reveals the potential of these materials in empowering the output of solar cell. Typical reason for this enhancement in performance owe to the betterment in light trapping, scattering, and absorption due to the surface plasmon and LSPR. Properties and advantages of metal nanoparticles, metal oxide nanoparticle and metal sulfide nanoparticles along with a few recently published (within a decade) research works were discussed in this chapter. According to the material property these nanomaterials were utilized as ETL, and HTL. All the mentioned work highlighted the potential and further possibility of these materials in the field of energy harvesting.

KEYWORDS

- **cadmium sulfide**
- **electron transport layer**
- **hole transport layer**
- **organic photovoltaics**
- **oxide nanoparticles**
- **semiconducting nanoparticles**
- **solar cells**
- **sulfide nanoparticles**

REFERENCES

1. Dieter, W., & Dieter, M., (1991). Adv. Muter., 3, No. 3.
2. Yang, F., Huang, Y., Li, Y., et al. (2021). Large-area flexible organic solar cells. *NPJ Flex Electron., 5,* 30. https://doi.org/10.1038/s41528-021-00128-6.
3. Santhosh, M. V., Deepu, D. R., Kartha, C. S., Kumar, K. R., & Vijayakumar, K. P., (2014). All sprayed ITO-free $CuInS_2/In_2S_3$ solar cells. *Solar Energy, 108*, 508–514.
4. Ninan, G. G., Kartha, C. S., & Vijayakumar, K. P. (2016). On the preparation of n-type SnS: Cu using chemical spray pyrolysis for photovoltaic application: Effect of annealing. *Solar Energy Materials and Solar Cells, 157*, 229–233.
5. Geethu, R., Kartha, C. S., & Vijayakumar, K. P., (2015). Improving the performance of ITO/ZnO/P3HT: PCBM/Ag solar cells by tuning the surface roughness of sprayed ZnO. *Solar Energy, 120*, 65–71.
6. Ninan, G. G., Kartha, C. S., & Vijayakumar, K. P., (2016). Spray pyrolyzed SnS thin films in n and p type: Optimization of deposition process and characterization of samples. *Journal of Analytical and Applied Pyrolysis, 120*, 121–125.
7. Rajeshmon, V. G., Kartha, C. S., Vijayakumar, K. P., Sanjeeviraja, C., & Abe, T., (2011). Role of precursor solution in controlling the opto-electronic properties of spray pyrolyzed Cu_2ZnSnS_4 thin films. *Solar Energy, 85*(2), 249–255.
8. John, T. T., Wilson, K. C., Ratheesh, K. P. M., & Sudha, K. C., (2005). $CuInS_2$ films using repeated chemical spray pyrolysis. *Physica Status Solidi (a), 202*(1), 79–84.
9. Sunny, G., Thomas, T., Deepu, D. R., Kartha, C. S., & Vijayakumar, K. P., (2017). Thin film solar cell using earth abundant Cu_2SnS_3 (CTS) fabricated through spray pyrolysis: Influence of precursors. *Optik, 144*, 263–270.
10. Santhosh, M. V., Deepu, D. R., Geethu, R., Rajeev, K. K., Sudha, K. C., & Vijayakumar, K. P., (2014). Spray pyrolyzed microporous TiO_2 thin films by optimization of substrate temperature for 'all sprayed' solar cells. *Semicond. Sci. Technol., 29*, 115026.

11. Kang, J. W., Kang, Y. J., Jung, S., Song, M., Kim, D. G., Kim, C. S., & Kim, S. H., (2012). Fully spray-coated inverted organic solar cells. *Solar Energy Materials & Solar Cells, 103*, 76–79.
12. Park, H. Y., Lim, D., Ohc, S. H., Kang, P. H., Kwak, G., & Jang, S. Y., (2014). Inverted-structure polymer solar cells fabricated by sequential spraying of electron-transport and photoactive layers. *Organic Electronics, 15*, 2337–2345.
13. La Notte, L., Cataldi, P., Ceseracciu, L., Bayer, I. S., Athanassiou, A., Marras, S., Villari, E., et al., (2018). Fully-sprayed flexible polymer solar cells with a cellulose-graphene electrode. *Materials Today Energy, 7*, 105–112.
14. Park, S. Y., Kang, Y. J., Lee, S., Kim, D. G., Kim, J. K., Kim, J. H., Kang, J. W., & Park, S. Y., (2011). Spray-coated organic solar cells with large-area of 12.25 cm^2. *Solar Energy Materials and Solar Cells, 95*, 852–855.
15. Peumans, P., Yakimov, A., & Forrest, S. R., (2003). Small molecular weight organic thin-film photodetectors and solar cells. *J. Appl. Phys., 93*, 3693.
16. https://eng.libretexts.org/Bookshelves/Materials_Science/Supplemental_Modules_(Materials_Science)/Materials_and_Devices/Organic_Solar_Cells (accessed on 06 September 2023).
17. Rostalski, J., & Meissner, D., (2000). Photocurrent spectroscopy for the investigation of charge carrier generation and transport mechanisms in organic p/n-junction solar cells. *Sol. Energy Mater. Sol. Cells, 63*, 37.
18. Zhigang, Y., Jiajun, W., & Qingdong, Z., (2016). *Adv. Sci., 3*, 1500362.
19. Aernouts, T., Geens, W., Poortmans, J., Heremans, P., Borghs, S., & Mertens, R., (2002). Extraction of bulk and contact components of the series resistance in organic bulk donor-acceptor-heterojunctions. *Thin Solid Films, 403*, 297–301.
20. Hu, G., Xu, T., Chen, X., James, T. D., & Xu, S., (2016). Solar-driven broad spectrum fungicides based on monodispersed Cu7S4 nanorods with strong near-infrared photothermal efficiency. *RSC Adv., 6*, 103930–103937.
21. Mbuyise, X. G., Arbab, E. A. A., Kaviyarasu, K., Pellicane, G., Maaza, M., & Mola, G. T., (2017). Zinc oxide doped single wall carbon nanotubes in hole transport buffer layer. *J. Alloy. Compd.* https://doi.org/10. 1016/j.jallcom.2017.02.249.
22. Yu, J. J. C., et al., (2018). Highly efficient and stable inverted perovskite solar cell employing PEDOT: GO composite layer as a hole transport layer. *Sci. Rep., 8*(1), 1070.
23. Hilal, M., & Han, J. I., (2018b). The effect of the functionalization of multiple carrier transporting interlayers on the performance and stability of bulk heterojunction organic solar cells. *J. Mater. Sci.: Mater. Electron.*, 1–16.
24. Mohammed, H. S. G., Saheed, O. O., Amit, K. D., Gaurav, S., D., & Genene, T. M., (2020). Nickel sulphide nano-composite assisted hole transport in thin film polymer solar cells *Solar Energy, 195*, 310–317.
25. Geethu, R., Santhosh, Sudha K. C., & Vijayakumar, K. P., (2019). Overcoming consistency constrains of ITO/ZnO/P3HT:PCBM/Ag solar cell by open air annealing and its systematic stability study under inborn conditions. *Journal of Materials Science: Materials in Electronics, 30,* 18981–18989.
26. Berson, S., De Bettignies, R., Bailly, S., Guillerez, S., & Jousselme, B., (2007). Elaboration of P3HT/CNT/PCBM composites for organic photovoltaic cells. *Adv. Funct. Mater., 17*, 3363e3370.

27. Li, X. M., Xie, D., Park, H. S., Zhu, M. T. H. Z., Wang, K. L., Wei, J. Q., Wu, D. H., & Zhu, J. K., (2013). Ion doping of graphene for high-efficiency heterojunction solar cells. *Nanoscale, 5*, 1945e1948.
28. Lee, J. M., Park, J. S., Lee, S. H., Kim, H., Yoo, S., & Kim, S. O., (2011). Selective electron-or holetransport enhancement in bulk-heterojunction organic solar cells with N-or B-doped carbon nanotubes. *Adv. Mater., 23*, 629–633.
29. Huang, L. L., Chen, X. Z., Bai, B. F., Tan, Q. F., Jin, G. F., Zentgraf, T., & Zhang, S., (2013). Helicity dependent directional surface plasmon polariton excitation using a metasurface with interfacial phase discontinuity. *Light Sci. Appl., 2,* 70.
30. Kosten, E. D., Atwater, J. H., Parsons, J., Polman, A., & Atwater, H. A., (2013). Highly efficient GaAs solar cells by limiting light emission angle. *Light Sci. Appl., 2*, e45.
31. Hwang, S. K., Lee, J. M., Kim, S., Park, J. S., Park, H. I., Ahn, C. W., Lee, K. J., Lee, T., & Kim, S. O., (2012). Flexible multilevel resistive memory with controlled charge trap band N-doped carbon nanotubes. *Nano Lett., 12*, 2217–2221.
32. Lee, D. H., Lee, W. J., Kim, S. O., & Kim, Y. H., (2011). Theory, synthesis, and oxygen reduction catalysis of Fe-porphyrin-like carbon nanotube. *Phys. Rev. Lett., 106*, 175502.
33. https://www.sciencedirect.com/topics/earth-and-planetary-sciences/surface-plasmon-resonance (accessed on 06 September 2023).
34. Ashraf, U., & Xiaohan, Y. (2014). *Surface Plasmonic Effects on Organic Solar Cells. 14*(2), 1099–1119.
35. Atwater, H. A., & Polman, A., (2010). Plasmonics for improved photovoltaic devices. *Nature, 9*, 205–213.
36. Qiaoqiang, G., Bartoli, F. J., & Kafafi, Z. H., (2012). Research highlights on organic photovoltaics and plasmonics. *IEEE Photonics J., 4*, 620–624.
37. Huang, X., Neretina, S., & El-Sayed, M. A., (2009). Gold nanorods: From synthesis and properties to biological and biomedical applications. *Adv. Mater., 21*, 4880–4910.
38. Muhammad, H., & Jeong, I. H., (2019). Preparation of hierarchical flower-like nickel sulfide as hole transporting material for organic solar cells via a one-step solvothermal method. *Solar Energy, 188*, 403–413. https://doi.org/10.1016/j.solener.2019.06.032.
39. Liu, K., Bi, Y., Qu, S., Tan, F., Chi, D., Lu, S., Li, Y., Kou, Y., & Wang, Z., (2014). Efficient hybrid plasmonic polymer solar cells with Ag nanoparticle decorated TiO_2 nanorods embedded in the active layer. *Nanoscale, 6*, 6180–6186.
40. Yu, J. C., et al., (2014). Highly efficient polymer-based optoelectronic devices using PEDOT: PSS and a GO composite layer as a hole transport layer. *ACS Appl. Mater. Interfaces, 6*(3), 2067–2073.
41. Sha, W. E. I., Choy, W. C. H., Chen, Y. P., & Chew, W. C., (2011). Optical design of organic solar cell with hybrid plasmonic system. *Opt. Express, 19*, 15908–15918.
42. Hamed, M. S. G., & Mola, G. T., (2019). Copper sulphide as a mechanism to improve energy harvesting in thin film solar cells. *J. Alloy, Compd., 802*, 252–258.
43. Tabitha, A. A., Genene, T. M., & Vincent, O. N., (2020). Reduced graphene oxide-germanium quantum dot nanocomposite. *Solar Energy, 203*, 83–90.
44. Ganeshan, D., Chen, S. C., Yin, Z. G., & Zheng, Q. D., (2015). *RSC Adv., 5*, 16153.
45. Krebs, F. C., (2010). *Polymeric Solar Cells: Materials, Design, Manufacturing.* https://www.amazon.com/Polymeric-Solar-Cells-Materials-Manufacture/dp/1605950173 (accessed on 06 September 2023).

46. Notarianni, M., et al., (2013). Plasmonic effect of gold nanoparticles in organic solar cells. *Sol. Energy.*
47. Morfa, A. J., Rowlen, K. L., Reilly, T. H., Romero, M. J., & Van De, L. J., (2008). Plasmon-enhanced solar energy conversion in organic bulk heterojunction photovoltaics. *Appl. Phys. Lett., 92*, 13504, 13505.
48. Wu, J. L., Chen, F. C., Hsiao, Y. S., Chien, F. C., Chen, P., Kuo, C. H., Huang, M. H., & Hsu, C. S., (2011). Surface plasmonic effects of metallic nanoparticles on the performance of polymer bulk heterojunction solar cells. *ACS Nano, 5*, 959–967.
49. Qiao, L., Wang, D., Zuo, L., Ye, Y., Qian, J., Chen, H., & He, S., (2011). Localized surface plasmon resonance enhanced organic solar cell with gold nanospheres. *Appl. Energy, 88*, 848–852.
50. Se-Woong, B., Garam, P., Jonghyeon, N., Changsoon, C., Chun-Ho, L., Min-Kyo, S., Hyunjoon, S., & Jung-Yong, L., (2014). Au@Ag coreshell nanocubes for efficient plasmonic light scattering effect in low bandgap organic solar cells. *ACS Nano, 8*, 3302–3312.
51. Wenxin, N., Yi, A. A. C., Weiqing, Z., Hejin, H., & Xianmao, L., (2015). *J. Am. Chem. Soc.* doi: 10.1021/jacs.5b05321.
52. Riski, T. G., Sandeep, K., Dong-Kwon, L., Jung-Mu, K., Joong-Hee, L., Seung, H. L., & Jae-Wook, K., (2017). *ACS Appl. Mater. Interfaces*. DOI: 10.1021/acsami.7b11084.
53. Sopit, P., Apichat, P., Chutiparn, L., Ryousuke, I., Kazunari, S., Keizo, K., Pitchaya, M., et al. (2019). Investigation of a gold quantum dot/plasmonic gold nanoparticle system for improvement of organic solar cells. *Nanoscale Advances*. doi: 10.1039/c8na00119g.
54. Kawasaki, H., Hamaguchi, K., Osaka, I., & Arakawa, R., (2011). *Adv. Funct. Mater., 21*, 3508.
55. Zhou, H., Zhang, Y., Mai, C. K., Collins, S. D., Bazan, G. C., Nguyen, T. Q., & Heeger, A. J., (2015). Polymer homo-tandem solar cells with best efficiency of 11.3%. *Adv. Mater., 27*, 1767−1773.
56. Günes, S., Neugebauer, H., & Sariciftci, N. S., (2007). Conjugated polymer-based organic solar cells. *Chem. Rev., 107*, 1324−1338.
57. Virginia, C., Manish, K. S., Gutierrez-Fernandez, E., Jaime, M., Rocío, D., Pilar De La, C., Ganesh, S. D., & Fernando, L., (2022). *ACS Appl. Mater. Interfaces, 14*, 11708−11717. https://doi.org/10.1021/acsami.1c22813.
58. Catchpole, K. R., & Albert, P., (2008). Plasmonic solar cells. *Optics Express, 16*(26), 21793–21800. https:// doi. org/ 10. 1364/ OE. 16. 021793.
59. Gupta, R., Dyer, M. J., Weimer, W. A. (2002). Preparation and characterization of surface plasmon resonance tunable gold and silver films. *J. Appl. Phys., 92*, 5264–5271.
60. Anthony, J. Morfa, Kathy L. Rowlen, Thomas H. Reilly, Manuel J. Romero, & Jao van de Lagemaat. (2008). Plasmon-enhanced solar energy conversion in organic bulk heterojunction photovoltaics, *Appl. Phys. Lett. 92*, 013504.
61. Baek, S. W., Noh, J., Lee, C. H., et al., (2013). Plasmonic forward scattering effect in organic solar cells: A powerful optical engineering method. *Sci. Rep., 3*, 1726. https://doi.org/10.1038/srep01726.
62. Koul, S., & Hakim, N. U. D., (2021). Analytical modeling of the effect of silver nanoparticle incorporation in active layer of an organic solar cell. In: *IEEE Transactions on Electron Devices* (Vol. 68, No. 9, pp. 4516–4521). doi: 10.1109/TED.2021.3097970.

63. Wang, D. H., et al., (2011). Enhancement of donor-acceptor polymer bulk heterojunction solar cell power conversion efficiencies by addition of Au nanoparticles. *Angew. Chem. Int. Ed., 50*(24), 5519–5523. doi: 10.1002/anie.201101021.
64. Rycenga, M., Cobley, C. M., Zeng, J., Li, W., Moran, C. H., Zhang, Q., Qin, D., & Xia, Y., (2011). *Chem. Rev., 111*, 3669.
65. Kai, Y., Michael, S., Chu-Chen, C., Xu-Kai, X., Yun-Xiang, X., Dane, W. De. Q., Ting, H., et al., (2014). *Adv. Energy Mater.*, 1400206.
66. Han, Y. W., Jeon, S. J., Lee, H. S., Park, H., Kim, K. S., Lee, H., & Moon, D. K., (2019). *Adv. Energy Mater., 9*, 1902065.
67. Kai, C. T., Hirotoshi, S., Philipp, M., Karen, F., Sarmad, F., Yutaka, H., Christoph, J. B., & Hans-Joachim, E., (2022). *Sol. RRL, 6*, 2100887.
68. Shabani, L., Mohammadi, A., & Jalali, T., (2022). Numerical study of plasmonic effects of Ag nanoparticles embedded in the active layer on performance polymer organic solar cells. *Plasmonics, 17*, 491–504. https://doi.org/10.1007/s11468-021-01539-x.
69. Vladimir, K., Lauren, N. N., Sungsool, W., Chih-Yu, J., & Hans, D. R., (2011). Aluminum nanoparticles for improved OPV devices. *Proceedings Volume 8111, Next Generation (Nano)Photonic and Cell Technologies for Solar Energy Conversion, II*, 811117. https://doi.org/10.1117/12.893532.
70. Maria, S., George, K., Barbara, P., Amanda, G., Emmanuel, K., & Emmanuel, S., (2015). Enhanced stability of aluminum nanoparticles doped organic solar cells. *ACS Appl. Mater. Interfaces, 7*(32), 17756–17764. https://doi.org/10.1021/acsami.5b03970.
71. George, K., Emmanuel, S., & Emmanuel, K., (2013). Aluminum nanoparticles for efficient and stable organic photovoltaics. *RSC Advances, 3*(37), 16288. doi: 10.1039/c3ra42792g.
72. Shao, Y. T., Chen, C. Y., Su, C. J., Jeng, U. S., Chen, Y. F., & Su, W. F., (2012). Nanoparticle-tuned self-organization of a bulk heterojunction hybrid solar cell with enhanced performance. *ACS Nano, 6*, 1657e1666.
73. Khan, M. T., Bhargav, R., Kaur, A., Dhawan, S. K., & Chand, S., (2010). Effect of cadmium sulphide quantum dot processing and post thermal annealing on P3HT/PCBM photovoltaic device. *Thin Solid Films, 519*, 1007e1011.
74. Oh, S. H., Heo, S. J., Yang, J. S., & Kim, H. J., (2013). Effects of ZnO Nanoparticles on P3HT: PCBM organic solar cells with DMF-modulated PEDOT: PSS buffer layers. *ACS Appl. Mater. Inter, 5*, 11530e11534.
75. Gollu, S. R., Sharma, R., Srinivas, G., Kundu, S., & Gupta, D., (2014). Effects of incorporation of copper sulfide nanocrystals on the performance of P3HT: PCBM based inverted solar cells. *Org. Electron., 15*, 2518e2525.
76. Itskos, G., Othonos, A., Rauch, T., Tedde, S. F., Hayden, O., Kovalenko, M. V., Heiss, W., & Choulis, S. A., (2011). Optical properties of organic semiconductor blends with near-infrared quantum-dot sensitizers for light harvesting applications. *Adv. Energy Mater., 1*, 802–812.
77. Mcpeak, K. M., & Baxter, J. B., (2009). ZnO nanowires grown by chemical bath deposition in a continuous flow microreactor. *Cryst. Growth Des., 9*, 4538–4545.
78. Olavo, C., Sajid, F., Patricia, F. M. A., Naum, F., Andreas, S., & Renato, E. De. A., (2022). Zinc oxide nanodiffusers to enhance p3ht:pcbm organic solar cells performance. *J. Mater. Sci., Mater. Electron, 33*, 3225–3236.

79. Yangdan, T., Haoran, L., Di, W., Feng, Z., Zeng, C., Haiming, Z., Hong, Z. C., & Chang-Zhi, L., (2022). Healing the degradable organic–inorganic heterointerface for highly efficient and stable organic solar cells. *InfoMat., 4*, e12276, https://doi.org/10.1002/inf2.12276.
80. Oyeshola, H., (2022). *Evaluation of Photovoltaic Properties of Green Synthesized Zinc Oxide Nanoparticles from Extract of Carica Papaya for Anode Buffer Film Layer of Polymer Solar* Cells. http://dx.doi.org/10.2139/ssrn.4070949/.
81. Ruiz, E., Alvarez, S., Alemany, P., & Evarestov, R. A., (1997). Electronic structure and properties of Cu_2O. *Phys. Rev. B, 56*, 7189/.
82. Tahir, D., & Tougaard, S., (2012). Electronic and optical properties of Cu, CuO and Cu_2O studied by electron spectroscopy. *J. Phys.: Condens. Matter., 24*, 175002.
83. Kidowaki, H., Oku, T., & Akiyama, T., (2012). Fabrication and characterization of CuO/ZnO solar cells. *J. Phys.: Conf. Ser., 352*, 012022.
84. Hafsa, S., Mohammad, R. P., Padmini, P., Qureshi, M. S., & Fozia, Z. H., (2020). Utility of copper oxide nanoparticles (CuO-NPs) as efficient electron donor material in bulk-heterojunction solar cells with enhanced power conversion efficiency. *Journal of Science: Advanced Materials and Devices, 5*, 104–110.
85. Aruna, P. W., Subhashini, G., Shengyi, L., Benjamin, C. C., & Abu-Zahra, N., (2015). Performance enhancement of polymer solar cells using copper oxide nanoparticles. *Semicond. Sci. Technol., 30*, 064004. doi: 10.1088/0268-1242/30/6/064004.
86. Prerna, M., Anoop, S., & Sandeep, A., (2020). Improved performance of solution processed organic solar cells with an additive layer of sol-gel synthesized ZnO/CuO core/shell nanoparticles. *Journal of Alloys and Compounds, 814*, 152292.
87. Sang, H. O., Seung, J. H., & Hyun, J. K., (2014). Organic solar cells with CuO nanoparticles mixed PEDOT: PSS buffer layer, regular paper 121. *J. KIEEME, 27*(2), 21–125. doi: http://dx.doi.org/10.4313/JKEM.2014.27.2.121.
88. Meyer, J., Hamwi, S., Bülow, T., Johannes, H. H., Riedl, T., & Kowalsky, W., (2007). Highly efficient simplified organic light emitting diodes. *Appl. Phys. Lett., 91*, 113506.
89. Tao, C., Ruan, S., Xie, G., Kong, X., Shen, L., Meng, F., Liu, C., et al., (2009). Role of tungsten oxide in inverted polymer solar cells. *Appl. Phys. Lett., 94*, 043311.
90. Miyake, K., Kaneko, H., Sano, M., & Suedomi, N., (1984). Physical and electrochromic properties of the amorphous and crystalline tungsten oxide thick films prepared under reducing atmosphere. *J. Appl. Phys., 55*, 2747–2753.
91. Vida, G., Josepovits, V. K., Gy″or, M., & Deák, P., (2003). Characterization of tungsten surfaces by simultaneous work function and secondary electron emission measurements. *Microsc. Microanal., 9*, 337–342.
92. Brabec, C. J., (2004). Organic photovoltaics: Technology and market. *Sol. Energy Mater. Sol. Cells, 83*, 273–292.
93. Lin, Z., Yuhua, Z., Tapas, K. M., & Senthilarasu, S., (2019). Enhanced efficiency of carbon- based mesoscopic perovskite solar cells through a tungsten oxide nanoparticle additive in the carbon electrode. *Scientific Reports, 9*, 8778 | https://doi.org/10.1038/s41598-019-45374-x.
94. Vijayakumar, P., Senthil, P. M., Sumita, M., & Ramasamy, P., (2015). Synthesis and characterizations of large surface tungsten oxide nanoparticles as a novel counter

electrode for dye-sensitized solar cell. *J. Sol-Gel. Sci. Technol, 75*, 487–494. doi: 10.1007/s10971-015-3719-z.
95. Remya, R., Gayathri, P. T. G., & Biswapriya, D., (2020). Studies on solution-processed tungsten oxide nanostructures for efficient hole transport in the inverted polymer solar cells. *Materials Chemistry and Physics, 255*, 123584. https://doi.org/10.1016/j.matchemphys.2020.123584.
96. Lennart, B., Jens, C., Radian, P., Dagmar, G., Alexander, C., & Claus, F., (2017). Surfactant-free synthesis of sub-stoichiometry tungsten oxide nanoparticles and their use as anode buffer layers in organic solar cells. *Solid State Sciences, 69*, 50–55.
97. Wang, D. H., Im, S. H., Lee, H. K., Park, O. O., & Park, J. H., (2009). *J. Phys. Chem. C., 113*, 17268.
98. Jian, X., Bingchu, Y., Conghua, Z., Junliang, Y., Haichao, D., Wenlong, H., Xiang, Z., et al., (2014). Enhanced efficiency and stability of polymer solar cells with TiO_2 nanoparticles buffer layer. *Organic Electronics, 15*, 835–843.
99. Ikram, M., Ali, S., Murray, R., Hussain, A., Islah-U-Din, & Ismat, S. S., (2015). Influence of fullerene derivative replacement with TiO_2 nanoparticles in organic bulk heterojunction solar cells. *Current Applied Physics, 15*, 48–54.
100. Kyaw, A. K. K., Sun, X. W., Jiang, C. Y., Lo, G. Q., Zhao, D. W., & Kwong, D. L., (2008). An inverted organic solar cell employing a sol-gel derived ZnO electron selective layer and thermal evaporated MoO_3 hole selective layer. *Appl. Phys. Lett., 93*, 221107.
101. Lee, D., Kim, J., Park, G., Bae, H. W., An, M., & Kim, J. Y., (2020). Enhanced operating temperature stability of organic solar cells with metal oxide hole extraction layer. *Polymers, 12*, 992.
102. Parthiban, S., Kim, S., Tamilavan, V., Lee, J., Shin, I., Yuvaraj, D., Jung, Y. K., et al., (2017). Enhanced efficiency and stability of polymer solar cells using solution-processed nickel oxide as hole transport material. *Curr. Appl. Phys., 17*, 1232–1237.
103. Jiang, F., Choy, W. C. H., Li, X., Zhang, D., & Cheng, J., (2015). Post-treatment-free solution-processed non-stoichiometric NiOx nanoparticles for efficient hole-transport layers of organic optoelectronic devices. *Adv. Mater., 27*, 2930–2937.
104. Cong, H., Han, D., Sun, B., Zhou, D., Wang, C., Liu, P., & Feng, L., (2017). Facile approach to preparing a vanadium oxide hydrate layer as a hole-transport layer for high-performance polymer solar cells. *ACS Appl. Mater. Interfaces, 9*, 18087–18094.
105. Michael, A. A., Mohammed, H. S. G., & Genene, T. M. (2020). *Light Trapping Using Copper Decorated Nano-composite in the Hole Transport Layer of Organic Solar Cell*. https://doi.org/10.1016/j.solener.2020.04.005.
106. Ranoo, B., Asit, P., Dhawan, S. K., & Gairola, S. P., (2018). Solution processed hole transport layer towards efficient and cost effective organic solar cells. *Solar Energy, 165*, 131–135.
107. Jun, L., Tonggang, J., Guo-Hong, T., Guojie, W., Chunming, S., Pandeng, L., Junfeng, F., & Ling, H., (2014). Manipulating surface ligands of copper sulfide nanocrystals: Synthesis, characterization, and application to organic solar cells. *Journal of Colloid and Interface Science, 419*, 142–147.
108. Yohan, K., Benjamin, H., Ahed, A., Christopher, P., Christian, I., Christina, G., Andreas, T., & Armin, W., (2018). CuS nanoplates from ionic liquid precursors—Application

in organic photovoltaic. *Cells Citation: The Journal of Chemical Physics, 148*, 193818. View online: https://doi.org/10.1063/1.4991622.
109. Hongwei, L., Guojia, F., Fei, C., Weijun, K., Pingli, Q., Zengcai, S., Qiao, Z., et al., (2014). Enhanced efficiency in organic solar cells via in situ fabricated p-type copper sulfide as the hole transporting layer. *Solar Energy Materials & Solar Cells, 128*, 77–84. http://dx.doi.org/10.1016/j.solmat.2014.05.021.
110. Krylova, V., Milbrat, A., Embrechts, A., & Baltrusaitis, J., (2014). Ag_2S deposited on oxidized polypropylene as composite material for solar light absorption. *Appl. Surf. Sci., 301*, 134–141.
111. Meng, Z. D., Ghosh, T., Zhu, L., Choi, J. G., Park, C. Y., & Oh, W. C., (2012). Synthesis of fullerene modified with Ag_2S with high photocatalytic activity under visible light. *J. Mater. Chem., 22*, 16127–16135.
112. Chen, C., Xie, Y., Ali, G., Yoo, S. H., & Cho, S. O., (2011). Improved conversion efficiency of Ag_2S quantum dot-sensitized solar cells based on TiO_2 nanotubes with a ZnO recombination barrier layer. *Nanoscale Res. Lett., 6*, 462.
113. Emmanuel, S., & Emmanuel, K., (2013). Nanoparticle-based plasmonic organic photovoltaic devices. *Mater. Today, 16*(4), 134–146.
114. Mohammed, S. G. H., Michael, A. Ad., Yong, Z., & Genene, T. M., (2020). Silver sulphide nano-particles enhanced photo-current in polymer solar cells. *Applied Physics A, 126*, 207. https://doi.org/10.1007/s00339-020-3389-8.
115. Zhiqi, L., Chunyu, L., Xinyuan, Z., Zhihui, Z., Wenbin, G., Liang, S., Liu, Z., & Yongbing, L., (2017). An easily prepared Ag_8GeS_6 nanocrystal and its role on the performance enhancement of polymer solar cells. *Organic Electronics, 45*, 247e255, http://dx.doi.org/10.1016/j.orgel.2017.03.029.
116. Gratzel, M., (2001). Photoelectrochemical cells. *Nature, 414*, 338–344.
117. Varga, A., Endrodi, B., Hornok, V., Visy, C., & Janaky, C., (2015). Controlled photocatalytic deposition of CdS nanoparticles on poly(3-hexylthiophene) nanofibers: A versatile approach to obtain organic/inorganic hybrid semiconductor assemblies. *J. Phys. Chem. C, 119*(50), 28020–28027.
118. (2015). Deposition of CdS nanoparticles on poly(3-hexylthiophene) nanofibers: A versatile approach to obtain organic/inorganic hybrid semiconductor assemblies. *J. Phys. Chem. C 119*, 28020–28027.
119. Ren, S. Q., Chang, L. Y., Lim, S. K., Zhao, J., Smith, M., Zhao, N., Bulovic, V., et al., (2011). Inorganic-organic hybrid solar cell: Bridging quantum dots to conjugated polymer nanowires. *Nano Lett., 11*, 3998–4002.
120. Martínez-Alonso, C., Rodríguez-Castañeda, A. C., Moreno-Romero, P., Coria-Monroy, C. S., & Hailin, H., (2014). Cadmium sulfide nanoparticles synthesized by microwave heating for hybrid solar cell applications. *International Journal of Photoenergy*. Hindawi Publishing Corporation. http://dx.doi.org/10.1155/2014/453747.
121. Dong-Bum, S., Songhee, K., Rajesh, G., Kiran, K. C., Kihyon, H., & Eui-Tae, K., (2020). Synthesis and organic solar cell application of RNA-nucleobase-complexed CdS nanowires. *Solar Energy, 206*, 287–293.
122. Ramakant, S., Sagar, B., & Dipti, G., (2016). Effect of incorporation of CdS NPs on performance of PTB7: PCBM organic solar cells. *Organic Electronics, 33*, 274–280.
123. Kristl, M., Dojer, B., Gyergyek, S., & Kristl, J., (2017). Synthesis of nickel and cobalt sulfide nanoparticles using a low cost sonochemical method. *Heliyon, 3*, 255–274.

124. Pitchaiya, S., Natarajan, M., Santhanam, A., Ramakrishnan, V. M., Asokan, V., Palanichamy, P., Rangasamy, B., et al., (2018). Nickel sulphide-carbon composite hole transporting material for (CH3NH3PbI3) planar heterojunction perovskite solar cell. *Mater. Lett., 221*, 283–288.
125. Min, H., (2016). Characterization of N-type and P-type thin films. *Chem. Sci. J., 7*, e113.
126. Son, M. Y., Choi, J. H., & Kang, Y. C., (2014). Electrochemical properties of bare nickel sulfide and nickel sulfide–carbon composites prepared by one-pot spray pyrolysis as anode materials for lithium secondary batteries. *J. Power Sources, 251*, 480–487.
127. Yang, H., et al., (2012). Organic solar cells employing electrodeposited nickel oxide nanostructures as the anode buffer layer. *Sol. Energy Mater. Sol. Cells, 101*, 256–261.

PART III

Selected Topics and Current Trends

CHAPTER 9

Metal Carbon Mesocomposites Synthesis in Polymeric Matrixes: Uniquely Structured Composites

V. I. KODOLOV,[1,2] V. V. KODOLOVA-CHUKHONTZEVA,[1,3] and YU. M. VASIL'CHENKO[1,2]

[1]*Basic Research–High Educational Center of Chemical Physics and Mesoscopy, Udmurt Scientific Center, Ural Division, Russian Academy of Sciences, Russia*

[2]*M.T. Kalashnikov Izhevsk State Technical University, Izhevsk, Russia*

[3]*Peter Great St. Petersburg State Polytechnic University, St. Petersburg, Russia*

ABSTRACT

This chapter presents the theoretical ideas about redox processes in polymeric matrixes nanoreactors. According to these ideas, the redox process is the main process preceding the nanostructures formation in which the work for charge transport corresponds to the energy of mesocomposite formation process in the reacting layer (mesoreactor).

The computational experiment is carried out with the reagents placed in the mesoreactor with the corresponding geometry and energy parameters. Avrami equations are used for such processes which usually reflect the share of a new phase produced. The selection of the modified Avrami equation depends on the mesoreactor shape and size and defines the mesocomposite growth within the mesoreactor. The examples of modified Avrami equation

New Advances in Materials Technologies. Hossein Hariri Asli, Ali Pourhashemi, Ann Rose Abraham, & A. K. Haghi (Eds.)

variants for the calculating experiments for the redox synthesis and the self-organization processes are brought.

9.1 INTRODUCTION

Previously [7, 8]. the models of nanoreactors in which carbon/metal containing nanostructures can be obtained were considered. The synthesis takes place due to the redox reaction between metal containing substances and mesoreactor walls which are the macromolecules of polymer matrix with functional groups participating in the interaction with metal containing compound or metal ion. If nanoreactors are microspores or cavities in polymer gels which are formed in the process of solvent removal from gels and their transformation into xerogels or during the formation of crazes in the process of mechanical-chemical processing of polymers and inorganic phase in the presence of active medium, the process essence is as follows.

The mesoreactor is formed in the polymeric matrix which, by geometry and energy parameters, corresponds to the transition state of the reagents participating in the reaction. Then the mesoreactor is filled with reactive mass comprising reagents and solvent. The latter is removed and only the reagents oriented in a particular way stay in the mesoreactor and, if the sufficient energy impulse is available, for instance, energy isolated at the formation of coordination bonds between the fragments of reagents and functional groups in mesoreactor walls, interact with the formation of necessary products. The initially appeared coordination bonds destruct together with the nanostructured product formation.

9.2 THEORETICAL IDEAS ABOUT THE INTERACTIONS IN MESOREACTORS OF POLYMERIC MATRIXES

The difference of potentials between the interacting particles and object walls stimulating these interactions is the driving force of self-organization processes (formation of mesoparticles with definite shapes).

The potential jump at the boundary "mesoreactor wall-reacting particles" is defined by the wall surface charge and reacting layer size. If we consider the redox process as the main process preceding the nanostructure formation, the work for charge transport corresponds to the energy of mesoparticle formation process in the reacting layer. Then the equation of energy conservation for mesoreactor during the formation of mesoparticle will be as follows:

$$nF\Delta\phi = RT\ln N_p/N_r \tag{1}$$

where; n is the number reflecting the charge of chemical particles moving inside "the mesoreactor;" F is the faraday number; $\Delta\phi$ is the difference of potentials between the mesoreactor walls and flow of chemical particles; R is the gas constant; T is the process temperature; N_p is the mol share of mesoparticles obtained; and N_r is the share of initial reagents from which the mesoparticles are obtained.

Using the above equation, we can determine the values of equilibrium constants when reaching the certain output of mesoparticles, sizes of mesoparticles and shapes of mesoparticles formed with the appropriate equation modification. The internal cavity sizes or mesoreactor reaction zone and its geometry significantly influence the sizes and shapes of nanostructures.

The sequence of the processes is conditioned by the composition and parameters (energy and geometry) of nanoreactors. To accomplish such processes, it is advisable to preliminarily select the polymeric matrix containing the nanoreactors in the form of nano-pores or crazes as process appropriate. Such selection can be realized with the help of computer chemistry. Further the computational experiment is carried out with the reagents placed in the mesoreactor with the corresponding geometry and energy parameters. Examples of such computations were given in Ref. [18]. The experimental confirmation of polymer matrix and mesoreactor selection to obtain carbon/metal containing nanostructures was given in Refs. [19, 20]. Avrami equations are widely used for such processes which usually reflect the share of a new phase produced.

As follows from Avrami equation:

$$1 - \upsilon = \exp[-k\tau^{n}] \tag{2}$$

where; υ is the crystalline degree; τ is the duration; k is the value corresponding to specific process rate; and n is the number of degrees of freedom changing from 1 to 6, the factor under the exponential is connected with the process rate with the duration (time) of the process. Under the conditions of the isothermal growth of the ordered system "embryo," it can be accepted that the mesoreactor activity will be proportional to the process rate in relation to the flowing process.

It was proposed to use the thermodynamics of small systems and Avrami equations to describe the formation processes of carbon nanostructures during re-crystallization (graphitization) [37, 38]. These equations are successfully applied [34] to forecast per molecular structures and prognosticate the conditions on the level of parameters resulting in the obtaining of nanostructures of definite size and shape. The equation was also used to forecast the formation of fibers [39]. The application of Avrami equations in the processes of nanostructure formation: (a) embryo formation and crystal growth in polymers [34]:

$$(1 - \upsilon) = \exp[-k\tau^n] \quad (3)$$

where; υ is the crystalline degree; τ is the duration; *k* is the value corresponding to the process specific rate; *n* is the number of the degrees of freedom changing from 1 to 6; (b) graphitization process with the formation of carbon nanostructures [37, 38]:

$$\upsilon = 1 - \exp[-B\tau^n] \quad (4)$$

where; υ is the volume share that was changed; τ is the duration; *B* is the index connected with the process rate; and *n* is the value determining the process directedness; (c) process of fiber formation [39]:

$$\omega = 1 - \exp[-z\tau^n] \quad (5)$$

where; ω is the share of the fiber formed; τ is the process duration; *z* is the statistic sum connected with the process rate constant; and *n* is the number of the degrees of freedom (for the fiber n equals 1).

Instead of k, B or z, based on the previous considerations, the activity of nanoreactors can be used (*a*).

At the same time, the metal orientation proceeds in interface regions and nano-pores of polymeric phase which conditions further direction of the process to the formation of metal/carbon mesocomposite. In other words, the birth and growth of nanosized structures occur during the process in the same way as known from the macromolecule physics [6], in which Avrami equations are successfully used. The application of Avrami equations to the processes of nanostructure formation was previously discussed in the papers dedicated to the formation of ordered shapes of macromolecules [34], formation of carbon nanostructures by electric arc method [37], obtaining of fiber materials [39].

9.3 MODIFIED AVRAMI EQUATION FOR PROCESSES IN MESOREACTORS

Then the share of the product being formed (W) in mesoreactor will be expressed by the following equation:

$$W = 1 - \exp(-a\tau^n) = 1 - \exp[-(\varepsilon_S/\varepsilon_V)\tau^n]$$
$$= 1 - \exp\{-[(\varepsilon^0_S d/\varepsilon^0_V)S/V]\,\tau^n\} \quad (6)$$

where; a is the mesoreactor activity; $a = \varepsilon_S/\varepsilon_V$; ε_S is the surface energy reflecting the energy of interaction of reagents with mesoreactor walls, $\varepsilon_S = \varepsilon^0_S dS$; ε_V is the mesoreactor volume energy, $\varepsilon_V = \varepsilon^0_V V$; $\varepsilon^0_S d$ is the multiplication of surface layer energy by its thickness; ε^0_V is the energy of mesoreactor volume unit; S is the surface of mesoreactor walls; and V is the mesoreactor volume.

When the metal ion moves inside the mesoreactor with redox interaction of ion (mol) with mesoreactor walls, the balance setting in the pair "metal containing – polymeric phase" can apparently be described with the following equation:

$$zF\Delta\varphi = RT\ln K = RT\ln(N_p/N_r) = RT\ln(1 - W) \quad (7)$$

where; z is the number of electrons participating in the process; $\Delta\varphi$ is the difference of potentials at the boundary "mesoreactor wall–reactive mixture"; F is the Faraday number; R is the universal gas constant; T is the process temperature; K is the process balance constant; N_p is the number of moles of the product produced in mesoreactor; N_r is the number of moles of reagents or atoms (ions) participating in the process which filled the mesoreactor; and W is the share of nanostructured product obtained in mesoreactor.

In turn, the share of the transformed components participating in phase interaction can be expressed with the equation which can be considered as a modified Avrami equation:

$$W = 1 - \exp[-\tau^n \exp(zF\Delta\varphi/RT)] \quad (8)$$

Where; τ is the duration of the process in mesoreactor; n is the number of degrees of freedom changing from 1 to 6. When "n" equals 1, one-dimensional nanostructures are obtained (linear nanostructures, nanofibers). If "n" equals 2 or changes from 1 to 2, flat nanostructures are formed (nano-films, circles, petals, wide nano-bands). If "n" changes from 2 to 3 and more, spatial nanostructures are formed as "n" also indicates the number of degrees of freedom. The selection of the corresponding equation recording form depends on the mesoreactor (nanostructure) shape and sizes and defines the nanostructure growth in the mesoreactor.

In case of nanostructure interaction with the medium molecules, the medium self-organization is effective with the proximity or concordance of the oscillations of separate fragments of nanostructures and chemical bonds of the medium molecules. This hypothesis of nanostructure influence on media self-organization was already discussed [21–24]. At the same time, it is possible to use quantum-chemical computational experiment with the known software products.

According to the chapter [25], the process vibration nature is discovered. This fact corroborates by IR spectroscopic investigations of nanostructures aqueous soles. Self-organization processes in media and compositions can be compared with the processes of crystalline phase origin and growth. At the same time, the growth can be one-, two-, and three-dimensional. For such processes Avrami equations are widely used which usually reflect the share of the new phase appearing. In this case, the degree of nanostructure influence on active media and compositions is defined by the number of nanostructures, their activity in this composition and interaction duration. The temperature growth during the formation of new phases in self-organizing medium prevents the process development.

Different shapes of nanostructures appear during the formation of lamellar or linear substances. For instance, graphite, clay, mica, asbestos, many silicates have lamellar structure and, consequently, they contain nanostructures with the shape of rotation bodies. It can be explained by the facts that bands of lamellar structure are rolled into clews and spirals with further sewing between them and formation of spheres, cylinders, ellipsoids, cones, etc. Nanostructures formed can be classified based on complexity. Under certain energy actions the distortion of coplanar (flat) aromatic rings is known, when π-electrons are shifted and charges are separated on the ring, the ring polarity goes up. For polymeric chains and bands formed from macromolecules the possibilities of formation of super molecular structures, corresponding to the possibility of taking a definite shape, can be predicted with the help of Avrami equation [34].

If the mesoreactor internal walls become the shells of nanostructures during the process, the nanostructures so obtained are the mesoreactor mirror reflections.

The formation processes of metal-containing nanostructures in carbon or carbon-polymeric shells in nanoreactors can be related to one type of reaction series using the terminology of the theory of linear dependencies of free energies (LFE) [40]. Then it is useful to introduce definite critical values for the volume, surface energy of mesoreactor internal walls, as well as the for one-, two-, and three-dimensional crystallization of macromolecules. Here, energy exchange during the formation of ordered nanostructures and without energy exchange with the surroundings, i.e., under thermal embryo-formation, is considered.

As mentioned before, at two-dimensional growth circles that form petals and furthermore complex nanostructures resembling flowers can be formed [36]. When being rolled, the circles formed can produce distorted semi-spheres or can serve as a basis for the transformation into "a beady,"

During the redox process connected with the coordination process, the character of chemical bonds changes. Therefore, correlations of wave numbers of the changing chemical bonds can be applied as the characteristic of the nanostructure formation process in mesoreactor:

$$W = 1 - \exp\,[-\tau^{n}(\nu_{\text{нс}}/\nu_{\text{кс}})] \tag{9}$$

where; $\nu_{\text{нс}}$ corresponds to wave numbers of initial state of chemical bonds; and $\nu_{\text{кс}}$ is the wave numbers of chemical bonds changing during the process.

The share of nanostructures (W) during the redox process can depend on the potential of mesoreactor walls interaction with the reagents, as well as the number of electrons participating in the process. At the same time, metal ions in the mesoreactor are reduced and its internal walls are partially oxidized (transformation of hydrocarbon fragments into carbon ones). Then the isochoric-isothermal potential (ΔF) is proportional to the product $zF\Delta\phi$ and Avrami equation will look as follows:

$$W = 1 - k_1 \exp[-\tau^{n} \exp(zF\Delta\phi/RT)] \tag{10}$$

where; k_1 is the proportionality coefficient taking into account the temperature factor; n is the index of the process directedness to the formation of nanostructures with certain shapes; z is the number of electrons participating in the process; $\Delta\phi$ is the difference of potentials at the boundary "mesoreactor wall–reactive mixture"; *F* is the Faraday number; and *R* is the universal gas constant.

Then the equation for defining the share (W) of nanostructures formed can be written down as follows by analogy with the aforesaid equations:

$$1 - W = \exp[-\beta a\tau^n] \tag{11}$$

where; β is the coefficient taking into account the changes in the activity in the process of nanostructured product formation.

Substituting the value of *a*, we get the dependence of the share of nanostructures formed upon the ratio of the energy of mesoreactor internal surface to its volume energy:

$$W = 1 - \exp\,[-\beta(\varepsilon_S/\varepsilon_V)\tau^n] = 1 - \exp\,[-\beta(\varepsilon_S^{\,0}/\varepsilon_V^{\,0}\cdot S/V)\tau^n] \tag{12}$$

When the ratio lg k/k_c is proportional $-\Delta\Delta F/RT$, the ratio W/W_c can be transformed into the following expression:

$$W/W_c = b\cdot\exp\,\{-(k/k_c)\cdot(\tau/\tau_c)^n\} = b\cdot\exp\,\{-(\tau/\tau_c)^n$$
$$\exp(-\Delta\Delta F/RT)\} = b\cdot\exp\{(\tau/\tau_c)^n[\exp k_T k_{VS}(\varepsilon_V/\varepsilon_{Vk} - \varepsilon_S/\varepsilon_{Sk})\theta/T]\} \tag{13}$$

where; b is the proportionality coefficient considering the temperature factor; k_{VS} is the coefficient considering correlations $\varepsilon_V/\varepsilon_{Vk}$ and $\varepsilon_S/\varepsilon_{Sk}$; ε_V and ε_{Vk} is the volume energies of mesoreactor and "equilibrium" mesoreactor calculated via the ratios of their volumes; ε_S and ε_{Sk} is the surface energy and its equilibrium value; T and θ is the temperature of the process and temperature of the equilibrium process; τ is the time required to develop the process of nanostructure formation; n is the index of the process directedness to the formation of nanostructures of definite shapes.

The values of volume and surface energies are given after the transformation of ΔΔF in accordance with Ref. [41], in which the physical sense of Taft constants is substantiated using the indicated energies.

At the same time, the share of nanostructures (W) during the redox process can depend upon the potential of mesoreactor wall interaction with reagents, as well as the number of electrons participating in the process. The metal ions in mesoreactor are reduced and its internal walls are partially oxidized (transformation of hydrocarbon fragments into carbon ones).

Then the Helmholtz thermodynamic potential (ΔF) is proportional to the product of $zF\Delta\phi$ and Avrami equations will be expressed by the following formulae in accordance with the above models, one of them can be as follows:

$$W = 1 - k\,\exp[-\tau^n\,\exp(zF\Delta\phi/RT)] \tag{14}$$

where; k is the proportionality coefficient considering the temperature factor; n is the index of the process directedness to the formation of nanostructures of definite shapes; z is the number of electrons participating in the process; Δϕ is the difference of potentials on the border " mesoreactor wall–reaction mixture;" F is the Faraday number; R is the universal gas constant. When *n* equals 1, one-dimensional nanostructures are obtained (linear nanostructured systems and narrow bands). If *n* equals 2 or changes from 1 to 2, narrow flat nanostructures are formed (nano-films, circles, petals, broad nano-bands). If *n* changes from 2 to 3 and over, spatial nanostructures are formed, since *n* also means the number of degrees of freedom. If in this equation we take *k* as 1 and consider the process in which copper is reduced with simultaneous formation of nanostructures of a definite shape, the share of such formations or transformation degree can be connected with the process duration.

9.4 EXAMPLES OF THE MODIFIED AVRAMI EQUATIONS APPLICATION

The experimental modeling of obtaining nano-films after the alignment of copper compounds with polyvinyl alcohol at 200°C revealed that optimal duration when the share of nano-films approaches 100% equals 2.5 hours. This corresponds to the calculated value based on the aforesaid Avrami equation. The calculations are made supposing the formation of copper nano-crystals on the nano-films. It is pointed out that copper ions are predominantly reduced to metal. Therefore, it was accepted for the calculations that *n* equals 2 (two-dimensional growth), potential of redox process during the ion reduction to metal ($\Delta\phi$) equals 0.34 V, temperature (T) equals 473 K, Faraday number (F) corresponds to 26.81 (A·hour/mol), gas constant R equals 2.31 (W·hour/mol·degree). The analysis of the dimensionality shows the zero dimension of the ratio $\frac{zF\Delta\varphi}{RT}$. The calculations are made when changing the process duration with a half-hour increment:

Duration (hours)	0.5	1.0	1.5	2.0	2.5
Content of Nano-Films (%)	22.5	63.8	89.4	98.3	99.8

If nano-films are scrolled together with copper nano-wires β is taken as equaled to 3, the temperature increases up to 400°C, the optimal time when the transformation degree reaches 99.97%, corresponds to the duration of 2 hours, thus also coinciding with the experiment. According to the calculation results if following the definite conditions of the system exposure, the duration of the exposure has the greatest influence on the value of nanostructure share. The selection of the corresponding equation form depends upon the shape and sizes of mesoreactor (nanostructure) and defines the nanostructure growth in mesoreactor or the influence distribution of the nanostructure on the structurally changing medium. With one-dimensional growth and when the activation zero is nearly zero, the equation for the specific rate of the influence distribution via the oscillations of one bond can be written down as follows:

$$W = 1 - \exp\left[-\beta \nu \tau^{n}\right] \tag{15}$$

where; ν is the oscillation frequency of the bond through which the nano-structure influences upon the medium; β is the coefficient considering the changes in the bond oscillation frequency in the process. In the case discussed the parameter βν can be represented as the ratio of frequencies of bond oscillations v_{is}/v_{fs}, that are changing during the process. At the same time v_{is} corresponds to the frequency of skeleton oscillations of C–C bond at 1,100 cm^{-1}, v_{fs} is the symmetrical skeleton oscillations of C=C bond at 1,050 cm^{-1}. In this case the equation looks as follows:

$$W = 1 - \exp\left\{-\tau^{n} \cdot \frac{\nu_{is}}{\nu_{fs}}\right\} \tag{16}$$

For the example discussed the content of nano-films in percentage will be changing together with the changes in the duration as follows:

Duration (hours)	0.5	1.0	1.5	2.0	2.5
Content of Nano-Films (%)	23.0	64.9	90.5	98.5	99.9

By the analogy with the above calculations the parameters *a* in Eqn. (5) should be considered as a value that reflects the transition from the initial to final state of the system and represents the ratios of activities of system states. Under the aforesaid conditions the linear sizes of copper (from ion radius to atom radius) and carbon-carbon bond (from C–C to C=C) are changing during the process. Apparently, the structure of copper ion and electron interacts with electrons of the corresponding bonds forming the layer with linear sizes $r_i + l_{C\text{–}C}$ in the initial condition and the layer with the size $r_a + l_{C=C}$ in the final condition. Then the equation for the content of nano-films can be written down as follows:

$$W = 1 - \exp\left\{-\tau^n \cdot \frac{r_\alpha + l_{C=C}}{r_i + l_{C-C}}\right\} \tag{17}$$

At the same time r_i for Cu^{2+} equals 0.082 nm, r_a for four-coordinated copper atom corresponds to 0.113 nm, bond energy C–C equals 0.154 nm, and C=C bond – 0.142 nm. Representing the ratio of activities as the ratio of corresponding linear sizes and taking the value n as equaled to 2, at the same time changing τ in the same intervals as before, we get the following change in the transformation degree based on the process duration:

Duration (hours)	0.5	1.0	1.5	2.0	2.5
Content of Nano-Films (%)	23.7	66.0	91.2	98.7	99.9

Thus, with the help of Avrami equations or their modified analogs we can determine the optimal duration of the process to obtain the required result. It opens up the possibility of defining other parameters of the process and characteristics of nanostructures obtained (by shape and sizes).

The influence of nanostructures on the media and compositions can be assessed with the help of quantum-chemical experiment and Avrami equations.

Modified Avrami equations were tested to prognosticate the duration of the processes of obtaining metal/carbon nano-films in the system "Cu–PVA" at 200°C [42]. The calculated time (2.5 hours) correspond to the experimental duration of obtaining carbon nano-films on copper clusters.

The influence of nanostructures on the media and compositions was discussed based on quantum-chemical modeling [9]. After comparing the energies of interaction of fullerene derivatives with water clusters, it was found that the increase in the interactions in water medium under the nanostructure influence is achieved only with the participation of hydroxyl fullerene in the interaction. The energy changes reflect the oscillatory process with periodic boosts and attenuations of interactions. The modeling results can identify that the transfer of nanostructure influence onto the molecules in water medium is possible with the proximity or concordance of oscillations of chemical bonds in nanostructure and medium. The process of nanostructure influence onto media has an oscillatory character and is connected with a definite orientation of particles in the medium in the same way as reagents orientate in nanoreactors of polymeric matrixes. To describe this process, it is advisable to introduce such critical parameters as critical content of nanoparticles, critical time and critical temperature [44]. The growth of the number of nanoparticles (n) usually leads to the increase in the number of interaction (N). Also, such situation is possible when with the increase of n critical value, N value gets much greater than the number of active mesoparticles. If the temperature exceeds the critical value, this results in the distortion of self-organization processes in the composition being modified and decrease in nanostructure influence onto media.

KEYWORDS

- **Avrami equations**
- **computational experiment**
- **mesoreactor**
- **metal/carbon nanocomposites**
- **polymeric matrixes**
- **redox process**
- **self-organization**

REFERENCES

1. Kodolov, V. I., Shabanova, I. N., Kuznetsov, A. P., Nikolaeva, O. A., et al., (1999). *Anal. and Control, 4.*
2. Stepanov, N. F., & Pupyshev, V. I., (1991). *Quantum Mechanics of Molecules and Quantum Chemistry* (p. 384). Moscow. MGU.
3. Dunning, T. H., (1971). *J. Chem. Phys. 55*, 716–723.
4. Schmidt, M. W., et al., (1993). *J. Comput. Chem., 14*, 1347–1363.
5. Molina, J. M., & Melchor, S. F. (1989). *Private Communication, 24*, 35.
6. Sadlej, J., (1985). *Semi-Empirical Methods of Quantum Chemistry.* Ellis Horwood limited.
7. Goodwin, L., (1991). *J. Phys.: Condens. Matter., 3*, 3869.
8. Fischer, E. O., & Hafner, W. Z., (1955). *Naturforsch., 10b*, 665.

9. Wachters, A. J. H., (1970). *J. Chem. Phys.*, ***52***, 1033.
10. Dupuiis, M., et al., (1989). *Comput. Phys. Comm.*, ***52***, 415.
11. Bethune, D. S., Klang, C. H., De Vries, M. S., et al., (1993). *Nature, 363*, 605.
12. Prigozhin, I., & Defay, R., (1966). *Chemical Thermodynamics* (p. 510). Novosibirsk: Nauka.
13. Berezkin, V. I., (2000). Fullerenes as embryos of soot particles. *Physics of Solids, 42*, 567–572.
14. Khokhriakov, N. V., & Kodolov, V. I., (2005). Quantum-chemical modeling of nanostructure formation. *Nanotechnics,* (2), 108–112.
15. Kodolov, V. I., Khokhriakov, N. V., Nikolaeva, O. A., & Volkov, V. L., (2001). Quantum-chemical investigation of alcohols dehydration and dehydrogenization possibility in interface layers of vanadium oxide systems. *Chemical Physics and Mesoscopy, 3*(1), 53–65.
16. Kodolov, V. I., Didik, A. A., Volkov, A. Y., & Volkova, E. G., (2004). Low-temperature synthesis of copper nanoparticles in carbon shell. Bulletin of HEIs. *Chemistry and Chemical Engineering, 47*(1), 27–30.
17. Lipanov, A. M., Kodolov, V. I., Khokhriakov, N. V., et al., (2005). Challenges in the production of nanoreactors for the synthesis of metallic nanoparticles in carbon shells. *Alternative Energetic and Ecology (ISJAEE), 2*(22). 58–63.
18. Nikolaeva, O. A., Kodolov, V. I., Zakharova, G. S., et al., (2004). *Method of Obtaining Carbon-Metal-Containing Nanostructures*. Patent of the RF 2225835.
19. Rubin, A. B., (1987). Biophysics. *Book 1: Theoretical Biophysics* (p. 319). Vysshaya Shkola, M.
20. Blokhintsev, D. I., (1961). *Basics of Quantum Mechanics* (p. 5112). Vysshaya shkola, M.
21. Yavorsky, B. M., & Detlaf, A. A., (1968). *Reference-Book in Physics (*p. 939). Nauka, M.
22. Christy, P., & Pitti, A., (1969). *Substance Structure: Introduction to Modern Physics* (p. 596). Translated from English. Nauka, M.
23. Korablev, G. A., (2005). *Spatial-Energy Principles of Complex Structures Formation* (p. 426). Brill Academic Publishers and VSP, Netherlands. (Monograph).
24. Airing, G., Walter, J., & Kimbal, J., (1948). *Quantum Chemistry* (p. 528). I.L., M.
25. Fischer, C. F., (1972). Average-energy of configuration Hartree-Fock results for the atoms helium to radon. *Atomic Data,* (4), 301–399.
26. Waber, J. T., & Cromer, D. T., (1965). Orbital radii of atoms and ions. *J. Chem. Phys., 42*(12), 4116–4123.
27. Clementi, E., & Raimondi, D. L., (1963). Atomic screening constants from S.C.F. functions, 1. *J. Chem. Phys., 38*(11), 2686–2689.
28. Clementi, E., & Raimondi, D. L., (1967). Atomic screening constants from S.C.F. functions, 1. *J. Chem. Phys., 47*(14), 1300–1307.
29. Korablev, G. A., & Zaikov, G. E., (2006). Energy of chemical bond and spatial-energy principles of hybridization of atom orbitals. *J. of Applied Polymer Science, 101*(3), 2101–2107.
30. Korablev, G. A., & Zaikov, G. E., (2008). Spatial-energy interactions of free radicals. *Successes in Gerontology, 21*(4), 535.

31. Korablev, G. A., & Zaikov, G. E., (2009). Spatial-energy parameter as a materialized analog of wave function. *Progress on Chemistry and Biochemistry* (pp. 355–376). Nova Science Publishers, Inc. New York.
32. Korablev, G. A., & Zaikov, G. E., (2006). Chemical bond energy and spatial-energy principles of atom orbital hybridization. *Chemical Physics. RAS. M., 25*(7), 24.
33. Zigban, K., Norling, K., Valman, A., et al., (1971). *Electron Spectroscopy* (p. 439). M.: Mir.
34. Vunderlikh, B., (1979). *Physics of Macromolecules (in 3 Volumes)* (Vol. 2, p. 574). M: Mir.
35. Kargin, V. A., & Slonimsky, G. L., (1960). *Essay on Physic-Chemistry of Polymers* (p. 321). – M.: MSU.
36. Proceedings of Conference *NATO-ASI: "Synthesis, Functional Properties and Applications."* (2002).
37. Fedorov, V. B., Khakimova, D. K., Shipkov, N. N., & Avdeenko, M. A., (1974). To thermodynamics of carbon materials. *Doklady AS USSR, 219*(3), 596–599.
38. Fedorov, V. B., Khakimova, D. K., Shorshorov, M. H., et al., (1975). To kinetics of graphitation. *Doklady AS USSR, 222*(2), 399–402.
39. Serkov, A. T., (1975). *Theory of Chemical Fiber Formation* (p. 548). M.: Himiya.
40. Palm, V. A., (1967). *Basics of Quantitative Theory of Organic Reactions* (p. 356). L: Himiya.
41. Kodolov, V. I., (1965). *On Modeling Possibility in Organic Chemistry* (Vol. 2, No. 4, pp. 11–18). Organic reactivity. Tartu: TSU.
42. Kodolov, V. I., Khokhriakov, N. V., Trineeva, V. V., & Blagodatskikh, I. I., (2008). Activity of nanostructures and its display in nanoreactors of polymeric matrixes and in active media. *Chem. Phys.& Mesoscopy, 10*(4), 448–460.
43. Kodolov, V. I., (2009). Commentary to the paper (2008) by Kodolov, V. I., et al. *Chem. Phys. & Mesoscopy, 11*(1), 134–136.
44. Kodolov, V. I., & Trineeva, V. V., (2012). Perspectives of idea development about nanosystems self-organization in polymeric matrixes. In: *Book the Problems of Nanochemistry for the Creation of New Materials* (pp. 75–100). Torun, Poland: IEPMD.
45. McLean, A. D., & Chandler, G. S., (1980). *J. Chem. Phys., 72*, 5639–5648.
46. Rappe, A. K., Smedley, T. A., & Goddard, III. W. A., (1981). *J. Phys. Chem., 85*, 2607–2611.

CHAPTER 10

Optical Properties and Applications of Rare Earth Elements in Solid Materials

P. ARJUN SURESH,[1,2] GREESHMA SARA JOHN,[1,2]
ATHIRA MARIA JOHNSON,[1,2] N. V. UNNIKRISHNAN,[3] and
K. V. ARUN KUMAR[1,2]

[1]*Department of Physics, CMS College (Autonomous), Kottayam, Kerala, India*

[2]*Nanotechnology and Advanced Materials Research Center, CMS College (Autonomous), Kottayam, Kerala, India*

[3]*School of Pure and Applied Physics, Mahatma Gandhi University, Kottayam, Kerala, India*

ABSTRACT

Rare earth elements (RREs), which include 15 lanthanides, scandium, and yttrium, are extensively used in various fields of research. Due to its optical and luminescent properties, it is used in lasers, and optical fiber and display applications. In laser applications, the studies are going on to improve its efficiency, optical quality, and thermal conductivity. It is found that using different rare earth materials such as neodymium, samarium, dysprosium (Dy), thulium, ytterbium, erbium, and holmium can improve its properties. For optical fiber applications, the fiber must show the properties such as high optical transparency, low signal loss, and high thermal stability. RRE such as Dysprosium, Samarium, Cerium, and Erbium are proven to increase these properties. In-display applications OLED and LED displays

New Advances in Materials Technologies. Hossein Hariri Asli, Ali Pourhashemi, Ann Rose Abraham, & A. K. Haghi (Eds.)

are the most popular. The main drawback of OLED display is its less lifetime and high production cost. Various research based on different rare-earth elements are going on to solve these issues. RRE such as Scandium, samarium, Europium, and Cerium are commonly used for this purpose. This chapter summarizes the use of RRE in the above-mentioned fields.

10.1 INTRODUCTION

Rare earth elements (RREs) comprise the 15 lanthanides in addition to yttrium and scandium. They generally exist as minerals, oxides, or metals. They are further classified as light RRE and heavy RRE [1]. RRE hold importance as they are found as electropositive in a natural system and could be employed in many applications such as lasers, lightning displays, solar cells, etc. They are known for their excellence in optical as well as luminescent properties which owe them the above-specified applications. The properties that account for excellence in optical applications are optical properties of fluorescence and high refractive index, unique electron configuration, ability to store and release oxygen catalytically, unique electron configuration, ability to store and release oxygen catalytically, high conductivity, etc. The rare-earth is being extensively used in phosphors which in turn is a wide growing area responsible for visual displays [1]. The optical applications could be broadly divided into optical components, visual displays such as LED, rare-earth-based optical amplifiers, display phosphors, and fiber lasers [2]. Rare-earth elements such as Cerium, Yttrium, Neodymium, lanthanum, europium, and terbium are mostly used in optical applications such as displays, lenses, lasers, CRTs, etc., due to their efficient properties. The class of rare-earth, especially lanthanides, are the key candidates exhibiting photonic and optoelectronic applications [3]. Three of their major applications include solid-state lasers, color lamps and displays, and optical fiber telecommunications [3]. The electronic configuration of these RRE comprises partially filled 4f shells, and the f-f transitions are the major cause of their optical properties.

This chapter thus focuses on the applications that arise due to the optical properties of the RRE. The following sessions give us a detailed review of the effect of rare earth in the above-mentioned applications. The first session deals with the laser applications, followed by optical fiber application, and the concluding session focuses on display applications.

10.2 ROLE OF RARE-EARTH ELEMENTS IN LASER DEVICES AND ITS APPLICATIONS

The fundamental of laser depends on the population inversion which results in the stimulated emission. This phenomenon is basically occurring by exciting to a third level which in turn transfers the energy to a metastable upper laser level [4]. For any laser to operate, the terminal level should satisfy the Boltzmann thermal equilibrium population [4]. The most employed type of laser among all the lasers existing is the four-level laser which is usually used in the rare earth lasers [4]. The rare earth lasers majorly depend on optical pumping [4] (Figure 10.1).

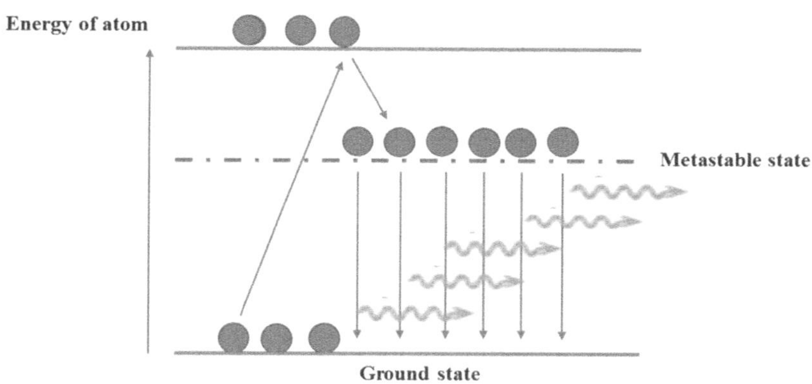

FIGURE 10.1 Working of basic laser.

We would be discussing the different trivalent rare-earth ions which are used in laser applications due to numerous fluorescence lines of high quantum efficiency and suitable absorption bands in the visible and NIR. Of which the majorly used rare earth includes Neodymium, Samarium, Europium, Terbium, Dysprosium, Holmium, and Erbium, and in addition to this Cerium and Praseodymium also show laser action under special conditions [4].

10.2.1 FIBER LASERS

Apart from the general rare earth lasers another type of laser that attracted research in the present world is rare earth lasers is the fiber laser. Fiber lasers are solid-state lasers that comprise light guided by total internal reflection [2]. Many rare-earth ions such as thulium, ytterbium, erbium, and holmium are commonly used in rare-earth-based solid-state lasers [2]. Ytterbium-based solid-state lasers hold their importance due to their higher efficiency which is the due effect of their small quantum defects [2]. Some of the Yb based lasers are Yb:SYS, Yb:YAG, Yb:KYW, and Yb:KGW [2]. Holmium is another important rare earth used in fiber lasers with appreciable efficiency that operates in the infrared wavelengths [2]. It could be used in medical applications either this is used for the production of optical fiber surgical devices [2]. Triply doped YAG laser with Holmium-chromium-thulium is another class of efficient fiber laser that has wide applications in different areas such as meteorology, medicine, and the military [2].

10.2.2 NEODYMIUM-BASED LASER AND ITS APPLICATIONS

Neodymium is a promising candidate that has higher efficiency when employed as a laser ion. The ion exists in faint yellow color discovered back in 1885 [5]. The ion accounts for higher optical pumping efficiency. Among all the Nd-based lasers, the most common and known solid-state laser is the Nd: YAG [4]. This laser exhibits higher efficiency, great optical quality, and increased thermal conductivity [3]. This laser has a wide range of applications in the medical field specifically in, podiatry, urology, dentistry, ophthalmology, neurosurgery obstetrics/gynecology, plastic surgery, dermatology, and gastroenterology. It is also used in the cosmetics industry for deep pigmentation and hair removal. It has a continuous wave that is capable of removing the hair as well as hair follicles. This laser when used for the purpose of hair removal employs the advantage of shortening the pulse duration when they are used as the pulse width lasers [5].

The functioning of the Nd: YAG could be explained as follows: Abbreviated as yttrium-aluminum-garnet crystal with neodymium ions replacing some of the aluminum atoms. It is capable of emitting the known intense lasing emission when irradiated with an optical source which could be a diode laser. The laser comprises a four-level system that could work in three different modes namely continuous modes, Q-switched, and long-pulse modes. The Q-switched mode work with its energy being stored in a closed cavity till the highest population inversion is attained. The continuous mode uses arc lamp-pumped lasers [5].

10.2.3 ERBIUM-BASED LASERS AND ITS APPLICATIONS

Erbium is another rare earth usually employed in YAG lasers. Erbium undergoes stimulated emission and these erbium-based lasers could be used in ophthalmology and dentistry [4]. The widely used erbium lasers are Er:YAG and Er,Cr:YSGG, respectively which differ from each other due to the difference in their laser wavelength [6]. These lasers possess the highest absorption among all infrared lasers and thus could be employed in dental treatment [6]. Erbium exists as the active material in both the solid-state lasers [6]. Pumping is employed using a pulsed broadband flashlamp [6]. The major difference among these Erbium lasers is in the technology employed in the flashlamp energization. There exist two types of pumping technologies, of which one is the traditional pulse forming network (PFN) and the variable square pulse (VSP). As the name indicates VSP pulses are square shaped with variable duration while the former one, either the PFN consists of bell-shaped pulses of fixed duration [6]. The effect of these pulses on medical applications differs due to fact that the average power and peak power of VSP pulses are almost the same which is quite different for the PFN pulses [6]. Erbium-based lasers are the safest among the commercially available lasers and also the most efficient one [6]. Comparing both the erbium-based lasers, it is well studied that Er:YAG, when utilized with VSP pumping technology, is the one with the highest absorption characteristics [6]. Thus, Er based lasers are found to be very important in the dentist's office [6].

10.2.4 SAMARIUM-BASED LASERS AND ITS APPLICATIONS

Rare-earth ion-doped glasses are finding wide applications in the area of optoelectronic devices which comprises solid-state lasers, optical fibers, etc. [7]. The properties of the rare earth ions depend on the host or glass matrix. The studies done in this field could find excellent optical properties on Sm doped alumino borate glasses [7]. The rare earth Sm holds its importance in laser technology because of its intense emissions in the visible regions which accounts for the reddish-orange emission [7]. The Sm ions possess the desirable properties for laser applications which are strong luminescence intensity, large-stimulated emission cross-section, high quantum efficiency, and very less probability of non-radiative decay [7]. The luminescence studies were carried out on Sm ions doped alumino borate glasses by employing the photoluminescence spectra which produced an intense peak that corresponds to the strong reddish orange emission thus making it capable for the application of solid-state lasers [7].

10.2.5 DYSPROSIUM-BASED LASERS AND ITS APPLICATIONS

The Dysprosium (Dy) ions are employed in laser applications specifically in the mid-infrared lasers [8]. The definition of mid-infrared (MIR) lasers, being a laser like beam emitted in the MIR region utilizing a light source. Dy ions with its characteristic f-f transition is a good candidate that produces the lasing action which is the consequence of their energy state distribution [8]. Studies shows that the MIR emission of the rare earth ion dysprosium is a host dependent property which has been studied on various host matrix. Out of all the matrix, chalcogenide glass was found to be the most efficient due to their high transparency and low phonon energy [8]. Another important result of the research done is that the co-doping of Dy ions and Prions showed good results giving they enhanced the MIR emission thus could be utilized for fabricating MIR laser materials [8]. These MIR laser materials could be used in optical fiber applications as well [8].

10.2.6 AN OVERVIEW OF OTHER RARE EARTH ELEMENTS (RRES) AND ITS APPLICATIONS

Holmium is a well-known rare earth which is regarded to be a widely used lanthanide laser ion [4]. It comprises of nine transitions and stimulated emission is exhibited for pulsed operation [4]. It could be also employed in stochiometric laser material and thin films as well [4]. It also shows phonon terminated laser action in some host. Thulium is also a rare earth candidate that exhibits laser action from two transitions. It does selective excitation to the upper laser level [4]. Pulsed laser action is observed in thulium-based lasers [4]. Ytterbium and Cerium are the rare earth that exhibits laser actions in special pumping conditions.

Apart from the rare-earth based lasers discussed above, recent works in this area includes Yb ions doped silica performs based laser technology which was a consequence of the silica based optical fiber technology [40].

Another recent study on laser technology is the Ho^{3+}/Yb^{3+} co-doped heavy metal oxide glass which finds applications in the MIR lasers due to their increased transparency in the MIR regions [41]. An interesting class of material used in lasers are tellurite glasses doped with rare earth such as Ho, Tm, Nd, Er. This rare earth possesses excellent optical properties and have absorption bands in the range of commercial laser diodes. Thus, lasers with rare earth ions are of high relevance in the present world [42]. Thus, we conclude by adding that rare-earth ion account for the laser applications in an accountable manner. Another innovative invention which could further be used for lasing is an optical fiber. The basics as well as the role of rare earth in optical fiber is well elaborated in the next session.

10.3 OPTICAL FIBERS

Optical fibers (Figure 10.2) are cylindrical-shaped structure that has an actual application in telecommunication. The optical fibers have an inner core and outer cladding under the condition that core glass has high refractive index than the cladding, thereby confining light in the core. But the difference in index between core and cladding need not be significant. This fiber is coated with protective sheaths of polymer to reduce the effects of bending and resist abrasion. Glass optical fibers operate under the principle of total internal reflection [9]. The fiber length exceeds its diameter, generally ranging from meters to kilometers. Conventionally, the outer cladding follows a diameter of 125 μm. The inner core diameter can vary depending on the mode of propagation of the waves. For a single waveguide mode, the core diameter is about 8 μm, and for the multiple modes, it is around 62.5 μm. An optical fiber doped with light-emissive species like RRE has a laser or optical amplifier application. An optical pump is required for the excitation of the active species [9]. Microstructured and photonic crystal optical fibers are other possibilities. The propagation characteristics of light are controlled by periodically modifying the cross-sectional geometry. This classification of optical fiber applies in low-latency communication, sensing, and supercontinuum generation. Optical fibers have got applications mainly in the areas of (i) information processing, communications, and data storage; (ii) energy; (iii) defense and national security; (iv) advanced manufacturing; and (v) health and medicine [9].

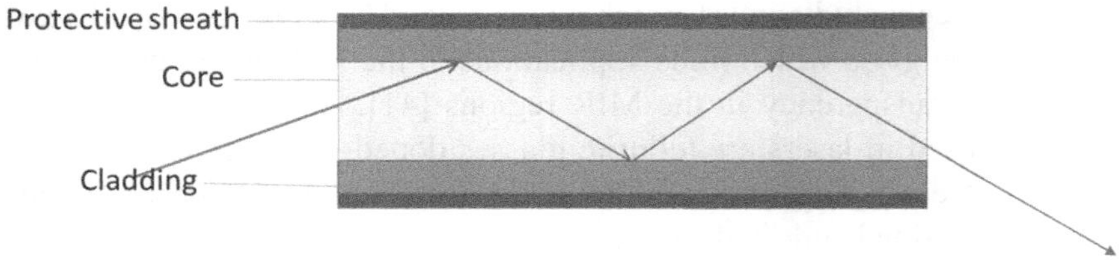

FIGURE 10.2 Diagram of optical fiber.

Various materials of glasses of optical fiber include silica and silicates, phosphates, infrared glasses, heavy metal oxides, fluorides, and chalcogenides. Usually, the leading glass former is silica on practical, large commercial scales due to its tensile strength and minimum losses. Amplifiers and lasers having high gain per unit length are the uses of phosphate glasses. Phosphate glasses can accommodate the higher concentration of rare earth up to the limit of onset of the concentration quenching. Infrared glasses do the sensing application, sensing chemical and biological species, identification of thermal signatures and their countermeasures, weather forecasting, and cosmology/astronomy [9].

10.3.1 RARE EARTH AS DOPANTS IN OPTICAL FIBER GLASSES

The lanthanides, the largest chemically identical group in the periodic table, are RRE. They include lanthanum (La), cerium (Ce), dysprosium (Dy), europium (Eu), samarium (Sm), ytterbium (Yb), etc. [11]. RREs exhibit exceptional optical properties [12] – absorption and luminescence [10]. They have specific electronic structure [10]. Since there is a high need for research in the development of optical communication and other fields where optical fibers are necessary, we need to focus on improving the properties of the optical fibers by doping them with the RRE [11]. The RREs occupy the voids of the glass network, acting as glass modifiers. They occupy the gaps such that their coordination number is high. The addition of RREs to glasses gives them particularly favorable properties [10]. They are selective absorption of radiation in visible, near-UV, and near-IR regions of the spectrum [10], increasing the bandwidth capacity of the optical fibers [11]; photoluminescence in various ranges of the energy spectra; inducing laser action; high refraction and low optical dispersion [10].

RRE having a large atomic size and high coordination number improve the structure of fiber glasses. More the concentration of RREs, the bonding becomes more ionic, and the glass structure becomes more rigid [11]. The doping concentration of the RREs must have an optimum value to avoid concentration quenching. Because increasing the RRE concentration decreases the ion spacing, thus hindering the energy transfer [11]. Rare earth doping in glasses has applications in the production of laser glasses, colorless optical glasses, including glasses for fiber optics, luminescent materials and optical filters [10], proton beam monitoring [13]. Let us concentrate on some rare-earth ions doped in different fiber optic glasses.

10.3.2 DYSPROSIUM (DY)-BASED OPTICAL FIBERS AND ITS APPLICATIONS

Ahmad et al. investigated and found that lithium strontium zinc borate glasses' thermal, physical, and absorption properties improved on Dy doping. Amorphous glasses with high optical transparency, very low signal loss, and high thermal stability were obtained in this study. Refractive indices and densities of the glasses increased with increase in the concentration of Dy. The optical bandgap energy of the glasses was reduced with increased Dy concentration. Also, the glass has got excellent stability against crystallization. These excellent traits make the glass suitable for drawing optical fibers [14]. Sun et al. studied Dy doped gallium-antimony-sulfur-iodine chalcohalide glasses and optical fiber for their potentiality in mid-infrared (MIR) luminescence. The glasses exhibited two intense emissions with a large-stimulated emission cross-section, a high quantum efficiency, and minimum losses. It also showed intense mid-infrared fluorescence making it a good candidate for MIR laser gain material [15].

10.3.3 SAMARIUM (SM)-BASED OPTICAL FIBERS AND ITS APPLICATIONS

Samarium doped glasses have possible applications in visible solid-state lasers, high-density optical storage, undersea communication, photodynamic therapy light sources, etc. [7]. Mohan et al. reported a samarium doped lead alumino borate glass suitable for photonic devices and lasers operating in the visible spectrum. The optimum concentration of Sm doping was found to be 1 mol%. The glass is of amorphous nature, having better strength, rigidity, and viscosity [7]. Singh et al. prepared lead borate glasses co-doped with samarium and gadolinium (Gd) suitable for lasing and high energy radiation sensing applications. The glasses with maximum rare-earth ions have the highest refractive index values, molar refraction, dielectric constant, absorption intensity and polarizability, and the lowest value for bandgap energy [16].

10.3.4 CERIUM (CE)-BASED OPTICAL FIBERS AND ITS APPLICATIONS

Francesca et al. studied the effect of doping cerium in two multimode optical fibers – germanosilicate and phosphosilicate glasses. They focused on the radiation response of the glasses. The dopant of cerium improves the resistance of the glasses to radiation-induced attenuation, which deteriorates the transmission of the glass. Comparing the effect of cerium on both the glasses shows that the radiation sensitivity of the germanosilicate glasses increased to the UV-Visible domain, whereas it decreased for phosphosilicate glasses. Information on the type and concentration of the defects induced by the radiation is another effect of cerium addition to the glasses. The cerium doped phosphosilicate glass has good application in radiation dosimetry [17].

10.3.5 ERBIUM (ER)-BASED OPTICAL FIBERS AND ITS APPLICATIONS

Galagan et al. reported a composite gain media for all optical fiber amplifiers with a composite of erbium-ytterbium co-doped phosphate core and double silica cladding. These fibers have a significant gain coefficient. The higher concentration of erbium helped reduce the length of the gain media [18].

The very recent and fast-growing area of research in the present era is display technologies. The rare earth ions play a major role in the development of display devices due to their very interesting luminescent and optical properties which is dealt in detail in this last session.

Sadeq et al. studied aluminoborate glass doped with copper (Cu) and samarium rare earth ion. An optimum value of samarium results in an optical fiber glass with least refractive index, least molar volume, minimum conductivity, uppermost density and optical band gap. This glass has an application to be a good band pass filter in the visible region [43]. Zhang et al worked on the impact of rare earth oxides Gd_2O_3, Pr_2O_3, Yb_2O_3, and La_2O_3 on alkali free aluminoborate glasses [44]. There is a decrement in the value of dielectric constant, viscosity, melting temperature and fiber drawing temperature of the glass due to the presence of the rare earth oxides. The working temperature of the glass is also reduced which is crucial for glass fiber drawing process and melting process [44].

Miluski et al investigated on Yb^{3+}/Tm^{3+} and Tm^{3+}/Ho^{3+} co-doped silica optical fibers synthesized by MCVD-CDS method. The rare earths ions provide the glasses with the property of wide range emission in the near IR region, eye safe emission. This glass is thermally robust, chemically, and mechanically stable. Also, it has high damage threshold and low attenuation. These optical fibers has application in lasing action and amplified spontaneous emission (ASE) [45].

Kaur et al studied barium tellurite and boro-tellurite glasses doped with varying concentrations of different rare earth ions (Dy^{3+}, Eu^{3+}, and Er^{3+}), alkali (Li^+, K^+, and Na^+) and Al^{3+} ions [46]. The barium tellurite and boro-tellurite glasses exhibit excellent thermal stability and luminescent properties. Increase in the rare earth doping concentration increases the glass transition temperature and decreases density of the glass. Optical transparency and thermal stability of the glass is enhanced due to the doping of the glass with B_2O_3. These optical fiber glasses serve the purpose of fluorescent devices and optical wave guides [46].

10.4 DISPLAYS

In the last few years, there is a massive improvement in the performance and resolution of displays. The technology has changed from old monochrome CRT display to vivid higher resolution display. LCD, LED, OLED are the most recent displays. LED displays are basically LCD unit, which uses LED as backlight. In these various display technology OLED displays is the most superior display and it have the capability to work even without backlight. Various research are taking place to make the OLED technology more superior.

10.4.1 ROLE OF RARE EARTH ELEMENTS (RRES) IN LED APPLICATIONS

LED is a commonly used device to provide backlight in LCD and TFT displays. For display applications, the backlight should be uniform and have the ability to emit high luminescence. The white light can be generated by mixing red, green, and blue, that is the RGB phosphors can be stimulated using UV LEDs [19]. LED are generally made of single-layered or multi-layered PN junction diodes. When it is biased with a suitable potential, the electrons and holes flow through the electrode and combines at the junction and emit light. The energy bandgap of the semiconductor material determines the wavelength of the light emitted.

10.4.1.1 MAJOR CRITERIA OF LED

LED phosphors are generally used to convert near blue or violet light into white light. It consists of a host and an activator. Generally, RREs are used as activators. The white light used as a backlight in modern televisions is achieved by doping phosphor with Ce^{3+} or Eu^{2+} which helps to emit longer-wavelength light and this light combine with light produced by the diode, forms white light [20]. Using Ce^{3+} doped garnet phosphor can produce better quality white light [21].

10.4.1.2 EMISSION SPECTRUM OF LED

The emission spectrum of phosphor is related to the pumping LED. The phosphor used should cover all the regions of the visible spectrum if the pumping LED is in the UV spectral range. In the case of visible LED, the emission of the phosphor material will be mixed with the emission from the pumping LED. In the case of a single phosphor, the wavelength of the emission spectra lies between 500 and 600 nm and for multiple phosphors, it can cover up to 700 nm range. Due to the spin-orbit split ground state of Ce^{3+}, Ce: YAG shows a wide emission range and has FWHM (full width at half maximum) of 100 nm [22].

10.4.1.3 EXCITATION SPECTRUM OF LED

Another important property that is required for phosphor is its spectral excitability, it is the matching between the emission of pumping LED and excitation spectrum. Thus, the phosphor material has spectral excitability near UV to the blue region is preferred. Also, the phosphors having wide emission spectra are preferred to compensate the variation in the optical properties while varying the temperature. Ce: YAG is a material that shows excitations in the region of 450 nm and it can be excited by a blue pumping LED [23].

10.4.1.4 LONG TERM STABILITY OF LED

The stability of LED is really important for the long-term life of LED. LED generally has a lifetime of nearly about 15,000 to 1,00,000 h, which is higher than that of incandescent and CFL lamps. Heat dissipation is the major problem that directly influences the lifetime of LED s. Even in the case of LEDs s 20% of energy is converted into heat and this will increase the temperature from 400 to 450 K near the chip area. It not only varies the optical properties of the material, but it will also decrease the overall lifetime of LED. It is found that phosphors such as CaS: Eu and SrS: Eu have several stability problems due to low stability with air. Ce: YAG is a material that shows high thermal stability and shows high photon emission [24].

But one of the disadvantages of Ce: YAG is that it does not show emissions above 600 nm and it decreases the color rendering and the CCT. To solve these problems researchers added different luminescent ions and also tried by varying the elements in the matrix. For high luminance applications, the number of red-emitting phosphors is small. Europium is the mostly used dopant (as (Ba, Sr)2Si5N8:Eu) and it shows emissions between 610 and 660 nm [25].

10.4.2 OLED (ORGANIC LIGHT EMITTING DIODE)

Organic light emitting diode (OLEDs) are one of the best displays available nowadays. The OLED displays work in a completely different way compared to LED and LCDs. In OLED displays the pixels are self-lighting compared to other displays which require a backlight for its working. Due to this property, it can produce more sharper and high-quality images. Even though it has very good display quality it has less lifetime and very expensive. Various research are going on to increase its lifetime and to make it more economical. An OLED device consists of different organic layers which include the HTL, emitting layer, ETL, cathode, anode (Figure 10.3).

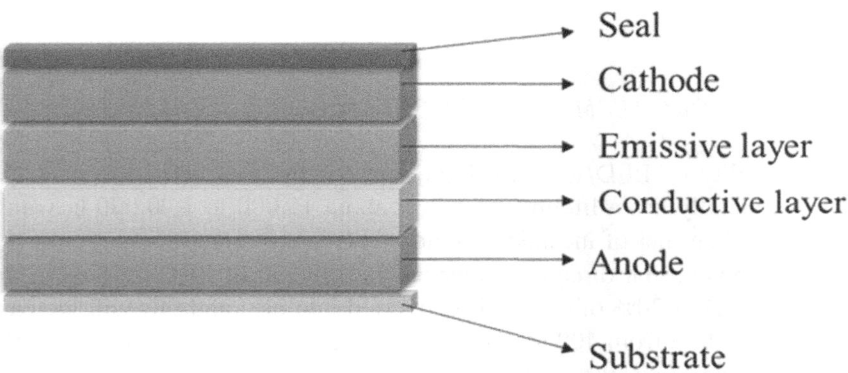

FIGURE 10.3 Layers of OLED.

10.4.2.1 RARE-EARTH DOPED OLED

By utilizing the different light extraction techniques, the electric photo conversion efficiency of OLED reached close to its theoretical limit [47, 48]. But now also it is a challenge to achieve high efficiency at high brightness levels due to the difference in carrier mobility and results in low excitations yield. It is found that these drawbacks of OLED can be eliminated to some extend by doping with RREs. Due to the interatomic transitions of 4f shell RE complexes emit line-like spectra and shows high color purity. Moreover, the narrow emission spectra of rare earth doped element will not get affected by the surrounding environment due to the shielding effect of 5s5p electrons [49, 50].

According to emission wavelength the RE complexes can be classified into two Visible emission and NIR emission. The important RE complexes which are capable of emitting in the visible region are Eu (III) (red emitting),

Tm (III) (blue emitting), Sm (III) (orange emitting), Dy (III) (yellow emitting), Tb (III) (green emitting). Nd (III), Yb (III) and Er (III) are the other RE complexes which emits radiations in the NIR region [51].

Lanthanides are mainly used as emitters in OLED s. The traditional material which is used as emitters will give an internal quantum efficiency (IQE) of 25% while using lanthanide compounds will increase the efficiency up to 100%. It also increases the color purity of the display and also decreases the cost of production [8]. But one of the drawbacks of RE doped complexes are its poor film forming property due to its special molecular structure.

10.4.2.2 ROLE OF SCANDIUM IN OLED DISPLAYS

Lanthanide compounds are generally used as active layers in OLED displays [26]. Even though trivalent elements like Eu, Er, Sm, Tb, and Yb exhibit brighter luminescence, other rare-earth materials which are less luminescent were also studied, carefully. Materials like Y, Gd, La, Lu, and Sc exhibit luminescence from fluorescence originating from $\pi^* \rightarrow \pi$ transitions in the organic ligand [27].

Alq_3 is the common material used as an emitting layer in OLED devices. The main drawback of Alq_3 is that its efficiency is low compared to other metalorganic emitters. Based on recent studies it is found that scandium 8-oxyquinolinate as emitting layer significantly improves the performance of OLED devices.

The turn-on voltage of Sc diode is 3.8 V compared to 4.8 V of Alq_3. Thus, Sc diode requires less turn-on voltage. In the case of the Sc diode, the power and current efficiency is almost double [28]. The performance raise is due to the high charge mobility of scandium 8-oxyquinolinate while using Yb metal as cathode (having work function 2.59 eV), instead of Al (having work function 4.31 eV) [29]. Thus, for increasing the performance of the device, we should decrease the energy barrier between the fermi level of the cathode and LUMO of the emitting layer [30].

10.4.2.3 ROLE OF SAMARIUM IN OLED DISPLAYS

It is found that using samarium in OLEDs, will emit light in the visible and NIR region. The majority of samarium derivatives used in OLED s are in the class of Beta-diketonates. Due to ${}^4G_{5/2} \rightarrow {}^6H_{5/2}$, ${}^4G_{5/}2 \rightarrow {}^6H_{7/2}$

and $^4G_{5/2} \rightarrow {}^6H_{9/2}$ transitions they result in orange transitions near the peak 563, 598, 643 nm [31].

10.4.2.4 ROLE OF EUROPIUM IN OLED DISPLAYS

In 1991 Kido and co-workers reported the electroluminescence study of the europium (III) complex. They used europium as the emitter in OLED. It emits a maximum brightness of about 0.3 cd m^{-2} and has a turn-on voltage of up to 12 V [32, 33]. Researchers create different europium (III) complexes by introducing conjugated pyrazine, thiadiazole, and selena-diazole rings.

Due to the strong metal-centered luminescence of Eu 3+, it emits light in the red region. The emission is due to electronic transitions from the lowest excited state $^5D_0$ to the ground state ^7F_J. 580, 590, 610, and 650 nm are the characteristic emission peaks of europium (III). The emission peak of 580, 590, 610, 650 belong, respectively to the transitions $^5D_0 \rightarrow {}^7F_0$, $^5D_0 \rightarrow {}^7F_1$, $^5D_0 \rightarrow {}^7F_2$, and $^5D_0 \rightarrow {}^7F_3$ transitions, respectively. The emission peak at 610 nm due to $^5D_0 \rightarrow {}^7F_2$ transition is the strongest transition compared to others [34]. Researchers were exploring different europium complexes and tries to increase their electroluminescent properties. Even though homoleptic β-diketonate europium gives high and efficient photoluminescence it can't be used in OLED applications due to its poor carrier transport ability [35, 36].

10.4.2.5 ROLE OF CERIUM IN OLED DISPLAYS

Even though the IQE of OLED has reached close to the theoretical value, the development of blue OLED with high efficiency and lifetime are still a challenge. The efficiency and stability of OLED decreases drastically due to the chemical reaction at high current density [52]. Research are going on to increase the stability of the device and to shorten the excited life time. The Cerium (III) is such a material which have short, excited state lifetime. And also, the cost of Cerium is much lesser compared to other material due to the abundance of cerium in the nature [53].

Like Europium (II) complex cerium (III) also exhibits a wide emission spectrum, tunable colors, and short excited-state lifetime. It generally emits in the blue or UV region and can be varied under different coordination environments. Yu et al. in 2007 [37] studied the electroluminescence and photoluminescence properties of Cerium complex. 225 and 313 nm are the excitation wavelength of Cerium complex in methanol solution and 360 nm is the emission wavelength. In comparison to Cerium complex, $CeCl_3$ exhibits an excitation wavelength of about 283 and 334 nm, an emission wavelength of about 376 nm.

Cerium complex shows higher luminescence intensity than $CeCl_3$ of the same concentration. The researchers studied the material by varying the concentration of Cerium complex in CBP and they came to the conclusion that when the doping concentration is above 3 wt.% the emission mainly came from CBP and when the doping concentration is more than 3 wt.%, due to poor carrier transportability the device won't work even at a higher voltage. Then the researchers fabricated the device by the structure ITO/CuPc/Ce-1:CBP/PBD/LiF/Al and the device shows a maximum wavelength of 376 nm and maximum radiance of 13 μW cm^{-2} [37, 38]. Ce^{3+}, Pr^{3+}, and Nd^{3+} are the three trivalent lanthanides for which f-d emissions can be observed. These three lanthanides can show a broad emission band in the UV-vis region, but among these Ce^{3+} only shows strong f-d transitions. Ce^{3+} normally exhibits emission in the UV or in the blue spectral region but it can be red shifted based on the ligand environment [39].

10.5 CONCLUSIONS

Rare-earth elements are a group of chemical elements in the periodic table. Most of the Rare-earth elements occur in nature but not in pure form, synthesized various methods. Currently RREs is most significant other than commonly exploited elements, due to its extreme properties in various fields and it is one of the inevitable parts in the modern technologies. In the modern era several modern devices we use RREs, applications like cell phones, televisions, LED light bulbs, laser sources, optical fiber amplifiers, etc. Here we discuss the role and importance of rare elements in various solid-state devices like lasers, optical amplifiers and LED/OLED display devices. Most solid-state lasers are made up of rare-earth ions such as thulium, ytterbium, erbium, and holmium. The most common and known solid-state laser is the Nd: YAG in which Neodymium is a promising candidate for higher efficiency and employed as a laser ion. Another commonly used laser is erbium doped YAG laser and erbium and chromium doped YSGG, respectively which differ from each other due to the difference in their laser wavelength. These lasers possess the highest absorption among all infrared lasers and thus could be employed in dental treatment. The reddish-orange emission of rare earth element samarium doped laser devices has held many applications in the visible regions.

An optical fiber doped with RREs has wide range of applications due to its light-emissive properties. Some of the RREs exhibit exceptional optical properties such as absorption and luminescence. They include lanthanum (La), cerium (Ce), dysprosium (Dy), europium (Eu), samarium (Sm), ytterbium (Yb), etc. We discussed here the importance of doping with the RREs and thereby improving the properties of the optical fibers. The RREs acting as glass modifiers in which it occupy the voids of the glass network. The improved the structure of fiber glasses depends upon the rage atomic size and high coordination number of the RREs. The optical fiber amplifiers with a composite of erbium-ytterbium co-doped phosphate core and double silica cladding have a significant gain coefficient. The higher concentration of erbium helped reduce the length of the gain media. Luminescence properties of dysprosium doped gallium-antimony-sulfur-iodine chalcohalide glasses and optical fiber are potentiality good in mid-infrared (MIR) region. The two intense emissions of these elements in glasses exhibited with a large, stimulated emission cross-section, a high quantum efficiency and minimum losses. It is a good candidate for MIR laser gain material.

Recently OLED displays improved the quality of the display devices and thereby improved the transmission quality of the pictures. The self-lighting properties of OLED displays are much better than the other display devices like CRT, LCD, LED, in which it require a backlight for its working. Due to this property, it can produce more sharper and high-quality images. OLED displays has very good display quality, but it has disadvantages like less lifetime and high cost. The RREs has its own role for improving its lifetime, the drawbacks of OLED could be eliminated to some extend by doping with RREs. The interatomic 4f transitions of RREs emit line-like spectra and it improve the color purity. The narrow emission spectra of rare earth doped element will not get affected by the surrounding environment due to the shielding effect of 5s5p electrons thereby improving the quality of display devices. Finally, we conclude that the role of RREs in solid state device is a promising research field in the future.

KEYWORDS

- **display device**
- **dysprosium**
- **fiber optics**
- **laser**
- **LED**
- **luminescence**
- **mid-infrared**
- **OLED**
- **rare earth elements**

REFERENCES

1. Haque, N., Hughes, A., Lim, S., & Vernon, C., (2014). *Resources, 3*(4), 614–635.
2. Song, X., Chang, M. H., & Pecht, M., (2013). *JOM, 65*(10), 1276–1282.
3. Steckl, A. J., Heikenfeld, J. C., Lee, D. S., Garter, M. J., Baker, C. C., Wang, Y., & Jones, R., (2002). *IEEE Journal of Selected Topics in Quantum Electronics, 8*(4), 749–766.
4. Weber, M. J., (1979). *Handbook on the Physics and Chemistry of Rare Earths, 4*, 275–315.
5. Littler, C. M., (1999). *Dermatologic Clinics, 17*(2), 401–430.
6. Diaci, J., &Gaspirc, B., (2012). *J. Laser Health Acad., 1*(1), 1–13.
7. Mohan, S., Kaur, S., Singh, D. P., & Kaur, P., (2017). *Optical Materials, 73*, 223–233.
8. Bünzli, J. C. G., Comby, S., Chauvin, A. S., & Vandevyver, C. D., (2007). *Journal of Rare Earths, 25*(3), 257–274.
9. Ballato, J., (2021). *Encyclopedia of Glass Science, Technology, History, and Culture, 1*, 705–719.
10. Łączka, M., & Stoch, L., (1990). *Journal of the Less Common Metals, 166*(1), 163–171.
11. Righini, G. C., & Ferrari, M., (2005). *La Rivista del Nuovo Cimento, 28*(12), 1–53.
12. Dousti, M. R., & Amjad, R. J., (2016). *In Reviews in Plasmonics 2015* (pp. 339–386). Springer, Cham.
13. Girard, S., Capoen, B., El Hamzaoui, H., Bouazaoui, M., Bouwmans, G., Morana, A., & Ouerdane, Y., (2016). *IEEE Transactions on Nuclear Science, 64*(1), 567–573.
14. Ahmad, A. U., Hashim, S., & Ghoshal, S. K., (2020). *Journal of Alloys and Compounds, 844*, 156176.
15. Sun, M., Yang, A., Ren, H., Qi, S., Lin, H., Feng, X., & Yang, Z., (2021). *Journal of Non-Crystalline Solids, 560*, 120718.

16. Singh, D., Singh, S., Singh, T., & Kaur, P., (2021). *Journal of Materials Science: Materials in Electronics, 32*(6), 6900–6911.
17. Di Francesca, D., Girard, S., Agnello, S., Alessi, A., Marcandella, C., Paillet, P., & Gelardi, F. M., (2015). In: *2015 15th European Conference on Radiation and Its Effects on Components and Systems (RADECS)* (pp. 1–4). IEEE.
18. Galagan, B. I., Denker, B. I. I., Egorova, O. G. N., Kamynin, V. A., Ponosova, A. A., Sverchkov, S. E. E., & Tsvetkov, V. B., (2018). *Quantum Electronics, 48*(6), 550.
19. Mikami, M., Watanabe, H., Uheda, K., Shimooka, S., Shimomura, Y., Kurushima, T., & Kijima, N., (2009). In: *IOP Conference Series: Materials Science and Engineering* (Vol. 1, No. 1, p. 012002). IOP Publishing.
20. Tsai, C. C., Wang, J., Chen, M. H., Hsu, Y. C., Lin, Y. J., Lee, C. W., & Cheng, W. H., (2009). *IEEE Transactions on Device and Materials Reliability, 9*(3), 367–371.
21. Mueller-Mach, R., Mueller, G. O., & Krames, M. R., (2004). In: *Third International Conference on Solid State Lighting* (Vol. 5187, pp. 115–122). SPIE.
22. Smet, P. F., Parmentier, A. B., & Poelman, D., (2011). *Journal of the Electrochemical Society, 158*(6), R37.
23. Lisitsyn, V., Lisitsyna, L., Tulegenova, A., Ju, Y., Polisadova, E., Lipatov, E., & Vaganov, V., (2019). *Crystals, 9*(9), 476.
24. Gang, S. R., Kim, D., Kim, S. M., Hwang, N., & Lee, K. C., (2012). *Microelectronics Reliability, 52*(9, 10), 2174–2179.
25. Lee, B., Lee, S., Jeong, H. G., & Sohn, K. S., (2011). *ACS Combinatorial Science, 13*(2).
26. Thejokalyani, N., & Dhoble, S. J., (2014). *Renewable and Sustainable Energy Reviews, 32*, 448–467.
27. Aleksanyan, D. V., Kozlov, V. A., Petrov, B. I., Balashova, T. V., Pushkarev, A. P., Dmitrienko, A. O., & Abakumov, G. A., (2013). *RSC Advances, 3*(46), 24484–24491.
28. Katkova, M. A., Ilichev, V. A., Konev, A. N., Bochkarev, M. N., Vitukhnovsky, A. G., Parshin, M. A., ... & Van der Auweraer, M. (2008). *Journal of Applied Physics, 104*(5).
29. Bochkarev, M. N., Katkova, M. A., Ilichev, V. A., & Konev, A. N., (2008). *Nanotechnologies in Russia, 3*(7), 470–473.
30. Katkova, M. A., Kurskii, Y. A., Fukin, G. K., Averyushkin, A. S., Artamonov, A. N., Vitukhnovsky, A. G., & Bochkarev, M. N., (2005). *Inorganicachimica Acta, 358*(13), 3625–3632.
31. Chu, B., Li, W. L., Hong, Z. R., Zang, F. X., Wei, H. Z., Wang, D. Y., & Lee, S. T., (2006). *Journal of Physics D: Applied Physics, 39*(21), 4549.
32. Kido, J., Nagai, K., Okamoto, Y., & Skotheim, T., (1991). *Chemistry Letters, 20*(7), 1267–1270.
33. Kido, J., Nagai, K., & Okamoto, Y., (1993). *Journal of Alloys and Compounds, 192* (1, 2), 30–33.
34. Fangfang, C., Zuqiang, B., & HUANG, C., (2009). *Journal of Rare Earths, 27*(3), 345–355.
35. Uekawa, M., Miyamoto, Y., Ikeda, H., Kaifu, K., & Nakaya, T., (1998). *Bulletin of the Chemical Society of Japan, 71*(9), 2253–2258.
36. Braga, S. S., Coelho, A. C., Gonçalves, I. S., Santos, G., Fonseca, F. J., Andrade, A. M. D., & Pereira, L., (2008). *Journal of Non-Crystalline Solids, 354*(19–25), 2736–2739.

37. Yu, T., Su, W., Li, W., Hua, R., Chu, B., & Li, B., (2007). *Solid-State Electronics, 51* (6), 894–899.
38. Zheng, X. L., Liu, Y., Pan, M., Lü, X. Q., Zhang, J. Y., Zhao, C. Y., &Su, C. Y., (2007). *Angewandte Chemie International Edition, 46*(39), 7399–7403.
39. Vogler, A., & Kunkely, H., (2006). *Inorganica Chimica Acta, 359*(12), 4130–4138.
40. Zheng, B., Yang, J., Qi, F., Wang, J., Zhang, X., & Wang, P., (2021). *Journal of Non-Crystalline Solids, 573*, 121129.
41. Ragin, T., Baranowska, A., Kochanowicz, M., Zmojda, J., Miluski, P., & Dorosz, D., (2019). *Materials, 12*(8), 1238.
42. Anashkina, E. A., (2020). *Fibers, 8*(5), 30.
43. Sadeq, M. S., & Morshidy, H. Y., (2020). *Journal of Rare Earths, 38*(7), 770–775.
44. Zhang, L., Qu, Y., Wan, X., Zhao, J., Zhao, J., Yue, Y., & Kang, J., (2020). *Journal of Non-Crystalline Solids, 532,* 119886.
45. Miluski, P., Kochanowicz, M., Żmojda, J., Dorosz, D., Łodziński, M., Baranowska, A., & Dorosz, J., (2020). *Optical Materials, 101*, 109711.
46. Kaur, A., Khanna, A., González-Barriuso, M., & González, F., (2021). *Journal of Materials Science: Materials in Electronics, 32*(13), 17266–17281.
47. Jou, J. H., et al., (2015). Approaches for fabricating high efficiency organic light emitting diodes. *J. Mater. Chem. C., 3*, 2974–3002.
48. Swayamprabha, S. S., et al., (2021). Approaches for long lifetime organic light emitting diodes. *Adv. Sci., 8*, 2002254.
49. Wei, C., et al., (2018). Advances in luminescent lanthanide complexes and applications. *Sci. China Technol. Sci., 61*, 1265–1285.
50. Zampetti, A., Minotto, A., & Cacialli, F., (2019). Near-infrared (NIR) organic light-emitting diodes (OLEDs): Challenges and opportunities. *Adv. Funct. Mater., 9*, 1807623.
51. Ma, C. G., et al., (2016). Energy level schemes of f(N) electronic configurations for the di-, tri-, and tetravalent lanthanides and actinides in a free state. *J. Lumin., 170*, 369–374.
52. Chan, C. Y., Tanaka, M., Nakanotani, H., et al., (2018). Efficient and stable sky-blue delayed fluorescence organic light-emitting diodes with CIEy below 0.4. *Nat. Commun., 9*, 5036–5044.
53. Lindqvist-Reis, P., R′eal, F., Janicki, R., et al., (2018). Unraveling the ground state and excited state structures and dynamics of hydrated Ce^{3+} ions by experiment and theory. *Inorg. Chem., 57*, 10111–10121.

CHAPTER 11

Smart Materials-Based E-Nose Technology: Fundamentals and Emerging Applications

JESNA SARA SHAJI[1] and RONY RAJAN PAUL[2]

[1]*Department of Chemistry, Mar Thoma College, Tiruvalla, Pathanamthitta, Kerala, India*

[2]*Department of Chemistry, CMS College, Kottayam, Kerala, India*

ABSTRACT

The world of material science is a challenging and exciting research field where so many innovations are taking place in a short span of time. Smart materials are a new group of materials that can respond to the external stimuli in its environment in a controlled and useful way. Some major applications of the smart materials include smart windows, smart fabric, thermochromic mugs, biomedical applications, etc. Electronic nose technology is the technical copy of mammalian olfactory system which is used for the assessment of volatile or vaporous organic compounds, odors, vapors, etc. E-nose has interesting applications in daily life including air quality monitoring, food quality and spoilage monitoring, for detection of diseases, etc. Since smart materials have the properties of sensing and actuating, some of them are widely used as gas sensors in the e-nose device. For more accuracy in sensitivity and specificity like human nose, bioelectronic nose is also used in which a biological element is used as an active component in the sensing system. Bioelectronic nose is a fascinating technology which employs the application of biotechnology

New Advances in Materials Technologies. Hossein Hariri Asli, Ali Pourhashemi, Ann Rose Abraham, & A. K. Haghi (Eds.)

and nanotechnology. In this chapter, we discuss about bioelectronic nose technology, smart materials, and the role of smart materials in e-nose applications.

11.1 INTRODUCTION

Most of the technologies are developed with the cause of a need or an inspiration except the accidental discoveries. The human olfactory system or the sense of detecting odors is an inspiring and exciting sensory system by which they can identify and distinguish the smell of different products and thereby analyzing and detecting their flavors. Most of the cases, the flavor is completely analyzed with the help of smell and taste.

With the popular notion created from the question of scientist Alexander Graham Bell in 1914, that whether we can measure smell and from the inspiration of mammalian olfactory system, for the purpose of detecting and distinguishing simple or complex odors from different volatile chemical products to the need of air monitoring [1], food quality monitoring [2], etc., as a quick, cheap method, a technology for artificial olfaction was developed. First attempt to develop an instrument for the detection of odors was done in 1961 by Moncrieff [3]. But it was a mechanical nose. Then, in 1964, Wilkens and Hartman developed an electronic nose [4] and in 1965, Dravnieks and Trotter reported one [5] and also, Buck et al. developed an electronic nose [6]. In 1982, Persaud and Dodd proposed the concept of using chemical sensor array system for distinguish between odors by developing an electronic nose by using a gas multisensor array [7]. In the late 1980s, the term electronic nose was introduced around the world. The definition of electronic nose was first given by Gardner and Bartlett in 1988 as, "An electronic nose is an instrument, which comprises an array of electronic chemical sensors with partial specificity and an appropriate pattern-recognition system capable of recognizing simple or complex odors."

Electronic nose or E-nose is a device fabricated to discriminate or distinguish among complex odors by a bunch of sensors and pattern recognition method by taking the mammalian olfactory system as the paradigm (Figure 11.1). It consists of mainly three parts: (i) A sample handling system; (ii) detection system; and (iii) a data processing system [8], which in detail discussed later. E-nose has been used for multidisciplinary applications. Some of its applications include categorization and recognition of fungal species [9], classifying, and distinguishing essential oils from fruits and herbs [10], in meat quality monitoring [11], for the analysis of volatile organic compounds in rice aging [12], for agricultural applications [13], for the detection of spoilage of vegetables [14], for cancer detection [15], for food adulteration monitoring, for stability determination of biodiesel [16], biomedical applications [17], for environmental monitoring [17], etc.

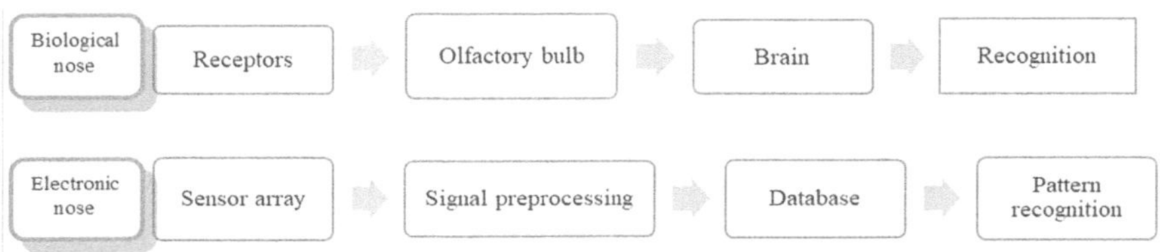

FIGURE 11.1 Comparison of E-nose and biological nose.

From the era of different civilizations, human beings were using materials to enhance their standard of living, thereby classification of civilizations were done according to the most advanced material at that respective period, for example, Stone age, Bronze age, Iron age. Bronze age was the dawn of a new period in which the field of metallurgy has flourished by the extraction of several materials that occupied a place in our daily lives. For the past two centuries, so many developments occurred in science and technology to synthesize new materials. There are mainly four groups of materials, Metals, Polymers, Ceramics, and Advanced materials/Smart, intelligent or responsive materials [18]. The concept of smart materials was originated in the 1980s [19].

Smart materials are materials which can alter their one or more properties in a reversible, controlled, and advantageous way when they come in exposure to external stimuli like stress, temperature, moisture, etc. [20]. The controllability and their ability to adapt to the external environmental conditions to activate their functions are the reasons for the unique and novel properties of these materials [21, 22]. Smart materials are most often utilized as sensors and actuators [23]. Based on the environmental or external stimuli, smart materials are classified as piezoelectric materials, chromic materials, shape memory alloys (SMA), electrorheological fluids (ERFs), electrostrictive materials, magnetostrictive materials, magnetorheological fluids [24]. Smart materials are embedding the whole world due to its advanced, intelligent abilities to act smart in response to different types of external conditions or stimuli. They are used for environmental safety applications such as bioremediation [25], for efficient detection of ozone [26], etc., for medical applications like drug delivery and cancer therapy [27], in orthodontics [28], etc., for robotics applications [29], in the field of manufacturing [30], for civil engineering applications [31], etc. In this chapter, we discuss about the smart materials for electronic nose applications.

11.2 ELECTRONIC NOSE

Electronic nose or E-nose is an instrument which is structured to duplicate the human olfactory system [32]. The term electronic sensing or E-sensing refers to the ability of enabling human-like senses by employing sensor clusters and pattern identification or recognition systems [33]. E-noses are made for specific applications, and one made for a particular application for example, an e-nose used for food quality monitoring, can't be used as a universal one, i.e., it can't be used for environmental monitoring [34]. The sensing system and recognition method of e-nose is not highly specific, i.e., it consists of sensors capable of identifying a wide class of odorant compounds which is smart. In mammalian olfaction, one odor is not detected by just one sensor, but is detected by a bunch of various sensors and each combination of signals is unique for each different odor, thereby producing "odor fingerprint." So, partial selective or cross-selective sensors are used in e-nose. One odor is detected by multiple sensors and one sensor can identify multiple odors, which can be said when olfactory receptors is replaced by sensor array in e-nose [8]. Odorant molecules are light weight (molecular mass approximately 300 Da), small, often hydrophobic, and polar [6]. A simple odor contains only one type of chemical component, but a complex odor constitutes many types of components each in different concentrations that may vary with time [35]. The simple or complex mixture of aroma is detected and identified as a unique odor signature pattern like fingerprint, without analyzing its individual chemical components just as traditional analytical techniques like gas chromatography, mass spectrometry did which is a time-consuming and extensive process. Like mammalian olfactory system, non-selective sensors are used in e-nose [36]. The e-nose technology can't completely mimic the biological nose but can help with monotonous application fields like food-quality monitoring, environmental monitoring and in security where using human panels for odor analysis is at risk [8] and the results can't be completely accurate because of the occurrence of individual variations that is possible depending upon one's physical and mental well-being [37]. The e-nose machine is an advanced technology where one can quantify or determine the aroma and quality of a compound. This automated technology gives an efficient, qualitative monitoring system that can be used instead of human sensory panels. In addition to these properties, E-nose can detect some odorless VOCs such as explosives, etc., also which is not practically possible by biological nose [38, 39]. Electronic nose employs a group of chemical sensors to obtain the odor fingerprint and pattern recognition software to identify and classify odors [40]. Also, it possesses an information processing unit such as artificial neural network system (ANN) and reference library database to compare the identified odors and distinguish it. In e-nose, there is technical equivalent of every part in mammalian nose. Corresponding to olfactory receptors in mammalian nose, there is multisensor array which detects the VOCs responsible for different types of odors. The input signal created from olfactory receptors is sent to olfactory bulb and processed and characterized. Then, the processed signal is sent to the brain where the smell is recognized and classified by olfactory cortex is the last part of biological olfaction. In e-nose, the signal generated from the sensors is preprocessed and using data analysis, pattern recognition and machine learning algorithms, the input scent is identified and classified based upon the extracted digital signatures of senses chemicals which are present in the database. Electronic nose consists of both hardware and software components whereas, software components being the signal conditioning, data processing unit and pattern recognition system where the input scent is identified, characterized, and classified based on comparing the digital signatures obtained from the sensed chemicals present in the database with that of the unknown sample. Hardware component is the electrochemical multisensory array used in this [41] (Figure 11.2).

In the above block diagram, basic working principle of an e-nose is depicted. First, the gas molecules with aroma are drawn to the surface of sensor array by various sampling methods such as headspace sampling, diffusion techniques, etc., and binds with the surface through different binding methods such as absorption, adsorption, coordination chemistry or chemisorption, etc. This odor molecules induces certain reversible physical or chemical changes in the sensing material such as change in resistance, voltage, current, and generates an input signal. The sensor array contains number of sensors and each sensor respond to a particular odor distinctly in varying amounts, also each sensor array contains different types of sensors, so the input signals must be preprocessed to understand the induced changes clearly. Then the signals are converted to electrical signals by a transducer. Then they are again processed to digitalize the signals so that they can form a digital dataset. In data gathering stage, the processed signals are analyzed based on their specific properties and after acquiring sufficient data, the collected information is preprocessed, respectively upon the requirement of the employed pattern recognition algorithm. The final stage is identification and classification of the unknown sample by comparing it with the digital data of sensed odors stored in the database. The response of e-nose to odorant molecules is considered as a first order time response.

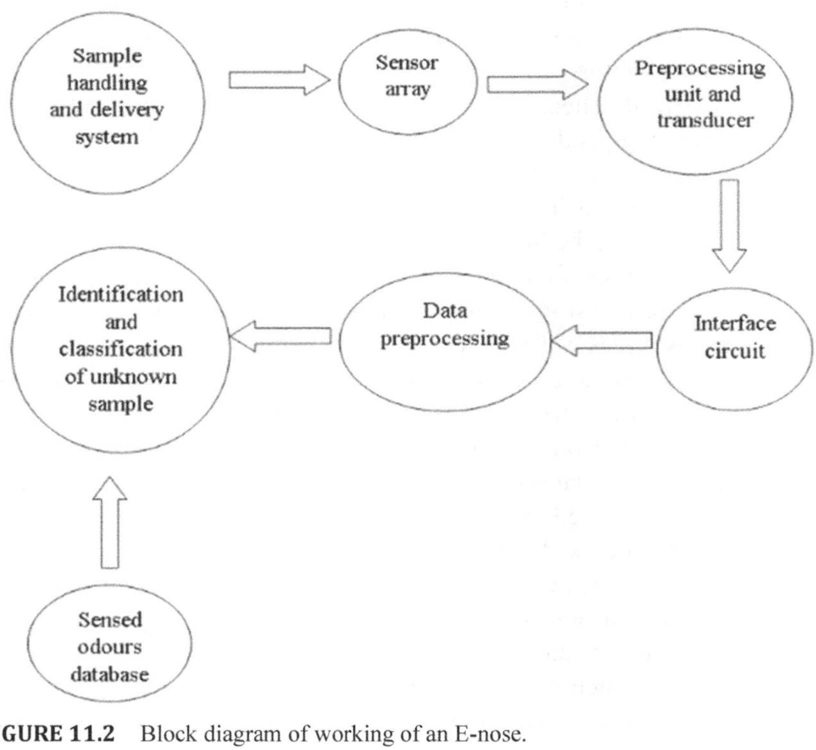

FIGURE 11.2 Block diagram of working of an E-nose.

The Electronic nose consists of mainly three components: (i) sample handling system; (ii) detection system; and (iii) data processing system:

1. **Sample Handling System:** Sample handling is a critical component of this instrument as if the sample is not properly conditioned, the results obtaining from the analyzer is of no value whether the analyzer is of good quality and proper maintenance or not. By using different types of sampling techniques, the analyzing standards can be improved. The methodology used to introduce the volatile chemicals present in the headspace of the sample into the detection system are:

    i. Static headspace sampling method can be used for the qualitative and quantitative analysis of odorant molecules. In this technique, the sample is positioned in a tightly sealed vial and when equilibrium is achieved between the matrix and gaseous phase, the volatile compounds are carried to the sensors. This is a more popular method which is also a cheap method. The transferred components into the chamber containing sensor array are saturated with odor, so that the first response of the sensors is large and after some time, the headspace is removed and clean air is filled in the chamber, thereby, sensor responses are returned to their baseline values. The important parameters that might be optimized are analyte's temperature, equilibration time, vial size and amount of analyte. Automated Headspace sampler is more preferred than manual Headspace injection because of its poor repeatability. Inert gas is used to carry the volatile components to the chamber of sensor array.

    ii. Purge and trap (P&T) and dynamic headspace (DHS) extraction methods are employed for high sensitivity applications as it is depending on the vaporizing capability of the target molecules to enable extraction from the matrix. They provide pre-concentration of volatile compounds. The volatile components are ejected by a flow of inert gas and captured onto the adsorbent. In the P&T technique, the gas flow is applied through the sample and in DHS technique, the headspace only is sweeped out with gas. The detachment of the adsorbed molecules from the matrix is favored by displacement of the equilibrium because of the constant depletion of the Headspace. The trapped molecules are passed into the detection system after desorption which is caused by heating. The main parameters to be considered are sample temperature, the equilibration time, the rate of flow of the purging gas and the purge time of the Headspace.

iii. Solid-phase-microextraction (SPME) method uses a silica fiber coated with an adsorbent in the HS of the sample. The volatile components in the sample are adsorbed onto the fiber. Then the adsorbed components undergo desorption by heating and will be entered into the detection system. The parameters to be considered are nature of adsorbent on the fiber, equilibration time, sample temperature and time of extraction. This is a user friendly and pre-concentration method. This method has a preferrable concentration capacity and because it does not need any additional equipment, this is a simple method.

iv. Flow injection analysis (FIA) method is usually computer automated and, in this method, the background gas is constantly being pumped into the chamber of sensory array. Gas containing the odor or the volatile components is applied in this background gas before it reaches the sensor array chamber. There is an accurate regulation on the ratio of the mixture of background gas to odor gas.

v. Stir bar sorptive extraction (SBSE) is a polymer coated magnetic bar, which is held in the headspace for sampling. Compared to SPME, SBSE has high loading capacity and SBSE is a more preferrable method for extraction when high sensitivity is needed.

vi. Inside-needle dynamic extraction (INDEX) method consists of a needle in which there is an adsorbing polymer phase like a fixed bed. The volatile components are ejected through the needle by continuous aspiration/ejection motions by syringe plunger. The advantage of INDEX method over SPME is in its mechanical robustness and in the possibility of increasing the amount of adsorbing polymer and surface area available for adsorbing odorant molecules.

2. **Detection System:** In electronic nose, a multisensor array is employed as the detection system. This sensor array is preferred as the main component of this instrumentation as it is the technical equivalent of human olfactory receptors and human olfactory

nerve cells. The heart of the e-nose is this sensor array, which is used for detecting chemicals in the medium, each individual sensor will be detecting a distinct type of aroma. The main aim of e-nose is to sense two or more chemicals at the same time. So, this is possible when various distinct sensors are used together in this array. A chemical sensor is a transducer which involves a chemical detection layer and transform the chemical interaction into an electrical signal. They can measure continuously and is not much expensive [42]. All chemical sensors consist of required sensitive materials which are interlinked to a transducer. Gas molecules or odorant molecules interact with solid-state sensors by absorption, adsorption, chemical reactions, etc., with thin or thick film of the sensor material. The type of binding mechanism is of relevance because it has dependence on the selectivity and reversibility of the sensing system. High selectivity with high reversibility is difficult to achieve. Therefore, E-noses consist of independently semi-selective and reversible gas sensors and the output is analyzed by the pattern recognition software. These interactions cause physical or chemical reversible changes, and these changes are converted to electrical signal and measured. Based upon the changes produced during the interaction between the odorant molecules and sensor materials, different types of sensing devices are used.

A good sensor should possess the following characteristics, first, the sensor must possess quick response to the analyte molecules for detection and it should have a threshold of detection like that of human nose, second, for being capable of diverse detection abilities, it should have less selectivity so that it can respond to an entire scale of different chemical compounds. The sensor should be of small size and easy to handle, inexpensive, can be reused, short response and recovery time, high resistance for different mediums, insensitive to temperature and humidity, high stability for the employed application.

A device that converts information including the quantity of a simple odorant molecule to complex mixture of odorant molecules, into a useful analytical, electrical signal is defined as chemical sensor. It consists of two components connected in series, a chemical identification system and physiochemical transducer. Biochemically derived recognition entity such as tissues, organelles, etc., is incorporated into the chemical sensor to develop Biosensors.

Multisensor array consists of an array of multiple sensors which can transduce chemical and physical changes or interactions into measurable output signals. The chemical sensors can be divided into 4 main categories, Optical, Thermal, Electrochemical, and Gravimetric based upon the detection principles used (Figure 11.3).

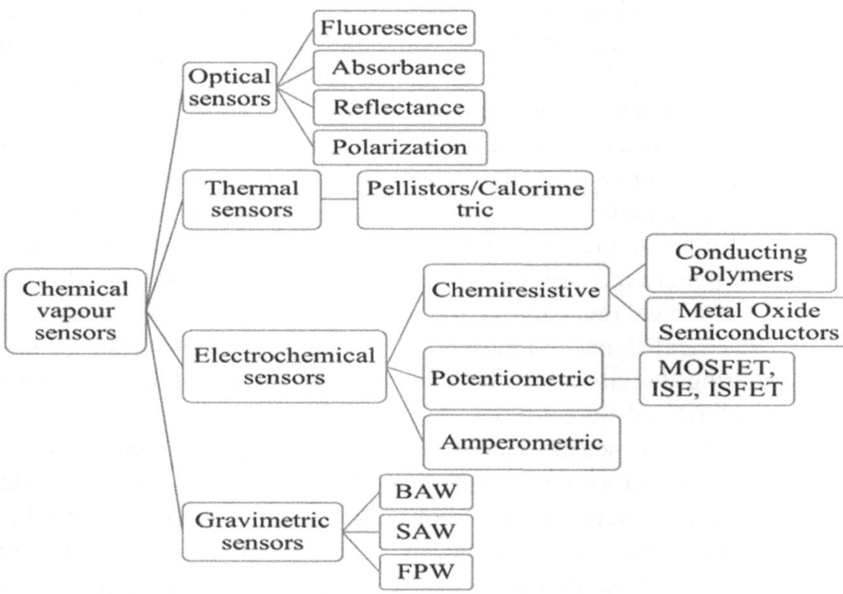

FIGURE 11.3 Types of detection principles used for sensor classification in e-nose.

The sensor types are determined based on the type of analyte that is needed to be detected. The multisensor array must be a neutral mixture of same and distinct types of sensors. For cross-selective sensing and precision improvement, identical, and different types of sensors are used. The number of sensors employed in this array depends upon the purpose for which the e-nose is used. When the background is unknown, different types of mixtures are present or the concentration of background gases are varying, a greater number of sensors are needed to eliminate the uncertainty in the interpretation of output signal pattern and for avoiding false alarm signatures. When the purpose of e-nose is to distinguish among a bunch of pure substances at known concentrations without any interfering background, limited number of sensors is used.

In odor analysis we have to purge a reference gas through the sensor array to obtain a baseline. Then the analyte gas is passed to the chamber of sensor array which causes variations in the output signal of sensors and the sensor is exposed to the odorant molecules until the sensor signal reaches the steady state. Then the odorant molecules are replaced by the reference gas and the sensor comes back to the baseline. The time during which sensor is exposed to odorant is called the response time and the time duration required by sensor to come back to its baseline resistance is called recovery time. The ability of a sensor for quick response is determined as measure of change in the output signal of the sensor when there is a change in the input signal.

There are different types of sensors are employed in e-noses depending on the detection principle used.

i. **Optical Sensors:** In this sensor technique, the measured parameter is the modulation of light properties. The optical sensor technology has its applications spreading from light sources with optical fibers to different types of photodiode and light-sensitive photodetectors. Different operation modes are employed based on the changes in absorbance, transmission, fluorescence, refractive index, optical layer thickness, colorimetric dye response and polarization. The simplest optical sensors use color-changing indicators like metalloporphyrins, to quantify the absorbance by the assistance of a LED and photodetector coming in contact with analyte vapors. Colorimetric sensors employ thin films of chemically responsive dye as its sensor array. Fluorescence based chemosensor contains optical fibers embedded with fluorescent indicator Nile Red dye in polymer matrices of changing attributes like polarity, hydrophobicity, etc., to develop unique sensing regions that interact differently with odorant molecules. The sensitivity depends upon the type of fluorescent dye, or its mixture and the type of polymer used to support the dye. The response time is controlled by the nature of the polymer used. At lower wavelengths, analyte vapors emit fluorescence that is identified by fluorescence sensors. Moisture related energy absorption of near Infrared light can be employed for quantification of the moisture present in the odorant mixture. The optical sensor can be used for determination of the composition of medium with the help of precisely quantified absorption of light by the sample medium. Reflective optical sensor contains a light source and detector. The light is radiated from the source to outwards and will be reflected if the material of interest became exposed to the sensor, then the reflected light will be detected by the detector. The fluorescence sensors have fast responses than colorimetric sensor clusters.

The fundamental structure of an optical fiber sensor array is that sides or tips of the optical fiber are coated with a fluorescent dye encapsulated in a polymer matrix. The polarity changes in the fluorescent dye on exposure to vapor changes the dye's optical properties such as intensity change, spectrum change, wavelength shift in fluorescence, etc. These optical changes are the response mechanism used for odor detection. These compact, lightweight optical sensors can be blend together on a single optical fiber network, which can resist the Electromagnetic interference and can operate in high radiation areas due to Bragg and other grating based optical sensors.

The advantages of the optical sensors are very high sensitivity, multi-parameter detection capabilities, able to identify individual components in mixtures, very fast response time, etc. Disadvantages are the electronics and software associated with it is very complex thereby making it expensive and short lifetime because of photobleaching, poor transportability due to fragile optoelectrical counterparts.

ii. **Thermal Sensors:** Pellistors or calorimetric sensors have the basic design which is a catalytic surface; Platinum and Palladium are the prevailing examples included with a heater to manage the temperature conditions required for sensor operation and that of the temperature probe. Based on the type of volatile chemical and its concentration, temperature changes are detected. Thermistors, thermopiles, and pellistors are different types of sensors employed for the measurement of heat change that occurs during the chemical reactions such as combustion, adsorption, and enzymatic reactions. Any chemical reaction or even adsorption or desorption process releases or absorbs a certain amount of heat from its surroundings.

iii. **Electrochemical Sensors:** They can transform chemical energy into a useful, reliable electrical energy or signal. Their responses are detected by the change in any electrical characteristics. Chemiresistive, Potentiometric, and Amperometric sensors belong to this group of sensors.

Chemiresistive sensors works on the principle that whenever the odorant molecule interacts with the sensor, resulting in variations in any property of the material which leads to change in resistance of the sensor material. The mechanisms that lead to change in resistance is different for each material. Common examples of chemiresistive sensors are CP, intrinsically CP and metal oxides (MOS).

Potentiometric Sensors employ a technique where a zero-current potential is formed at a suitable membrane or electrode interface which is in exposure with the analyte in accordance with the analyte concentration. Examples are metal-oxide-semiconductor field effect transistor (MOSFET), Ion-selective electrodes (ISE), ion-sensitive field effective transistor (ISFET), etc.

Amperometric sensors were one of the sensors used in E-nose initially. Amperometry is a former electroanalytical method where coulometry, voltammetry, and constant potential techniques are joined together to identify and measure electroactive moieties in liquid and gas phases. Amperometry can be employed in gas phase analytes by involving a gas-liquid/solid interfacial transport process. The measurements are taken in this technique by recording the current in the electrochemical cell between the working and counter electrodes as a function of the analyte concentration. This sensor consists of a working electrode, counter electrode and reference electrode that are immersed in an electrolyte. The analyte odorant molecule diffuses into the electrochemical cell and to the working electrode surface via a porous membrane. Then the analyte molecule becomes oxidized/reduced and by governing the Faraday's law, resulting in the production or consumption of electrons at the working electrode.

iv. **Gravimetric Sensor:** They are also called Piezoelectric sensors which are often used for e-noses. This category comprises of bulk acoustic wave (BAW), surface acoustic wave (SAW) or Sound acoustic wave (SAW), flexural plate wave (FPW), or shear horizontal acoustic plate mode sensors (SH-APM) sensors. BAW and SH-APM are commonly referred to thickness shear mode (TSM) or quartz crystal microbalance (QMB or QCM). Here, the operating principle is identification of alterations in the propagation of acoustic waves as a result of binding of analyte molecules onto the sensor surface. The SAW device and the QCM device are predominantly employed in gas sensing system. The SAW device produces a surface wave that travels along the surface of the sensor whereas QCM develops a wave that travels through the bulk of the sensor. Due to adsorption of odorant gas molecules, variations in the thickness of sensing layer are occurred which leads to a change in the resonant frequency, thereby detecting the odor.

v. **Mass Spectrometry:** Mass spectrometers in combination with Gas chromatograph are often used for the identification of pure chemicals. After interaction of the analyte molecules with reagent ions or ionization of compounds through therm-ionic emitted electrons, the molecular ion and their fragment ions are separated according to their mass to charge ratio in the presence of electric/magnetic field. Variety of mass analyzers are available such as sector instrument employs classical approach with tunable static fields, quadrupole mass analyzer contains four parallel metal rods and filters the different ions by oscillating electric fields. Finally, the ions collide at electron multiplier and the current is measured.

vi. **Ion Mobility Spectrometry:** Here, the target molecules are not only separated based on mass to charge ration, but also on the difference in the mobilities of ions. This means that the reduced mass, the charge and the collision cross section which is determined by size and shape has a direct impact on the separability of ions. Therefore, the collisions between the ions and the ambient air molecules are utilized and the measurement can be performed under normal pressure. After the air sample is ionized, the ions are pulsed through a shutter into a drift tube, which is isolated from atmospheric air. The drift tube contains a poor steady electric field, which accelerates the ions along the tube. The movement is hindered by collisions until the ions reach the detector at the end. Depending on the ion impact, a current is generated and measured over the time of flight. This gives information about the identity and concentration of the analyte.

vii. **Gas Chromatography:** Most widely used technique in analytical chemistry for the separation of mixtures. In the case of volatiles, Gas-liquid or gas-solid chromatography are possible. The sample is transported by the mobile gas phase and is moved towards the stationary solid or liquid phase and reacts with it. Based on the physical properties such as H-bonding, polarity, boiling point, etc., the affinity of each component towards the stationary phase is different. Based on the partition behavior, retention time is analyzed and the order of elution. In the e-nose sensor technology, GC is employed in the fast or ultrafast mode. To increase the separation speed, the carrier gas flow rate can be increased, reduction of column length, reduction of thickness of stationary phase, use of faster carrier gas in the case of gas-liquid chromatography. Depending on the sample, it is important to avoid using all possibilities at once, because this always results in a decrease of the resolution, the sample capacity or both.

viii. **Infrared Spectroscopy:** In between the range of 4,000 and 200 cm^{-1}, molecular vibrations and higher energy levels are excited. Through characteristic absorption bands, the type of chemical bonds can be determined, and pure chemicals can be identified by their unique fingerprint spectrum. Classical electronic nose algorithms are used to understand the spectrum of mixtures. In photoacoustic IR spectroscopy, a modulation of intensity of an IR source result in a temperature alteration and the resulting expansion and contraction of the gas will be quantified as audible frequencies with a microphone. Commercially available devices are mostly used for absolute measurements of concentration either in detection of a single species which has a distinct absorption wavelength or by analysis at multiple wavelengths for a known gas mixture. The instrument can be combined with pattern recognition and can be used as an e-nose.

3. **Data Acquisition System:** This part of the e-nose is responsible for recording the signals given by the sensor cluster and delivering them to the processing or computing system which has the required software for analyzing them. The control and data acquisition systems can be conformed into an exclusive device which can be a data acquisition card, a microcontroller, a Digital Signal Processor, or a computer. Authentic connections between electrodes and measuring device are maintained for consistent and valuable data. It must contain enough power stage for managing the components which needed more power consumption. This part of the system must be able to handle and saving data from different sensors instantaneously.
4. **Pattern Recognition Algorithm:** The job is to identify, classify, and quantify the analyte odor relying on the information from the data-storage. Digital signature patterns of known chemicals are stored in the database prior to the analysis of unknown odor and training of the Pattern recognition system will ensure that unknown odors can be distinguished and detected. Multiple procedures are available to facilitate this process and are divided into three groups.

11.3 GRAPHICAL ANALYSIS

This is the simplest form of data reduction for comparison between the odorant molecules of the unknown analytes with those of known sources in database. For each type of analyte, the sensor array produces a unique response pattern in which the pattern of changes is distinguishable of vapor whereas the amplitude of patterns denotes the amount of the vapor present. Various types of graphs are used for this process such as polar, bar graph, etc.

11.4 MULTIVARIATE DATA ANALYSIS (MDA)

This method consists of a group of methods for examining data sets with more than one variable by decreasing high dimensionality in a multivariate problem when variables are slightly correlated, so they can be exhibited in two or three dimensions. This procedure is depended on the multivariate statistics that comprises monitoring and examining of multiple variables at a time. The different types of multivariate data analysis (MDA) methods are, Cluster analysis, featured within, Principal component analysis and Canonical discriminate analysis. MDA is useful when sensors have partial selectivity towards the individual components in the analyte mixture. Multivariate analysis could be carried out as both untrained and trained techniques. Untrained methods are employed when there is no database of known samples, so in this case, comparisons between different unknown samples are taking place to distinguish between them. Most common example of untrained MDA technique is PCA. PCA is useful when there is no dataset of known sample is available or concealed relationships between analyte components are doubted. Trained learning techniques classify and identify unknown samples on the basis of datasets of the known samples maintained in a reference library.

11.5 NETWORK ANALYSIS

This analysis method gives important associations between various components of analyte mixture. The various techniques used for this are, ANN and radial basis function (RBF). Among ANN and RBF, ANN is the widely known and most evolved analysis technique. An ANN is a technical copy/mathematical model of biological neural network. A neural network comprises an interlinked bunch of artificial neurons and during each learning phase it changes its structure from which we can say it is an adaptive system. The training process requires a certain amount of data of known samples to train the system and is effective for comparison between unknown components in analyte mixture to known references in the data library. The result of ANN data analysis is in the form of percentage similarity of detection components in the analyte with those from known samples in the library.

Among the different analysis techniques, precision comes up at the rate of escalating complications. Depending upon the type of available input and the type of information sought, each analysis methods are used.

11.6 SMART MATERIALS

Smart materials are a group of materials that possess a smart behavior, that means, when a material perceives certain stimulus from its surroundings and provides a dependable, beneficial, controlled response in a reproducible manner [43]. Also, smart materials provide a way of getting an effective intelligent response in a material which is insufficient to do that otherwise and the potential to yield a variety of enhanced capabilities and functionalities [19]. The controllability of the materials considered is another important point because it is necessary to know that up to which level these materials can be affected by these external stimuli. The different types of stimuli to which smart materials are sensitive are stress, strain, light, electrical field, pressure, pH, temperature, etc.

A smart structure consists of a structure that incorporates smart materials in it to perform functions like sensing, actuation, transmitting, and recording of data. A smart structure consists of five important parts, actuators, sensors, control strategies, power, and signal conditioning electronic and a computer. Smart structure has the potential to react to an altering external and internal environment. It contains smart material actuators that enable the variation of system properties and system response in a controlled manner. The three basic components of a smart system are sensor, processor, and actuator which are embedded to the system containing central control and command unit to form an integral part of it. The smart system consists of sensing, processing, actuating, feedback, self-diagnosing, and self-recovering subunits. The system utilizes the functional properties of advanced materials to achieve high performances with capabilities of recognition, discrimination, and adjustification in response to a change of its environment. Each component of smart system should have functionality and the entire system is integrated to perform a self-controlled smart action which is like a living being that can think, judge, and take actions [44]. Smart structures are classified as Adaptive structures, Sensory structures, Controlled structures, Active structures, and intelligent structures. Adaptive structures have no sensors but possess actuators to alter the system characteristics in a desired and controlled manner. A sensory structure consists of sensors but no actuators for monitoring the structure properties. Sensory materials may identify strain, displacement, acceleration, etc. The combination of sensory and adaptive structure comprises the whole controlled structure. This employs sensors, actuators, and a feedback control system is integrated in a closed loop for the active control of characteristics of the structure. An active structure belongs to the subset of controlled structures. It consists of integrated sensors and actuators which fulfill both structural and control functions which include load-carrying capability, etc. Intelligent or smart structures are a subunit of active structures that has highly integrated control logic and electronics in addition to distributed actuators and sensors.

The smart materials possess some significant properties, that makes them distinguishable from other materials. They are: (i) Transiency, which is they can respond to different types of external impulses or factors and can be in different states; (ii) Immediacy, that is they can respond to external changes without wasting time or fast response time; (iii) Self-actuation, the ability which is inside the matter to change the appearance and shape; (iv) selectivity, the response is predictable and distinguishable; (v) both stimuli and response are occurred in the same place; (vi) shape changing with respect to external stimuli; (vii) self-diagnostic, automatic determination of damages on material; and (viii) self-healing, capable of autonomous repair when damaged.

Smart materials are divided into two distinct groups, Passive and Active. Passive smart materials are materials that can transfer some types of energy, for example, optic fibers can transfer electromagnetic waves. They lack inherent capability to transduce energy. Active smart materials possess the capacity to modify their geometric or material characteristics over the influence of electromagnetic or thermal fields, thus acquiring an inherent capacity for transduction of energy. Active smart materials are then classified to two groups, first type being the materials which can change their characteristics on exposure to external effects and the second type is capable of energy conversion, examples are Piezoelectric materials that can generate electricity upon the action of external strain, Solar cells that can transform solar energy to electrical energy, etc. Examples of active smart materials include SMA, shape memory polymers, magnetic shape memory alloys (MSMAs), magnetostrictive materials, electrostrictive materials, piezoelectric materials, dielectric elastomers, photomechanical materials. Examples of passive smart materials are optic fibers, etc. Examples of semi-active smart materials are magnetorheological fluids, ERFs and magnetorheological elastomers.

Smart materials are divided into several groups based on the type of stimulus and their corresponding response:

1. **Shape Memory Alloys (SMA):** They are metallic materials which changes their mechanical properties because of change in temperature and reform to their actual structure when heated due to a change in the stable crystalline phase either martensitic or austenitic phase. They have the ability to memorize their inherent structure. SMAs' memory behavior arises from the temperature induced austenite to martensite phase transformation, sometimes referred as thermoelastic martensitic transformation. Martensite phase and austenite phase is stable at lower and higher temperatures, respectively. This thermos mechanical property of SMAs is due to the reversible solid-solid phase conversion which is controlled by temperature and mechanical stress. They are made of Nickel and Titanium alloys, or Copper based ternary alloys. Disadvantages are slow response time, need of a backup system. They can withstand large strain levels, shows high specific force and power density with low absolute forces. They can be used as an energy harvesting structure as well as actuators. Nitinol is an acronym for Ni-Ti alloy which reflects the constituent elements of the alloy and location of the original discovery, US Naval Ordnance Laboratory. Ni-Ti metal combination is an efficient and familiar type of SMA which consists of 45–50% of Titanium. SMAs' are widely available and about 20 varieties of alloys got recognition for displaying the phenomenon of remembering and recovering their initial structure even after undergoing changes due to an external stimulus. The shape recovery conversion occurs due to the need of the crystal lattice to stay at the least possible energy at a given temperature. Certain SMA displays two-way shape memory effect, in which, after regaining the initial shape due to the heating above transformation temperature, they come back to a different structure on lowering the temperature. In addition to thermal stimuli, some SMAs can change shape when exposed to magnetic fields, that comprises ferromagnetic shape-memory alloys which is a class of shape memory materials (SMMs). MSMAs are ferromagnetic materials which exhibit large strains on exposure to an applied magnetic field due to martensitic phase transformation. MSMAs with near-stoichiometric Ni_2MnGa are widely studied and differing from magnetostrictive materials, they produce much larger strains by twinning, sometimes as large as 9%, under relatively low bias

magnetic fields. Their mechanism is based on the magnetic inhomogeneity of the material. Shape-memory polymers (SMPs) are another type which are smart materials based on polymers that can be actuated by heat, light or magnetic field and they can retain 2 or even 3 shapes. The SMPs' mechanism is based on transition of the polymer above or below a transition temperature, always the T_g. Many polymers exhibit these properties which generates lower forces compared to SMAs, so they are less applicable. A group in MIT in 2013, accounted for shape memory ceramics.

2. **Magnetostrictive Materials:** They are materials that can be magnetized on the application of stress or deform on exposure to the magnetic field such as change in length can take place upon magnetization. They are always referred to as a transducer. These materials can be classified as positive and negative magnetostrictive. By varying the magnetic field, the material can contract or relax. These materials exhibit frequency dependent hysteresis which is a non-linear characteristic due to the several challenges occurring on accurately capturing the complex behavior. The mechanical properties are workability, moderate saturation magnetization, high coercivity, high cubic magneto crystalline anisotropy, high Curie temperature, and high chemical stability. Cobalt ferrite is used for applications like sensors and actuators. This is because of high saturation magnetostriction. It is a good substitute for Terfenol-D or Galfenol (alloy of terbium, iron) due to no RREs. The magnetostrictive materials contains RREs so they are very expensive, and they can reach moderate strain levels, high elastic modulus and they have narrow hysteresis loop so low loss.
3. **Electrostrictive Materials:** They are like magnetostrictive materials, but the activation is due to the exposure of an electric field. Their stimulus is an electric field and their response is exhibited as a variation in their size. This is because, when electric field is applied on the material, the ions are displaced from their original sites thus the size is increased. There exists a non-linear correlation between material size and applied electric field. In both directions, by the application of electric field, the size of the material is only increased, so it can't be used in bipolar mode applications. They have electrical capacitance more than piezoelectric materials. Mainly used materials are Lead manganese niobate-lead titanate and lead

lanthanum zirconate titanate. These materials are mostly used for actuation purposes, and they are very fast, very stiff, very brittle especially in traction, so they need a precompression. Because of the fast response time, narrow hysteresis loop is generated and low material loss. These materials have multiple material properties such as ferroelectricity, pyroelectricity, and piezoelectricity. An induced strain in the material is not dependable on the direction of the applied electric field based on the electrostriction effect (induced strain is in proportion with square of applied electric field) and the same deformation occurs when the field is reversed. Materials that have high dielectric constant can exhibit very large electrostrictive strain.

4. **Piezoelectric Materials:** When a mechanical stress is applied on a material, particularly in the form of crystals and ceramics, an electric potential is generated as a response. This phenomenon is called direct piezoelectric effect. The direct piezoelectric effect is also called generator effect or sensor effect. This effect is mainly used in sensors. The inverse piezoelectric effect is whenever an electric field is applied, the materials produce a stress/strain. This effect is also known as minor effect or actuator effect and is utilized in piezoelectric actuators. At very high electric intensity, there will be change in width of crystals that can be of better than micrometer precision, which makes piezocrystals, the most accurate tool for positioning objects. Since both effects are reversible, these materials can be used as both sensors and actuators. Berlinite, cane sugar, topaz, and quartz are examples of natural crystal materials that are piezoelectric, commonly used polymer is polyvinylidene fluoride, most used for engineering applications are artificial ceramics like barium zirconate titanate and lead zirconate titanate, lithium niobate and lithium tantalate. Piezoelectric actuators have large variety of applications such as adjusting mirrors, various automotive parts, etc.
5. **Magneto-Rheological Fluids (MRFs):** On the application of magnetic field, the rheological properties of materials like stress and viscosity changes, such materials are called MRFs, also known as magneto-sensitive smart materials. They are smart materials composed of a carrier fluid, a distinct oil in which there is a dispersion of micron-sized ferromagnetic particles plus some thixotropic

additives. When came in contact with a magnetic field, the fluid viscosity elevated to higher values until it resembles a viscoplastic solid. This behavior arises when the magnetized particles start attracting each other to the direction of the field, hence forming chains. The chains then form a skeleton within the fluid, which contrast the movement of fluid itself. There is a great control over the yield stress of this material which is dependent on the applied magnetic fluid only. Magneto-rheological elastomers (MREs) are a rubber-like delicate material whose mechanical characteristics can be altered on the exposure to a magnetic field and comprises of magnetic and non-magnetic particles and additives. MRFs can be used in damper, clutch, buffer, etc. Carbonyl iron, due to its high saturation magnetization, is used to prepare MRFs. Some characteristics of MRFs are visco-elastic and magnetic, delicate, regulatory modulus, and good sound-absorption property.

6. **Electrorheological Fluids (ERFs):** They are like MRFs, but the viscosity change arises from the application of electric field. In the neutral state, very small particles are randomly and uniformly distributed in an electrical insulation fluid, but under electric field, the large dielectric constants redistribute the particles changing the viscous properties, obtaining also in this case a non-Newtonian fluid, similar to a solid-like structure with a controllable yield stress and along the direction of electric field. The ERFs were discovered before the MRFs but, they have less engineering applications because the usable yield stress is lower and it is very much easier to create and control a high magnetic field rather than a large electric field. Although, ERFs are used in automobile industry, building base-isolation, electro-active actuators, vibration isolators, shock absorber, clutch, etc. The properties of ERFs are stiff and dielectric in nature, damping coefficient is changed in electric field, strong interfacial bonds, large dielectric constant.
7. **Optical Fibers:** A flexible and transparent fiber which is produced by extracting glass/plastic to a diameter which is obscurely greater than that of a human hair is called Optical fiber. They are widely used in examining civil structures for strain alterations resulting from cracks or damages in buildings, mechanical blades, shafts, and in fiber-optic communications. Light can be transmitted between the ends of the fiber. One of the most used smart materials for structural

health monitoring are fiber-optic sensors. Fiber Bragg grating (FBG) and fiber-optic polarimetric sensor (FOPS) are the most studied fiber-optic sensors in the last two decades. FBGs are used for localized strain determination; it can be easily embedded into composites for organized health supervision and its small diameter are employed for measuring the non-homogeneous internal strain fields. FOPS are used for the measurement of strain in whole length of a structure. The increase in strain due to the presence of crack/damage in a structure are detected by FOPS irrespective of the location of damage in structure. Optical fibers can be used to transmit energy by the utilization of a solar cell, can be used in medical applications as illuminating sources, and also for toys, signs, etc.

8. **Smart Nanomaterials:** Under stimulating agents, the certain structure of a nanomaterial can be varied (Figure 11.4).

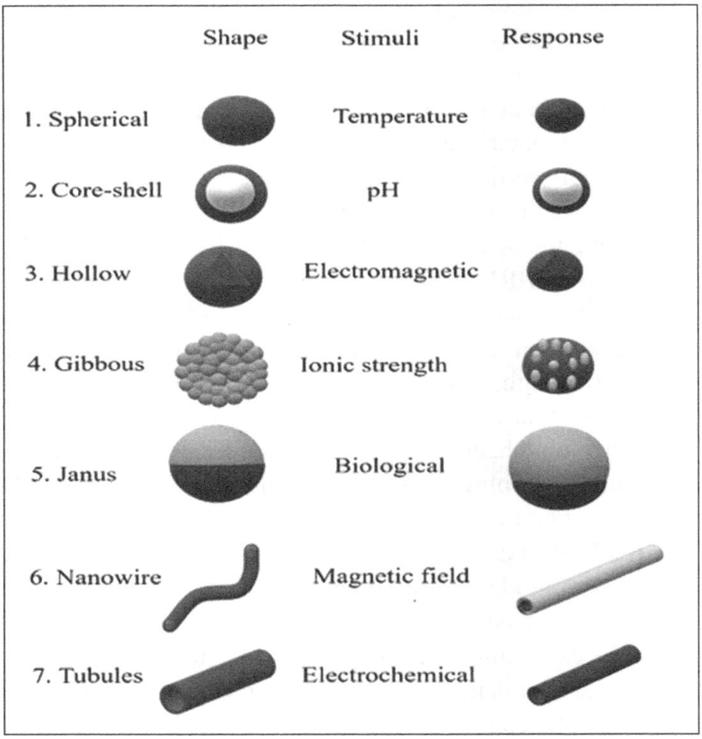

FIGURE 11.4 The changes in morphology of nanomaterials upon specific stimuli response.

Various sorts of classifications are possible for smart nanomaterials. One such classification based on the specific stimuli response includes piezoelectric smart materials, electrochromic, and photochromic smart materials, thermoresponsive smart materials, pH dependent smart materials, photoresponsive smart materials, electroresponsive smart materials, and magnetoresponsive smart materials.

i. **Piezoelectric Smart Nanomaterials:** Piezoelectric materials can transform the mechanical energy to electrical energy under the influence of mechanical stress and vice versa since this effect is reversible. Most materials exhibiting piezoelectric effects are non-conductive materials such as SiO_2 (Quartz), along with ceramics that also exhibit excellent piezoelectric effects, for example Lead zirconium titanate (PZT). The PZT materials possess high piezoelectric coefficient, and it is very cheap to manufacture. When a force or stress is applied on a piezoelectric material, a dipole moment is originated in the crystal structure because of a variation in the ion balance and the net dipole moment should not be zero. To obtain this, the piezoelectric crystal should possess an asymmetric atomic structure. Piezoelectricity smart nanomaterials can be used for various applications such as vibrational generation and actuation, and for commercial applications such as production of microphones, radio antenna oscillators, fuel injection, time keeping using Quartz resonance. Piezoelectricity smart nanomaterials are developed broadly and manufactured as delicate coating layers, discs, or stacked sheets upon the development of advanced nanotechnology. The serious drawback and the main limitation so far of these materials is degradation because of the application of repeated forces on them. Piezoelectricity smart nanomaterials are widely employed for administration of human health including the field of nanomedicine. The cells and tissues are extremely receptive to the employed electric fields so that the piezoelectric materials can be employed for clinical use. Stretchable, flexible, and cost-effective e-skins based on Piezoelectric smart materials are an advanced tool for avoiding illness, health monitoring, human physiological monitoring, early prediction of diseases.

ii. **Electrochromic and Photochromic Smart Materials:** When a potential is applied, color or opacity of a material varies, and this process is called electrochromism resulting in reversible variations in the optical properties of material observed under applied voltage. The principle mainly involves the insertion/extraction of ions into/out of the Electrochromic materials. In nanostructured materials, the ion insertion/extraction takes place with good efficiency as they possess large surface area. Therefore, various materials display electrochromic properties. The applications of EC materials include anti-glare mirrors, smart windows, displays, etc. Smart window application is a potential application through which energy conservation and indoor comfort is achieved by reverse color changes. In a similar way, Photochromism is the color change of materials that occur under photon energy and such materials exhibit reversible color changes with respect to the specific photon energy. Photochromism occurs in organic, inorganic, and biological systems. Because of the inherent properties of the photochromic material, its applications are extended in sensors, and fast optical shutters. In human health applications, such materials are widely used in ophthalmic sun screening applications and in UV light protection glasses. In this UV protection/cooling glasses, photochromic materials turn to dark when exposed to sunlight and reverse back to colorless otherwise for personal comfort and safety.

iii. **Thermoresponsive Smart Materials:** They are materials that can respond to a change in temperature and such properties are predominantly seen in polymers. Polymers are of two types which include lower critical solution temperature (LCST) and upper critical solution temperature (UCST). Various attributes like temperature, molecular weight, etc., determines the solubility of polymer in aqueous solution. From the phase diagram of polymer mixture against temperature, we can effortlessly identify the critical solution temperature. In the case of both LCST and UCST, at the critical temperature, the solvent and polymer are entirely miscible. So, if the polymeric solution below LCST is a clear and homogeneous solution whereas above the LCST, it is cloudy. This happens because it is energetically favorable. The main reason for favorable phase separation by increasing temperature is due to the entropy of the systemin accordance with Gibbs equation, $\Delta G = \Delta H - T\Delta S$, G is Gibbs free energy, H is enthalpy and S is entropy. Various kinds of synthetic methodologies such as free radical copolymerization, self-assembly method, photopolymerization, sol-gel transition phase method, copolymerization, end-group functionalization, etc., are employed to produce different kinds of thermoresponsive smart nanomaterials. The thermoresponsive polymer smart nanomaterials can be used for biomedical applications like drug delivery, tissue engineering and gene delivery.

iv. **pH Dependent Smart Materials:** These materials can respond to pH and display advanced effective characteristics. Various efforts took place for the development of pH responsive biomaterials along with other stimuli responsive materials. In human body, different parts have varied pH levels, as, pH of saliva is 6.5–7.5, pH of upper and lower stomach have 4–6.5 and 1.5–4.0, respectively. Moreover, pathogen infected stage displays irregular pH value compared to the physiological state pH. Based on these pH variations, various pH responsive materials have been developed. Several pH responsive materials can be prepared by the addition of: (i) protonatable groups (amino, sulfonates, etc.) comprising polymers like polysaccharides and polypeptides that undergo pH responsive solubility and conformational changes; (ii) polymers carrying acid labile bonds (hydrazine, imine, etc.). The pH responsive smart nanomaterials like Poly-*l*-histidine (PLH), chitosan, and its derivative based biomaterials can be used to target and provide site specific therapeutic efficiency that includes tumor-specific drug delivery also and other biomedical applications. In the case of polymers carrying acid labile bonds, the cleavage of chemical bonds under pathological environment has gained much popularity in the field of targeted and site-specific drug delivery. The rate-determining step for pH-responsive bond breaking is acid-catalyzed, which can be modulated by choosing suitable connectors (imine, hydrazone, etc.). Some of challenges they face are lower invading efficiency in high-depth tumors, unpredictable tumor environments due to rapid mutation, nano-carrier's accumulation in tumor environment, etc.

v. **Photoresponsive Smart Materials:** These materials can react when come in exposure to external light. Controlled properties and non-invasiveness of phototriggered smart nanomaterials are playing a major part for obtaining success in effective therapeutics, while other internal stimuli such as temperature, etc., have some serious restrictions in tumor biology. To overcome these, external stimuli such as ultrasound, magnetic field, light responsive drug delivery systems become an efficient solution. Among all the external stimuli, light is contemplated as a potential one because of its tunable and controllable properties. Light responsive smart materials can be used as a biomarker which can be employed for tracking the targeting capability, site of drugs and capturing images of the tumors by optical imaging methods. They also can be applied for delivering medicines to site of non-curable diseases for their cure, photodynamic, and photothermal activation in which utilization of photoabsorbing materials leading to the formation of reactive oxygen intermediates and local hyperthermia for the destruction of cancer or bacteria cells, respectively. To overcome the challenges in these therapeutic systems, Near Infrared active smart nanomaterials/polymers which can efficiently produce the reactive oxygen moieties, have good water solubility, targeting ability are utilized.

vi. **Electroresponsive Smart Materials:** These materials vary their structure when an electric field is imparted. They have applications in sensors, optical systems, actuators, robotics, energy harvesting, etc. They are divided into two types: (i) ionic; and (ii) dielectric elastomers and electrostrictive polymers. In the first type, an electric field is employed resulting in mobile ions thereby generating a change in the number of ions in materials and is applicable in conductive polymers, ion polymers and polymer-metal composites. The dielectric elastomers and electrostrictive polymers have applications in biomedical field based on unique redo reactions, which is the regulated delivery of drugs and aggregation on electroresponsive nanoparticle drug carriers such as polymers and microgels. Despite the benefits, large voltage and a prolonged electrical potential treatment restrict their wide usage.

vii. **Magnetoresponsive Smart Materials:** They respond to an applied magnetic field as a stimuli agent and have applications in photonic, electronic devices, etc. More attention is given to the magnetic nanoparticles (MNPs) and their response due to its distinct physiochemical characteristics and possess applications in biomedical field such as magnetic hyperthermia, magnetic resonance imaging (MRI), etc. They are successfully synthesized by various techniques that involves chemical vapor deposition, hydrothermal process, combustion, carbon arc, electrochemical synthesis, co-precipitation method, laser pyrolysis, etc. In magnetic hyperthermia, traditionally thermotherapy and thermal ablation is used to destroy the cancer cells by elevating the local region/whole body temperature up to 42–45°C using microwaves or ultrasounds and by exerting a temperature above 45°C to the infected area, respectively. But these methods have low accuracy in targeting and deep tissue invasion problems. To overcome these challenges, magnetic NPs mediated hyperthermia for certain cancers have been evolved where the magnetic properties of MNPs in the fluid efficiently convert into heat under the application of magnetic field. This technique is still in its beginning stage and more clinical developments needed before the practical use. MRI is a versatile tool which is used in clinical diagnostics for obtaining a real time spatial resolution and higher contrast of soft tissues without side effects. The functioning is based on alignment of protons within a sample under applied external magnetic field. In order to replace existing contrast agents with economic, high contrast ability agents, and for making better signal to background noise ratio, several MNPs such as gadolinium, iron, and non-iron based NPs systems were widely applied as the MRI contrasting agent. Despite the advantages, retarded rate of imaging and accuracy, sensitivity, and toxicity in some contrast media are still some major issues. In magnetic induced drug delivery, several Magnetic NPs are efficiently utilized in the controlled release of drugs to target tissue.

9. Self-Actuation is closely in resemblance to shape changing materials and materials employed will create a certain amount of strain or displacement in reaction to an external stimulus. Examples are SMAs, ERFs, magnetostrictive elements, piezoelectric elements. It is hoped that in future, such structures will be produced which can imply greater displacements as a response.
10. Self-healing materials are materials that respond to a damage by implementing some form of repair mechanism on its own. One approach among several self-healing mechanisms is to employ glass capillaries or fibers deposited in a composite material and filling the resulting porous network with a monomer. When a crack occurs, the monomer will be released and tubes containing a curing agent also crack and mix with the monomer, causing the defect to be healed. Similar approach involves in the technology of microcapsule in which a monomer is enclosed and whenever a crack reaches there, the monomer polymerises, and the structure is repaired. Another application is the development of 3D microvascular networks containing healing agents which provide the material a prolonged working life by incessant repairing.
11. Self-diagnostic materials are those materials that can detect and measure the stress, or other forces imparted on them. This is an application of self-sensing and most of this technology involves the addition of sensory elements into a material or structure. The usage of clusters of piezoelectric materials to identify the variables and a material embedded with a conductive component which can alter their resistance when external forces are imparted are widely studied. Most developments involve composites, although concrete has also got popularity due to its universal usage for construction.
12. Self-sensing unravels various bright and clever methodologies that includes applying sensing skills to those materials which don't possess this property inherently such as concrete, textiles, etc., and can employ them to enable self-diagnosis, self-healing, etc. This will be successful by the application of a cluster of sensing elements and sensing materials which are embedded into other materials, which is also called sensing skins, that can be used to coat a structure or a component. FBG sensors are a group of sensors based on fiber optics which gained so much attention that the fibers can be incorporated easily into composites, elastomers, and polymers.

11.7 SMART MATERIALS USED FOR E-NOSE APPLICATIONS

In the detection system of electronic nose, different smart materials are used as the gas sensors:

1. **Chemiresistive Sensor:** A chemiresistive sensor which is one among the commonly used gas sensors because of its understanding electrical properties and readout unit. The time taken for the resistance of the sensor to change from base resistance to 90% of the response after being in contact with an analyte vapor is known as the response time. The recovery time is the time taken for target gas to be desorbed, which is the time taken for resistance to change from 90% to base resistance. The operating principle of the chemiresistive gas sensor depends on the chemical reaction between the gas molecules and the surface of the sensing materials. During the temperature range, 100°C to 400C, the chemisorption of the O_2 gas molecules from the air, onto the surface of the sensing material generates surface receptor states (O_2^-, O^-, O^{2-}). The electrons near the surface are captured on the surface receptor resulting in an increased electron depletion region. Based on charge carriers, semiconducting materials can be divided into two, n-type, and p-type materials. Target gas species can also be divided as oxidizing gas and reducing gas. With an n-type semiconductor material, the electrons present in the conduction band of the n-type semiconductors are removed by adsorbed oxygen ions. This change in charge carrier concentration increases the resistance of the sensing material layer. When the sensing layer come in contact with a reducing (oxidizing) gas, the amount of adsorbed oxygen ions decreases (increases) and thus the resistance of the material decreases (increases). The change in resistance when exposed to the target gas defines the response of the semiconducting gas sensor, which changes with the concentration of the target gas. Gas adsorption and subsequent resistance changes are influenced by three basic factors, receptor function, transducer function and utility function. Chemiresistive sensors include mainly three types of sensing materials, CP, intrinsic CP and metal oxide semiconductors (MOS).

i. **Metal Oxide Semiconductor (MOS):** These based sensors relies on the working principle of whenever a volatile chemical is attached onto a semiconductor surface, there will be variations in its resistance. This sensor comprises a ceramic structure which is heated by a wire and encapsulated with metal oxide or ceramic gas sensors. They are made by putting down a compact, porous metal oxide layer over a ceramic pellet which is electrically heated and at large temperatures, annealing is done. The electrical resistance of the sensing material increases when the oxygen from air is attached to sensor surface thereby removing electrons from the conduction band of the semiconductor sensing material. There will be an increase in the conductance of the sensing material when electron capturing is reduced because of the contact of vapors with adsorbed oxygen on the surface of MOS sensing layer. To reduce the response and recovery time, MOS are operating at elevated temperatures (up to 400°C). At temperature level, 250–450°C, the semiconducting sensing layer oxidizes the analyte and resulting in the absorption of free electrons by sensing semiconducting material. As a result, the conductivity and resistance of the sensing material varies which is the basic principle for MOS sensor-based e-nose system. Advantages are small size, robust, good sensitivity, low cost, long lasting operating life, easily fabricated, short response time, electronic simplicity. Disadvantages are very much sensitive to moisture, sensing material damaging by poisoning from sulfur compounds, etc., large working temperature becomes unfitting in surroundings where highly inflammable chemicals are present, sensitivity loss due to sensor loss, difficulty in signal analysis because of the non-linear signals, requirement of high power, poor specificity and selectivity, slow baseline recovery when compounds of high molecular weight are analyzed (Figure 11.5).

Joarder et al. conducted a work based on the MOS sensor based electronic nose and its non-invasive biomedical applications [45]. There is a work reported on MOS sensor-based e-nose and its application for the identification of olfactory characteristics of edible oil during storage period. Through this work, Karami et al. concluded that MOS sensor-based e-nose can be used for determining shelf-life of edible oil, in combination with ANN to obtain satisfactory results [46]. Binson et al. did a research work on the discrimination of Chronic obstructive pulmonary disease and lung cancer by MOS sensor based electronic nose from healthy controls [47]. In this work they concluded that further examinations are essential to enhance the sensor array system, to investigate the long-run reproducibility, repeatability, and enlarge its relevancy. Deng et al. published a work based on a portable e-nose for the subjective evaluation function of air quality in vehicles, by using an array of MOS sensors [48]. Moufid et al. conducted a work on wastewater monitoring by means of e-nose, VE-tongue, TD-GC-MS, and SPME-GC-MS, in which MOS sensor is used in the e-nose [49].

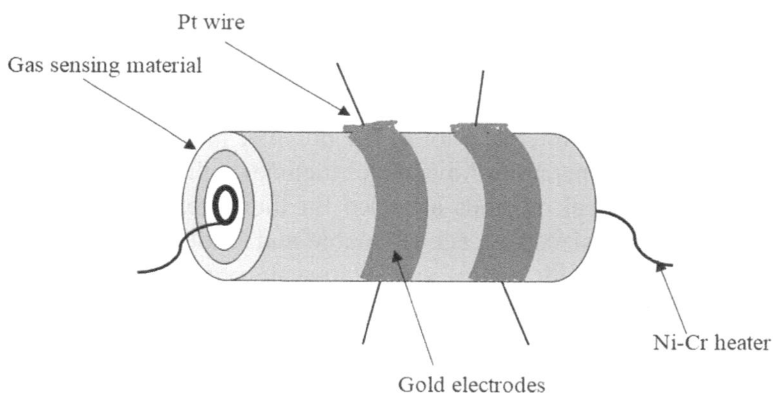

FIGURE 11.5 Structure of a metal oxide semiconductor sensor.

ii. **Conducting Polymers (CP):** Polymers possess good insulating properties and they are widely used among the society. CPs are polymer materials with the properties of both metals and semiconductors and possessing alternating single and double bonds in the polymeric backbone is an important characteristic of CP. With the addition of counter ions or functional groups to the polymeric backbone during its synthesis can enhance their molecular interactions such as analyte's affinity towards the polymer and thereby its conductivity. An advantage of CP is they can operate at ambient temperatures. The sensitivity for odor detection is 0.1 ppm but 10–100 ppm is more usual. Examples of CPs are polyacetylenes, polypyrroles, polythiophenes, polyanilines, polyfluorines, polynaphthalenes, etc. The functioning mechanism of CPs based e-nose system relies on the variations in the conductivity of polymers as a result of binding of different volatile chemicals with the polymer chain either as an ionic bond or a covalent bond. The conductivity is affected by the exposure to VOCs due to the variations in electron transfer along the polymeric backbone. The usage of CP nanocomposite/nanoparticles has greater surface area so its surface of exposure increases resulting in an enhanced dispersion. Conducting materials change resistance when chemicals/gases enter their environment. So, when the volatile analytes come in contact with the polymer composite, some of the chemical permeated into the polymer leads to swelling the polymer, thereby increasing the electrical resistance of polymer composite because the number of conducting pathways for charge carriers are reduced due to expansion. In the CP-based e-nose systems, a lot of the polymer materials are dependent on temperature. Advantages include, both electrochemical and chemical methods are used for their preparation and can be synthesized in a set of soluble and insoluble forms in accordance with micro- and nano-scale production methodology, possess distinct electromagnetic and optical characteristics, biomaterials such as enzymes, antibodies, etc., can be incorporated into the polymer matrix, diverse range of fabrication techniques involving electrochemical, optical, mass-based, etc., strong biomolecular interaction, low detection limits, enhanced sensitivity, cost-effectiveness, reversible response at ambient temperatures. Disadvantages are long response times or a quick shift of response with time, innate time- and temperature-related drift, poor reproducibility, high cost of sensor fabrication, extreme sensitivity to humidity because of the reduced working temperature, greater sensitivity to water vapors renders them susceptible to humidity (Figure 11.6).

Esteves et al. did a research work for the identification of tobacco types and cigarette brands using an electronic nose based on conductive polymer composite sensors [50]. Bonah et al. published a review about the application of e-nose as a non-invasive technique for odor fingerprinting and detection of bacterial foodborne pathogens, in which one of the sensors used was CP-based gas sensors [51]. Machungo et al. reported a comparison work on the performance of metal oxide and CP e-noses for detection of aflatoxin using artificially contaminated maize [52]. It is said in the work that an accuracy of 68%–94% was achieved for these different technologies. Seesaard et al. published a work on advances in gas sensors and e-nose technologies for agricultural cycle applications which is a detailed review about different gas sensors and their applications in the agricultural field [13]. Tan et al. published a review article based on the food quality determination applications of e-nose and e-tongue [53].

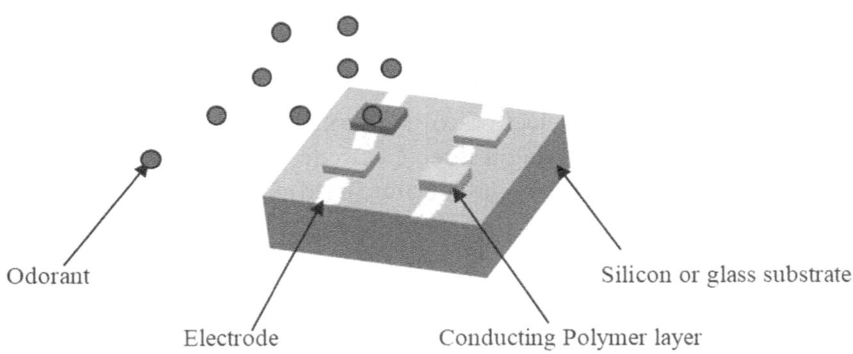

FIGURE 11.6 General structure of conducting polymer sensor.

Intrinsic or doped conducting polymers: The conductivity of polymers could be enhanced by doping. The intrinsically CPs have linear backbones consists of unsaturated monomers, that can be doped as semiconductors or conductors. They are described as π-electron conjugated polymers where the π symbol relates to the unsaturated structure of the monomer containing an unpaired carbon electron [54]. These CPs can be positively or negatively doped depending on the doping materials used. A lot of CPs are extensively used for e-nose sensing. The doping of these materials generates charge carriers and changes their band structure; as a result, both induce increased mobility of holes or electrons in the polymer depending on the type of doping used. The principle of operation for intrinsically CPs is that the odorant molecule is adsorbed into the polymer and alters the conductivity of the polymer. Three types of conductivity are affected in these materials, the intrachain conductivity in which the conductivity along the backbone is altered, the intermolecular conductivity which is due to the electron jumping to different chains because of odorant sorption, the ionic conductivity which is affected by proton tunneling induced by hydrogen bond interaction at the backbone and by ion migration through the polymer. Physical structure of polymer also has a major role on the conductivity. With the help of electrochemical methods or chemical polymerization, intrinsically CPs are usually embedded onto a substrate, with intertwined electrodes. Sensors with a simpler structure can be prepared where the polymer is deposited between the two conducting electrodes. Electrochemical polymerization is carried out using a three-electrode electrochemical cell with the electrodes on the substrate used as the working electrodes. The response of these sensors depends on the sorption of the vapor into the sensing material, causing swelling, and electron density on the polymeric chain backbone. The sorption properties of these materials depend on the diffusion rate of the permeant into the polymer matrix. The response time vary from seconds to minutes. Advantages can be increased discrimination when developing the sensor arrays can easily be achieved with these materials as a wide range of them are available, operate at room temperature thereby simplifying the electronics, good response to wide range of odorants, fast response, and recovery times especially for polar compounds. Disadvantages include poorly clarified signal transduction mechanisms, difficulties in resolving some types of analytes, high sensitivity to humidity and sensor response drift with time, time consuming and tough fabrication techniques, short lifetime of the sensors.

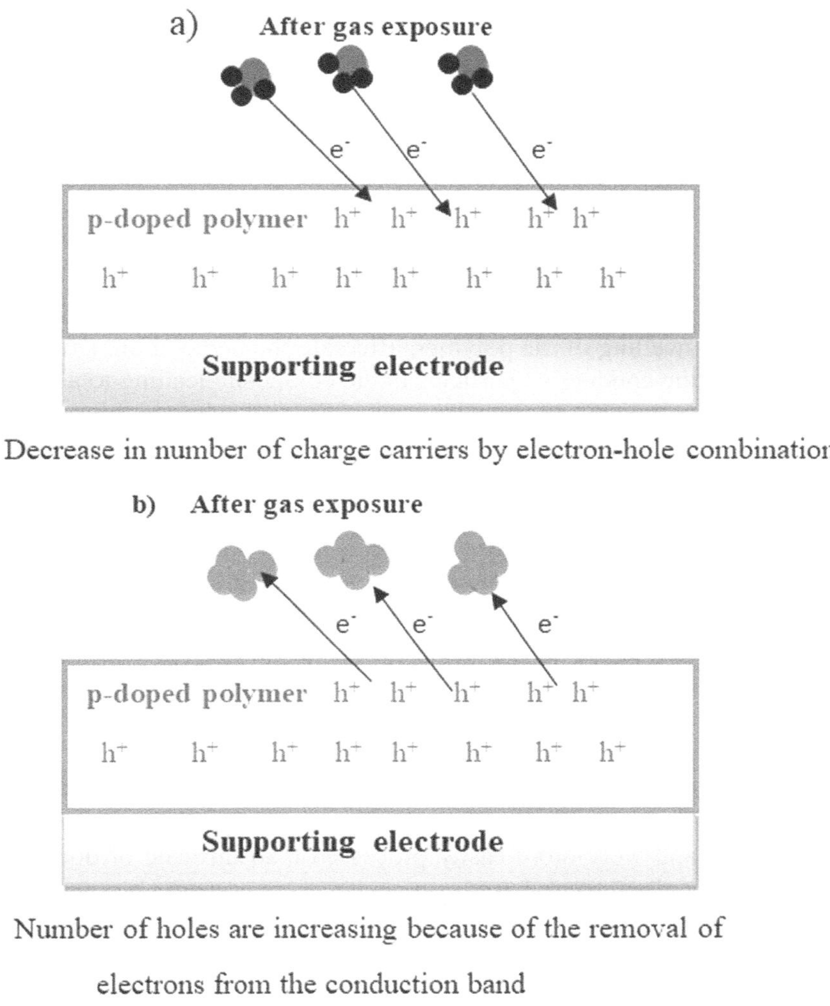

FIGURE 11.7 General structure of intrinsically conducting polymer.

iii. **Extrinsic or Composite Conducting Polymers:** The conductivity of conventional, insulating, polymers can be increased by combining them with conductive polymers or with fillers (metal powders, carbon black, or graphite) to create a composite or extrinsic. It is manufactured by covering or enfolding an electrode surface with a set of conducting and non-conducting or insulating materials. The polymer is the insulating material here and corresponds to a particular receptor material. It can attach and release the analyte compounds in vapor phase in the early and late vapor dispersion phases. The electrical conductivity of the sensing material is provided by the conducting materials and whenever the polymer is combined with an analyte or exposed to a vapor, it will start to swell. Sorption of analyte gases leads to swelling of the polymer, affecting the percolation network of the conductive particles in the composite leading to a change in the DC electrical resistance. Since they are highly sensitive to humidity, it is required to remove background humidity or moisture and regulate the shift in sensor baseline while working them. Although the sensitivity of the sensing film is also affected by vapor pressure, in open areas this relation is unusual. Various methods of fabrication of composite CP includes hot pressing, simple dissolution followed by sonication and evaporation, polymer grafting by gamma radiation as well as by reactive polymers. The dissolution method with carbon black filler is commonly applied to fabrication of gas sensors. Coating of the electrodes is relatively easy, soluble composite CP is applied using masks and spray coating. These gas sensors have higher sensitivity and reproducibility and were more easily processed than intrinsic or doped CP. Carbon black-based composites were prepared by suspending the carbon black in an insulating polymer solution made in an appropriate solvent. The overall composition of this solution is usually 80% insulating polymer and 20% carbon black by weight (Figure 11.8).

2. **Metal Oxide Semiconductor Field Effect Transistor (MOSFET):** It has a simpler structure due to which it gained popular attention. It is a transducer device employed in E-nose for the conversion of physical or chemical change into an electrical signal. This device comprises of a four-terminal device and the current flow is regulated by a vertical electric field applied externally. There will be no flow of current when there is no voltage applied because of the p-n

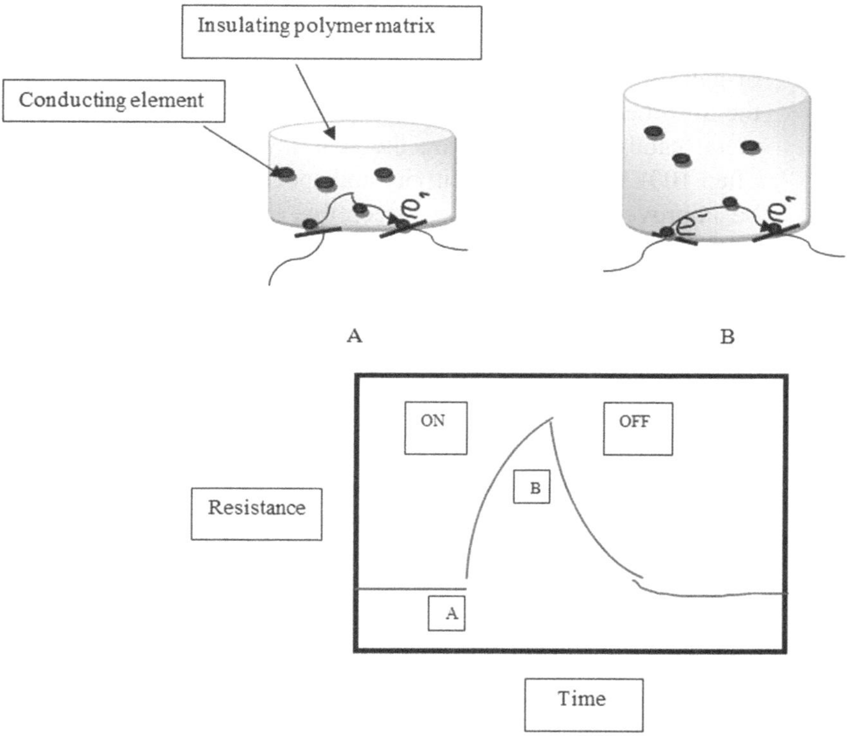

FIGURE 11.8 Structure of conductive polymer carbon-black composite which is an example of Conductive polymer composite.

junction between the drain and the source. Negative charge carriers form a conduction pathway between the source and the drain when a positive voltage is employed over the gate with respect to the source and the current is regulated by the vertical as well as the lateral field, hence the name Field effect transistor or FET. A transistor, a catalytic metal and an insulator is included in the general structure of MOSFET. The basic principle proceeds as the gaseous compounds react with catalytic metal and produce species that can disperse through the metal film and occupy onto a metal insulator. Consequently, the voltage will be altered and the whole current-voltage properties of the sensing material will be altered. Different catalytic metals can be used to analyze the sensor sensitivity. The working proceeds as when volatile chemicals encounter a catalytic metal, a reaction will occur and the resulting products will be disseminated through the gate due to which the surface potential of gate is altered. Usually, the change in potential depends upon the absorbed gas concentration. By altering the thickness/type of metal catalyst and changing the working temperature, which is often 100–200°C, the sensitivity and specificity of the sensors can be improved. They are too susceptible to drift like conductivity sensors. Advantages are they are robust sensors, having low sensitivity to humidity, good sensor reproducibility, and are of low-cost. Disadvantages include continuously changing temperature can vary the sensor selectivity and sensitivity, for good quality and reproducibility high quality production technology is required, poor response to gases such as ammonia, CO_2, etc., shift in baseline, poor moisture sensitivity. Badoussi et al. published a book chapter on a very smart monitoring technology which combines electronic nose that employs MOSFET sensors and image processing for quick determination of FAW pest in agriculture [55]. Zaukuu et al. did a review on evolving trends for quality determination of meat, fish based on modified and advanced sensor instruments [56]. Srivastava et al. published a review about electronic nose applications in medical field [57]. Aouadi et al. did a detailed review on the origin of near IR spectroscopy, e-nose, and e-tongue and their successful applications in food quality control [58]. Pulluri et al. published a paper on wine quality assessment using e-nose [59].

3. **Piezoelectric Sensors or Acoustic Wave Sensors:** In the case of piezoelectric sensors, they employ a mechanical wave as the detecting mechanism. As a result of adsorption of VOCs on the surface of sensor coating material, the variations in the properties of the propagation path of the wave which often propagates through or over the surface of the sensitive layer results in affecting the velocity or amplitude of wave or both. They contains a piezoelectric substrate, usually quartz (SiO_2), lithium niobate ($LiNbO_3$), lithium tantalite ($LiTaO_3$) or zinc oxide, doped with suitable sorptive material. These piezoelectric acoustic wave devices are compact, cheap, and respond quickly to almost every gas, in specific, reactivity is in direct relation with the weight of adsorbed compounds and molecular mass of the analyte is the fundamental parameter for its identification. Four types of such

devices are present, BAW, SH-APM, FPW and surface acoustic wave (SAW). The two main types used for gas sensing are those based on BAW devices and those based on SAW devices. In the former, the wave propagates through the bulk of substrate and in latter, the wave passes over the surface of the substrate. In both cases, waves are at ultrasonic frequencies, typically 1 to 500 MHz. The sensitivity to VOCs is defined by the kind of sorptive coatings applied on sensors. Various materials have been used for coating such as monolayer films, surface attached molecules and layers of different types of polymer films. BAW sensors are commonly called thickness shear mode sensors (TSM) or quartz crystal microbalances (QCM or QMB). It is the best known and oldest, simple piezoelectric acoustic wave device. It contains a piece of single crystal quartz, which is about 1 cm thick and metal electrodes, usually gold electrodes are deposited on the two large surfaces, both connected to lead wires. The sensor is used in an oscillator circuit, when desirable voltage is employed across the electrodes, excitation occurs and the oscillating velocity retards pointing out the mass accumulation on the surface of sensitive material. Commonly, 5–20 MHz is the range of resonance frequency of this device. The transducer for QCM sensor is mass sensitive like that of SAW sensor, only difference is in the sensor employment, that is, QCM uses BAW sensor whereas SAW uses SAW sensor. The oscillation frequency of QCM sensor decreases when mass is attached on the surface of sensing crystal, and hence it is identified as a sensitive mass measuring device. The shift in quartz crystal resonant frequency because of the binding of VOCs onto the sensing material is the sensing mechanism used in QCM. The sensor selectivity is maintained by the thickness of coatings. An appropriate electronic circuit unit is used for conversion of the measured signals. QCM sensor is an efficient sensitive detector of mass changes. Advantages are, for large dynamic range, signal is linear, ultra-sensitivity, and rapid response, and available for using in different environments and various locations as it is stable for a long term, can be used as a single sensor or group of sensors, detection of trace gases, low power utilization and miniature size. Disadvantages are temperature and humidity dependence, poor reproducibility in deposition of the coating material, tough, and time-consuming manufacturing methods, can't be used in multi-sensory clusters because of the requirement for a thin crystal in order to improve the sensitivity, low signal to noise performance because of size of crystal thickness. SAW sensors are most sensitive, miniature of e-nose piezoelectric acoustic wave type of sensors which is based on the SAW platform. SAW sensors are those sensors in which the resonance frequencies of oscillator circuits are regulated by resonator devices in the feedback path. The basic principle of the sensing mechanism is whenever a volatile chemical is attached onto the sensitive layer which can be the selective polymer thin film in the acoustic wave propagation region, there will be a shift in the oscillator frequency. The SAW sensors need waves to travel over the surface of the device. They function at large frequencies and thus generate a large change in frequency upon vapor sorption. Their limitations include decreased long-term stability and are prone to humidity. Important properties are its ongoing advancements in performance through improving working frequency, alteration in device framework, and improvement in polymer interfaces. Main benefits are high sensitivity and fast response time, easy fabrication, diverse sensor coatings, small size, cost-effective, responsive to almost all gases. Their disadvantages include poor signal to noise ratio because of requirement of large working frequency, operating circuit is tough to fabricate and of high cost, vulnerable to humidity and sensitive to temperature, reproducibility is difficult to achieve.

Sreelatha et al. published a conference paper on SAW e-nose sensor-based pattern generation and recognition of toxic gases using ANN techniques [60]. Kuchmenko et al. did a review on recent advances in piezoelectric chemical sensors for environmental monitoring and food stuff analysis [61]. Sianghio et al. reported the discrimination of different types of coffee beans by electronic nose based on piezoelectric quartz crystal array sensors [62]. Behera et al. reported a review on the e-nose based non-invasive technology for breath analysis of diabetes and lung cancer patients [63]. Matatagui et al. proposed portable low-cost electronic nose based on SAW sensors for the identification of BTX vapors in air in their research article [64]. Okur et al. published a work on the identification of mint scents using a QCM sensor-based e-nose [65].

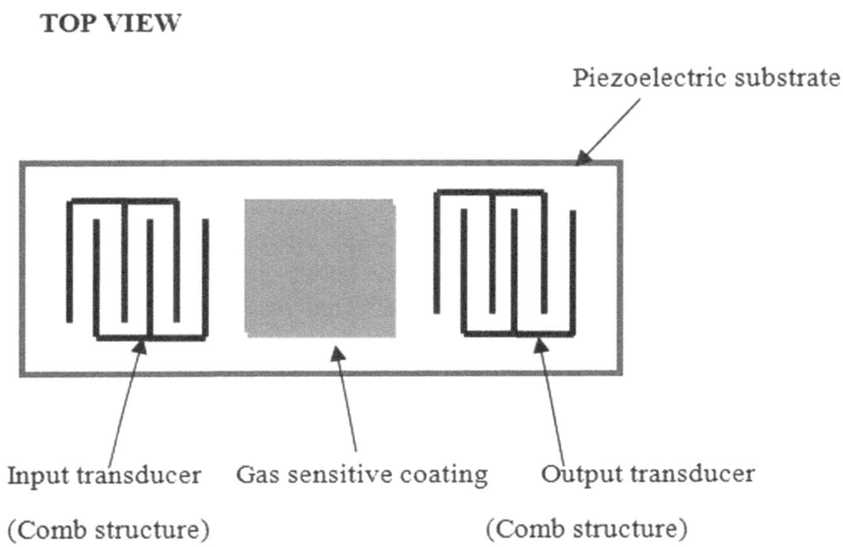

FIGURE 11.9 General structure of a piezoelectric sensor.

4. **Optical Sensors:** This technology is based on optical properties of the sensitive material like reflection, fluorescence, wavelength of maximum absorption, light polarization, optical layer thickness, refractive index, colorimetric dye response, etc. Similar to the other sensors, optical sensor also works by making contact between the sensitive layer and gaseous molecules, as a result, the volatile compounds are adsorbed on their surface which causes variations in their optical properties. Optical fiber sensors utilize glass fibers coated with thin chemically active materials on their sides or ends that contain immobilized fluorescent dyes in an organic polymer matrix. The changes in dye polarity because of the interaction of VOCs with a light source cause a shift in the emission spectrum. Wide sensitivities are obtainable because of the availability of different dyes, but because of the phenomenon like photobleaching, the lifetime will be minimized. The "SPR" or the "surface plasmon resonance" is an optical method that considers the evanescent wave development to estimate changes in the refractive index near the sensor surface. This is an optical development in which incident light excites a charge-density wave

at the interface between a highly conductive metal and a dielectric metal. The situation for excitation are observed using permittivity of the metal and the dielectric material. To obtain small changes in the refractive index of a thin region close to a metal surface, the transduction concepts of SPR is typically considered as an analytical tool. For sensitive detection of chemical species like odor, vapor, and liquid, the optical excitation of surface plasmon on a thin metallic film has been distinguished as a positive technique. In order to be able to better observe the excitation of SPR by estimating the light reflected from the sensor interface, many techniques have been used such as:

- Phase modulation;
- Intensity modulation;
- Wavelength modulation; and;
- Analysis of angle modulation.

Optical SPR sensors are sensitive to variations in refractive index of a sample surface. The change in refractive index of a sample surface sensitivity is determined using optical SPR sensors.

Gases such as:

- Ethylacetate;
- Xylene;
- Ammonia;
- Toluene, etc.

—can be recognized by determination of the optical SPR using angle modulation. Another approach is the optically based chemo-sensors such as:

- Hydrophobicity;
- Elasticity;
- Pore size;
- Optical fibers deposited with fluorescent indicator Nile Red dye in polymer matrices of varying polarity; and
- Swelling likelihood to generate distinctive sensing ranges that interact diversely with vapor molecules.

Fiber-optic sensors consist of an analyte sensing element deposited at the end of an optical fiber. The optical sensing element is usually composed of a reagent phase immobilized at the fiber tip by either physical entrapment or chemical binding. This reagent phase usually consists of a chemical indicator that experiences some change in optical properties such as intensity change, spectrum change, lifetime change and wavelength shift in fluorescence upon interaction with analyte gases or vapors. The responses depend upon the nature of organic vapor and the strength of its interaction with different polymer systems used. Advantages are excessive sensitivity, identification of individual components in mixtures, capabilities of finding multiparameter. Disadvantages include connected electronics and software are very complex, highly expensive, sensors have quite a short lifetime, complex sensor array systems, low portability on account of elegant optics and electrical components. Another technique is to use white light interferometry to measure the thickness changes in a polymer sensing layer that swells on exposure to odorant molecules. One simplest approach is to use color-changing indicators such as metalloporphyrins and measure with an LED and a photodetector system their absorbance upon analyte gas exposure. Colorimetric sensors use thin films of chemically responsive dyes as a colorimetric sensor cluster. Fluorescent sensors are more sensitive than colorimetric sensor arrays (Figure 11.10).

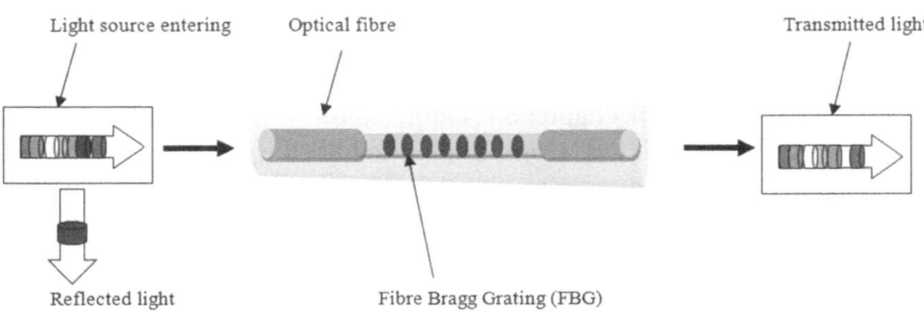

FIGURE 11.10 General structure of optical fiber.

Bieganowski et al. reported an analysis of hydrocarbon soil pollution by employing e-nose [66]. Zhu et al. conducted a work for the determination of VOCs related to urinary bladder cancer with the help of an optical sensor cluster [67]. Han et al. proposed a work for the qualitative and quantitative detection of beef adulterated with duck by fusion of low-cost e-nose and Fourier transform near IR spectroscopy [68]. Maw et al. presented an article on hybrid E-nose system about MOS gas sensors and compact colorimetric sensors [69]. Li et al. did a comprehensive study on the recognition of lung cancer using a homemade e-nose [70]. Hidayat et al. published an article about the e-nose combined with chemometric tools for distinguishing the grade of black tea samples *in situ* [71].

5. **Nanomaterials-based Sensors [34]:** With the progress in nanoscience, microelectromechanical systems (MEMS), nanoelectromechanical systems (NEMS), and very large-scale integration (VLSI), different types of novel nanostructures were applied in the fabrication of e-nose device. The micro-sized nanostructures like nanoparticles (NPs), nanospheres, and NCs are employed as a film for sensing applications. Higher dimensional nanostructures could be embedded singly, in multiples or can applied as a film for sensing applications. The elevated surface area to volume ratio is very advantageous for sensor applications because the majority of material is made accessible for being in contact with the target analytes. Moreover, some of the advantages of nanomaterials-based sensors are quick response time, high sensor response values revealing finer separation between the interfering responses, identification of very low concentrations and smaller training periods.

   Jian et al. did a detailed review on gas sensors based on chemi-resistive hybrid functional nanomaterials [73]. Ramgir et al. published a book chapter on Nanotechnology based E-nose for smart manufacturing [74]. Conti et al. conducted a work on discriminative detection of volatile organic compounds using an e-nose based on TiO_2 hybrid nanostructures [75]. Sinju et al. published a work on ZnO nanowires-based e-nose for detection of H_2S and NO_2 toxic gases [76]. Germanese et al. reported an article on E-nose for the monitoring of severe liver impairment [77]. Seesaard et al. published a work on hybrid electronic nose system for distinguishing bacterial volatile compounds which are pathogenic [78]. Thazin et al. did a work on Formalin adulteration detection in food using e-nose based on nanocomposite gas sensors [79]. Andre et al. published a work on nanocomposite based chemiresistive e-nose and its application in coffee analysis [80]. Yin et al. did a review on carbon-based nanomaterials for the detection of volatile organic compounds [81] (Figure 11.11).

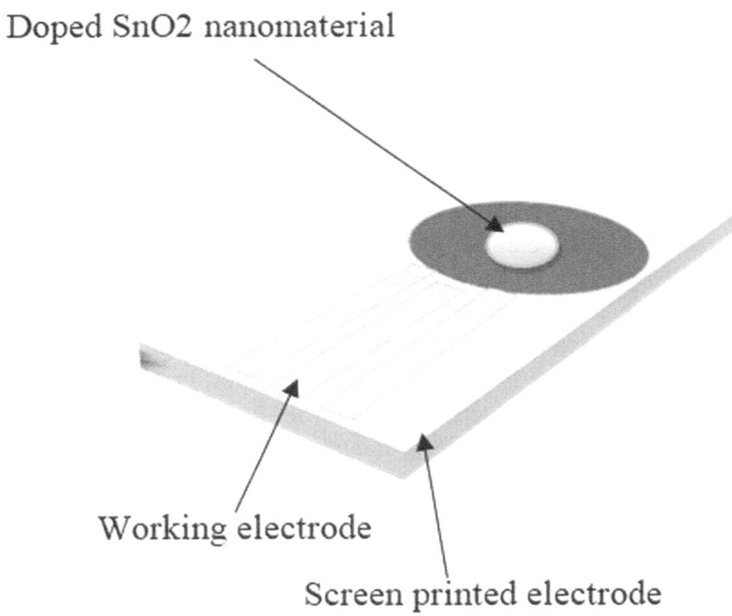

FIGURE 11.11 General structure of nanomaterial-based sensor.

i. **Isolated and Single Nanostructures:** One-dimensional structures are peculiarly beneficial as they can be productively employed for development of electronic nose based on single structures. The power required for single structured devices is less and hence one can envisage the e-nose and energy harvesting devices on a single unit. In spite of the high aspect ratio, changes along the length could be advantageously utilized as variables which can lead to individual and different sensing elements. A simple and excellent performing e-nose based on single SnO_2 nanobelt was made possible by employing a combination of bottom-up fabrication method and with the state-of-the art microfabrication methods. The electronic nose was exhibited to identify low concentration complex odorant chemicals such as brandy, vermouth, champagne, gluhwein under the strong background of ethanol. Use of various types of sensing elements in the sensor cluster would bring on better selectivity and an improvement in the detection limits of target molecules. The allotropes of carbon such as single walled carbon nanotube (SWCNT) and graphene are widely explored for practical chemical and biosensing applications. Graphene is a two-dimensional crystalline monolayer made of sp^2 hybridized carbon atoms assembled in a honeycomb lattice. Moreover, it is the fundamental structure required for the construction of various carbon nanomaterials. As an example, SWCNT can be made by rolling up a graphene sheet into a cylinder along a certain lattice vector. Both the structures, CNT and graphene are nanostructures possessing the simplest chemical composition and atomic bonding configuration with higher surface area to volume ratio. The limit of detection can be narrowed down to an individual molecule level if the nanostructures, CNT, and graphene are equipped with the size and surface similar to that of biomolecules. Because of the problems associated with the device-to-device heterogeneity originating from alterations in the morphology, alignment, and receptor loading, the application of these structures has not yet been in a deliverable position.

ii. **Multiple Nanomaterials:** An electronic nose based on SnO_2 multiple nanowire array using heterogeneous catalysis as the functionalization strategy was demonstrated [82]. SnO_2 NWs have been used as gradient microarray sensor to fabricate e-nose. The e-nose systems were examined to identify and discriminate between the reducing gases in air at ppb concentration level. The distinguishing power of the e-nose system is determined by the density and morphological anisotropy of the NW cluster. The distinctive inhomogeneous framework and the ample structure altering capabilities have led to a vast investigation of Si-NWs based sensing devices. It is the most ubiquitous electronic material amicable with most of the standard manufacturing and processing techniques. The tunability of its diameter is in the range of 10–100 nm which is analogous to various chemical and biological analyte molecules. It has been employed to understanding the optical/fluorescent sensing, noble metal-enabled surface enhanced Raman scattering (SERS) biosensing and FET-based electrochemical and biosensing. The surface of SiNWs aids as a gate in FET-based sensing, which is reformed with the receptors discerning towards a certain biomolecule. Attachment of biomolecule over the surface causes parallel alterations in the carrier density inside the NWs. The identification of protein, Ph, biomarkers, peptides, and viral peptides has been exhibited by utilizing this effortless approach.

iii. **Nanomaterials Film:** The ability to accurately regulate the size, shape, and functional groups on the surface of nanostructures resulted in a huge rise in their application as sensors. Chemically altered metal and alloyed nanoparticles have been utilized for the production of chemiresistive sensors. In particular, the sensing properties of metals such as Au, Pt, Pd, Ag, and alloys of these metals with several changes have been analyzed including functionalization with polymers, surfactants, ions, organic self-assembled monolayers and biomolecules based on the sensing application. There are various methods for sensor realization including drop casting, air blushing, spin coasting, inkjet printing, microdispensing, immersion, etc. Based on various metal NPs being investigated, gold nanoparticles have demonstrated the good potential to be an efficient sensor with better selectivity which is based on altering surface reactivity. NPs based sensors has the potential for doing careful diagnostics and medical care. Also, gold NPs were found to demonstrate its use in identification and creation of evaporative fingerprints of cancer specific genetic mutations. The fluorescence intensity of protein directed prepared Au NPs can be magnified remarkably on Ag plasmonic substrates [83]. Using this, an efficient sensor array capable of identifying 10 kinds of protein was developed. The analyte proteins upon the encounter with Au NPs influences the fluorescence process (increase/decrease), thereby providing distinct fluorescence image patterns that can be applied to detect and categorize various protein molecules. Chemiresistive FET based on other nanostructures, namely, CNTs, Si NWs and Pt NPs have demonstrated their efficiency in identification of VOCs delivered from the breath associated with various diseases which includes lung cancer, breast cancer, etc. Nanomaterials based chemiluminescence (CL) is a potential replacement for techniques that has been used widely for determination of volatile organic compounds, protein sensing, identification of sugars and artificial sweeteners. Distinct CL pattern is obtained as fingerprint for each analyte with each catalytic nanomaterial.

iv. **Conductive Polymer Nanocomposite:** Polymer nanocomposites embedded with nanomaterials are an interesting class of materials for the fabrication of room temperature sensors. Besides room temperature, other benefits like being delicate, low-cost, likelihood of regulating their properties based on the nanomaterial content and chemical nature of polymer matrix are offered by conductive polymer nanocomposites. Electronic nose based on them has been exhibited to be efficient and successful in the determination of lung cancer biomarkers, VOCs, etc. An e-nose based on Au NPs-fluorescent polymer conjugates that can determine, recognize, and assess the analyte proteins in a fast, general, and efficient [84] manner was demonstrated by You et al. Its mechanism is the presence of Au nanoparticles quenches the fluorescence of the polymer conjugate and fluorescence of protein disrupts the Au nanoparticles and polymer reciprocation, as a result of which forms a unique fluorescence feedback. This method does not need any special equipment, its sensitivity and speed make possible the protein identification.

v. **Microcantilever-based EN:** For high order analyzes, as in the cases of cells, DNA, etc., cantilever based on nanomechanics has been used as an option. The cantilever properties can be altered with receptor specific moieties and consequently, high selective and sensitive sensors are developed with rapid responses kinetics. The attachment of the target substance on the cantilever surface results in a change in the resonant frequency. Due to the improvements in MEMS and nanoelectromechanical systems, innovative, and compact cantilever-based sensors have been a promising advancement in e-nose technology, and it shows sensitivity down to zeptogram level

(1 zg = 10^{-21} g). Cantilever based e-nose have been used for applications like successful detection of explosive chemicals and for determination of hydrogen, primary alcohols, natural flavors and water vapor.

vi. A 2D material is defined as a material having a single layer. The two-dimensional materials can be classified as homogeneous materials consisting of unique chemical elements such as graphene, silicone, germanen, etc., and various chemical or compounded elements such as hexagonal boron nitride, transition metal dichalcogenides (TMDs), etc. The latter contains two or more elements forming a covalent bond. Each layer is glued via van der Waals force and can be stripped into a thin 2D layer. The combination of different 2D materials is commonly called the van der Waals heterostructure. Gas sensors are devices that can identify and measure concentrations of certain gases such as humidity, organic vapors and noxious gases. So much studies are occurring about gas sensors and its applications in environmental monitoring and emission control, personal and military safety, agricultural and industrial production management and medical diagnosis. Metal oxides are widely used material for gas sensing systems because they are highly sensitive and of low cost. But it has shortcomings such as need of high operation temperature, high power consumption, sensitive to environmental elements and low selectivity. A CP can be operated at room temperature, but it has long response and recovery time, it is vulnerable to heat and is difficult to process. So, carbon nanotubes (CNTs) are considered in its place. Gas sensing technology based on 2D materials can work efficiently at room temperature with high sensitivity. They have different active sites, such as defects, vacancies, and edge sites to increase the adsorption of gas molecules. They are efficient for sensing application because of its high surface to volume ratio and large surface area. The doping with noble metals such as Ag, Au, Pd, Pt, and elements with low work functions make them more suitable for sensing applications. The 2D nanostructures can provide benefits like more active sties, easier surface functionalization, better compatibility with device integration and possibility of being assembled in a three-dimensional structure and perform a major part in the making of high-performance gas sensors. Sensitivity and selectivity of gas sensors can be increased by fabricating a hybrid structure based on a 2D material. 2D materials can be easily fabricated with chemically resistive field effect transistors that work with less demand of power and ultra-safety. These advantages make 2D materials a highly sensitive platform and reduce the required power consumption, resulting in an ideal material for portable applications [85].

11.8 ADVANCEMENTS IN ELECTRONIC NOSE SYSTEMS

With the capability to be compact, rapid, and economical, electronic nose has great advantages which is required for a gas identification tool. Also, e-nose is suitable for any non-expert user and can be applied in daily life effortlessly. But commercial e-noses are large, non-portable, lab-type instruments and expensive also. So, e-noses can be included in the budgets of big companies, organizations, and research institutions only. Its bulky size and high price are restricting its applications in daily life. In 1994, Hatfield conducted a work about fabrication of components of an e-nose for bringing down the size and power demand by utilizing the recently developed integrated circuit (IC) technology [86]. An IC is a group of electronic circuits embedded on a tiny chip of a semiconductor material. In comparison to the size of its individual components, ICs can be built in a way which is more compact than them. The price of IC can be very cheap with huge production. Presently, IC technology is a necessary component of the contemporary world because of which the economical production of computers, smartphones, and numerous digital devices has been practical. An advantage of advanced IC technology is the incessant discussion and attention on the construction of electronic noses with the help of ICs. As an example, application-specific IC (ASIC) is specially designed for odor classification based on electronic nose data [87]. For a highly integrated e-nose, a conductive sensor like metal-oxide semiconductor, CP is highly appropriate for the unification of IC because of its understandable electrical properties and uncomplicated interface circuit.

Certain low-cost e-nose systems are designed for indoor air-quality monitoring [88], in addition to that, combining them with a wireless module enable online odor detection [89]. For individuals, transportable e-nose systems which includes the sensing module, data acquisition board and a personal computer for data analysis and the personal computer can be replaced with an individualized ultraportable electronic device, such as personal digital assistant (PDA). A support vector machine algorithm in MATLAB is fixed in the PDA. Gas delivery system, a chip containing the sensor cluster and signal conditioning electronics are comprised in the sensing module. By combining a handheld sensing module and a PDA, a lightweight and portable e-nose system was achieved. The PC can be replaced with a potent central microcontroller furnished with an implanted odor classification program. As a result, the volume and weight of the digital device, which is employed as a personalized, smart, portable e-nose, is reduced. A database building procedure should be accomplished in the odor recognition algorithm, and that is the most intricate phase in the algorithm. In addition to that, the microprocessor has to be examined based on its power consumption, system operation frequency, data capacity, instrument size restriction, production price and feasibility with other electronic devices such as through Ethernet, buses, ports, and display interfaces.

Another advancement is in the field of development of gas sensors. We know, different types of smart materials are used for e-nose applications. In order to attain a low-cost production and as an environmental friendly method, green synthesis methodology has been utilized in the production of smart materials, as a result, environmental friendly, inexpensive gas sensors are made available for e-nose device. Qiu et al. reported a valuable work on smart materials based on cellulose and its applications [90]. Lignin based smart materials have emerged as suitable options for advanced biomaterials because of their intrinsic functional properties and green carbon footprint. In a review published by Moreno et al., they have discussed the importance of them as stimuli-responsive polymers in sensor applications, biomedical applications, and SMMs [72].

11.9 CONCLUSION

The world of material science is always changing at an unexpected level of speed, because of the innovative research taking place in the field and its world-wide applicability. Smart materials can be considered as one of the jewels in the crown of material science advanced technologies. In this chapter, we have discussed about bioelectronic nose and its applications in daily life, smart materials and the smart materials employed for e-nose applications. There are different types of smart materials that can be used as gas sensors in e-nose device such as piezoelectric sensors, optical sensors, chemiresistive sensors, etc. Despite the advancements in e-nose technologies, still it is not a portable device and is highly expensive. Proposed works on cheap, portable e-noses for some particular applications are published nevertheless the practical potential has not been explored completely. Research is being carried out extensively to make a low-cost, portable e-nose that any non-expert users can use in daily life. Also, with the application of green synthesis in the production of smart materials, low-cost, and environmental friendly e-nose devices will be available in short time. So, in the coming years, let us hope that with further advancements in both smart materials and bioelectronic nose technologies, the shortcomings will be eliminated and have new exciting applications in daily life which makes life easier and smarter.

KEYWORDS

- **electronic nose**
- **gas sensors**
- **nanomaterials**
- **optical fibers**
- **piezoelectric sensors**
- **smart materials**

REFERENCES

1. Wilson, A. D., (2012). Review of electronic-nose technologies and algorithms to detect hazardous chemicals in the environment. *Procedia Technology, 1*, 453–463.
2. Roy, M., & Yadav, B. K., (2021). Electronic nose for detection of food adulteration: A review. *Journal of Food Science and Technology, 59*(3), 846–858.
3. Moncrieff, R. W., (1961). *An Instrument for Measuring and Classifying Odors, 16*, 742–749. https://doi.org/10.1152/jappl.1961.16.4.742.
4. Wilkens, W. F., & Hartman, J. D., (1964). An electronic analog for the olfactory processes. *Journal of Food Science, 29*, 372–378.
5. Dravnieks, A., & Trotter, P. J., (1965). Polar vapor detector based on thermal modulation of contact potential. *Journal of Scientific Instruments, 42*, 624.

6. Gardner, J. W., & Bartlett, P. N., (1994). A brief history of electronic noses. *Sensors and Actuators B: Chemical, 18*, 210–211.
7. Persaud, K., & Dodd, G., (1982). Analysis of discrimination mechanisms in the mammalian olfactory system using a model nose. *Nature, 299*, 352–355.
8. Gutiérrez, J., & Horrillo, M. C., (2014). Advances in artificial olfaction: Sensors and applications. *Talanta, 124*, 95–105.
9. Mota, I., Teixeira-Santos, R., & Rufo, J. C., (2021). Detection and identification of fungal species by electronic nose technology: A systematic review. *Fungal Biology Reviews, 37*, 59–70.
10. Rasekh, M., Karami, H., Wilson, A. D., & Gancarz, M., (2021). Classification and identification of essential oils from herbs and fruits based on a MOS electronic-nose technology. *Chemosensors, 9*, 142.
11. Wijaya, D. R., Sarno, R., & Zulaika, E., (2021). DWTLSTM for electronic nose signal processing in beef quality monitoring. *Sensors and Actuators B: Chemical, 326*, 128931.
12. Xu, J., Liu, K., & Zhang, C., (2021). Electronic nose for volatile organic compounds analysis in rice aging. *Trends in Food Science and Technology, 109*, 83–93.
13. Seesaard, T., Goel, N., Kumar, M., & Wongchoosuk, C., (2022). Advances in gas sensors and electronic nose technologies for agricultural cycle applications. *Computers and Electronics in Agriculture, 193*, 106673.
14. Ghosh, A., et al., (2022). Development of electronic nose for early spoilage detection of potato and onion during post-harvest storage. *Journal of Materials NanoScience, 9*, 101–114.
15. Tyagi, H., Daulton, E., Bannaga, A. S., Arasaradnam, R. P., & Covington, J. A., (2021). Electronic nose for bladder cancer detection. *Chemistry Proceedings, 5*, 22.
16. Vidigal, I. G., *et al., (*2021*).* Applications of an electronic nose in the prediction of oxidative stability of stored biodiesel derived from soybean and waste cooking oil. *Fuel, 284*, 119024.
17. Lozano, J., (2021). *Electronic Noses for Biomedical Applications and Environmental Monitoring*. MDPI-Multidisciplinary Digital Publishing Institute.
18. Bahl, S., Nagar, H., Singh, I., & Sehgal, S., (2020). Smart materials types, properties, and applications: A review. *Materials Today: Proceedings, 28*, 1302–1306.
19. Bogue, R., (2014). Smart materials: A review of capabilities and applications. *Assembly Automation, 34*, 16–22.
20. Materials which can think on their own and have Smart materials can respond to a change and are example: Smart concrete building (suitable to earth quake areas) processor: Smart wires (automatic shrink/expand) (fillers preventing crack propagation). https://www.tce.edu/sites/default/files/PDF/RV4-Smart-Materials.pdf.
21. Wieszczycka, K., et al., (2021). Surface functionalization – The way for advanced applications of smart materials. *Coordination Chemistry Reviews, 436*.
22. Spaggiari, A., Castagnetti, D., Golinelli, N., Dragoni, E., & Scirè, M. G., (2019). Smart materials: Properties, design, and mechatronic applications. *Proceedings of the Institution of Mechanical Engineers, Part L: Journal of Materials: Design and Applications, 233*, 734–762.
23. Ghareeb, N., & Farhat, M. (2018). Smart materials and structures: State of the art and applications. *Nanotechnology & Applications, 1*, 1–5.

24. Qader, I., Kök, M., Dagdelen, F., & Aydogdu, Y., (2019). A review of smart materials researches and applications. *El-Cezeri Fen ve Mühendislik Dergisi, 6*, 755–788.
25. Shrivastava, S., & Singh, T. R., (2020). Smart materials: An effective tool for bioremediation. *Materials Today: Proceedings, 29*, 508–511.
26. Yan, D., Wang, Z., Cheng, P., Chen, Y., & Zhang, Z., (2021). Rational fabrication of crystalline smart materials for rapid detection and efficient removal of ozone. *Angewandte Chemie, 133*, 6120–6125.
27. Yang, Y., Zeng, W., Huang, P., Zeng, X., & Mei, L., (2021). Smart materials for drug delivery and cancer therapy. *View, 2*, 20200042.
28. Hassan, R., Aslam, K. M. U., Abdullah, A. M., & Abd, R. S. I., (2021). A review on current trends of polymers in orthodontics: BPA-free and smart materials. *Polymers (Basel), 13*, 1409.
29. Yufei, H., et al., (2022). A review of smart materials for the boost of soft actuators, soft sensors, and robotics applications. *Chinese Journal of Mechanical Engineering= Ji Xie Gong Cheng Xue Bao, 35*.
30. Haleem, A., Javaid, M., Singh, R. P., & Suman, R., (2021). Significant roles of 4D printing using smart materials in the field of manufacturing. *Advanced Industrial and Engineering Polymer Research, 4*, 301–311.
31. Zhu, S., (2022). Study on smart materials using in civil engineering. *Proceedings of the 2021 International Conference on Social Development and Media Communication (SDMC 2021), 631*, 1462–1466.
32. Kalita, P., Saikia, M. P., & Singh, N. H., (2015). Electronic-nose technology and its application: A systematic survey. *IJIREEICE, 3*, 123–128.
33. Brand, G., (2019). The electronic nose. *Discovering Odors*, 167–169. doi: 10.1002/9781119687467.ch43.
34. Ramgir, N. S., (2013). Electronic nose based on nanomaterials: Issues, challenges, and prospects. *ISRN Nanomaterials, 2013*, 1–21.
35. Craven, M. A., Gardner, J. W., & Bartlett, P. N., (1996). Electronic noses – development and future prospects. *TrAC – Trends in Analytical Chemistry, 15*, 486–493.
36. Chiu, S. W., & Tang, K. T., (2013*)*. Towards a chemiresistive sensor-integrated electronic nose: A review. *Sensors (Switzerland), 13*.
37. Schiffman, S. S., & Pearce, T. C., (2002). Introduction to olfaction: Perception, anatomy, physiology, and molecular biology. *Handbook of Machine Olfaction,* 1–31 Preprint at https://doi.org/https://doi.org/10.1002/3527601597.ch1.
38. Wilson, A. D., & Baietto, M., (2009). Applications and advances in electronic-nose technologies. *Sensors, 9*, 5099–5148.
39. Röck, F., Barsan, N., & Weimar, U., (2008). Electronic nose: Current status and future trends. *Chemical Reviews, 108*, 705–725.
40. Arshak, K., Moore, E., Lyons, G. M., Harris, J., & Clifford, S., (2004). A review of gas sensors employed in electronic nose applications. *Sensor Review, 24*, 181–198.
41. Karakaya, D., Ulucan, O., & Turkan, M., (2020). Electronic nose and its applications: A survey. *International Journal of Automation and Computing, 17*, 179–209.
42. James, D., Scott, S. M., Ali, Z., & O'Hare, W. T., (2005). Chemical sensors for electronic nose systems. *Microchimica Acta, 149*, 1–17.
43. Bogue, R., (2012). Smart materials: A review of recent developments. *Assembly Automation, 32*, 3–7.

44. Wang, Z. L., Kang, Z. C., & Uchino, K., (1998). Functional and smart materials: Structural evolution and structure analysis. *Physics Today, 51*, 70, 71.
45. Joarder, M. S. H., Kulhari, L., Bin, A. B. H., & Ray, K., (2021). MOX based E-nose for non-invasive biomedical applications. *Przeglad Elektrotechniczny, 97*, 119–122.
46. Karami, H., Rasekh, M., & Mirzaee-Ghaleh, E., (2021). Identification of olfactory characteristics of edible oil during storage period using metal oxide semiconductor sensor signals and ANN methods. *Journal of Food Processing and Preservation, 45*, 1–3.
47. Binson, V. A., Subramoniam, M., & Mathew, L., (2021). Discrimination of COPD and lung cancer from controls through breath analysis using a self-developed e-nose. *Journal of Breath Research, 15*, 46003.
48. Deng, X., Meng, Q. H., Jing, T., & Hou, H. R., (2022). A portable E-nose endowed with subjective evaluation function of air quality in vehicles. *IEEE Transactions on Instrumentation and Measurement, 71*, 1–11.
49. Moufid, M., et al., (2021). Wastewater monitoring by means of e-nose, VE-tongue, TD-GC-MS, and SPME-GC-MS. *Talanta, 221*, 121450.
50. Esteves, C. H. A., *et al., (*2018*).* Identification of tobacco types and cigarette brands using an electronic nose based on conductive polymer/porphyrin composite sensors. *ACS Omega, 3*, 6476–6482.
51. Bonah, E., Huang, X., Aheto, J. H., & Osae, R., (2020). Application of electronic nose as a non-invasive technique for odor fingerprinting and detection of bacterial foodborne pathogens: A review. *Journal of Food Science and Technology, 57*, 1977–1990.
52. Machungo, C., Berna, A. Z., McNevin, D., Wang, R., & Trowell, S., (2022). Comparison of the performance of metal oxide and conducting polymer electronic noses for detection of aflatoxin using artificially contaminated maize. *Sensors and Actuators B: Chemical, 360*, 131681.
53. Tan, J., & Xu, J., (2020). Applications of electronic nose (e-nose) and electronic tongue (e-tongue) in food quality-related properties determination: A review. *Artificial Intelligence in Agriculture, 4*, 104–115.
54. Yasufuku, S., (2001). Electroconductive polymers and their applications in Japan. *Electrical Insulation Magazine*, *17*, 14–24. IEEE.
55. Badoussi, P. E. N. A., & Assogba, K. M., (2020). A multi-level smart monitoring system by combining an e-nose and image processing for early detection of FAW pest in agriculture. In: *Innovations and Interdisciplinary Solutions for Underserved Areas: 4th EAI International Conference, InterSol 2020, Nairobi, Kenya, Proceedings* (Vol. 321, p. 20). Springer Nature.
56. Zaukuu, J. L. Z., Bazar, G., Gillay, Z., & Kovacs, Z., (2020). Emerging trends of advanced sensor based instruments for meat, poultry and fish quality: A review. *Crit. Rev. Food Sci. Nutr., 60*, 3443–3460.
57. Article, R., (2015). *Electronic Nose Applications in Medical Field, 1*, 111–116.
58. Aouadi, B., et al., (2020). Historical evolution and food control achievements of near infrared spectroscopy, electronic nose, and electronic tongue—Critical overview. *Sensors, 20*. Preprint at https://doi.org/10.3390/s20195479.
59. Pulluri, K. K., & Kumar, V. N., (2021). Wine quality assessment using electronic nose. In: *2021 Asian Conference on Innovation in Technology (ASIANCON)* (pp. 1–5). IEEE.

60. Sreelatha, M., & Nasira, G. M., (2018). In: Panigrahi, B. K., Hoda, M. N., Sharma, V., & Goel, S., (eds.), *Surface Acoustic Wave e-Nose Sensor-based Pattern Generation and Recognition of Toxic Gases Using Artificial Neural Network Techniques BT – Nature Inspired Computing* (pp. 135–145). (Springer Singapore, 2018).
61. Kuchmenko, T. A., & Lvova, L. B., (2019). A perspective on recent advances in piezoelectric chemical sensors for environmental monitoring and foodstuffs analysis. *Chemosensors, 7*. Preprint at https://doi.org/10.3390/chemosensors7030039.
62. Sianghio, G. S., Albano, D. R. B., & Abarintos, V. R., (2018). AR2. 4-electronic nose based on piezoelectric quartz crystal array sensors for the discrimination of different types of coffee beans. *Proceedings IMCS 2018,* 268, 269.
63. Behera, B., Joshi, R., Vishnu, G. K. A., Bhalerao, S., & Pandya, H. J., (2019). Electronic nose: A non-invasive technology for breath analysis of diabetes and lung cancer patients. *J. Breath Res., 13*, 24001.
64. Matatagui, D., Bahos, F. A., Gràcia, I., & Horrillo, M. D., (2019). Portable low-cost electronic nose based on surface acoustic wave sensors for the detection of BTX vapors in air. *Sensors,* 19. Preprint at https://doi.org/10.3390/s19245406.
65. Okur, S., et al., (2021). Identification of mint scents using a QCM based e-nose. *Chemosensors, 9*, 31.
66. Bieganowski, A., et al., (2018). Evaluation of hydrocarbon soil pollution using e-nose. *Sensors, 18*, 2463.
67. Zhu, S., et al., (2019). Optical sensory arrays for the detection of urinary bladder cancer-related volatile organic compounds. *J. Biophotonics, 12*, e201800165.
68. Han, F., Huang, X., Aheto, J. H., Zhang, X., & Rashed, M. M. A., (2022). Fusion of a low-cost electronic nose and Fourier transform near-infrared spectroscopy for qualitative and quantitative detection of beef adulterated with duck. *Analytical Methods, 14*, 417–426.
69. Maw, A. K., Somboon, P., Srituravanich, W., & Teeramongkonrasmee, A., (2021). A hybrid e-nose system based on metal oxide semiconductor gas sensors and compact colorimetric sensors. In *2021 IEEE International Conference on Automatic Control & Intelligent Systems (I2CACIS)* (pp. 352–357). IEEE.
70. Li, W., et al., (2020). Recognizing lung cancer using a homemade e-nose: A comprehensive study. *Comput. Biol. Med., 120*, 103706.
71. Hidayat, S. N., et al., (2019). The electronic nose coupled with chemometric tools for discriminating the quality of black tea samples in situ. *Chemosensors, 7*, 29.
72. Moreno, A., & Sipponen, M. H., (2020). Lignin-based smart materials: A roadmap to processing and synthesis for current and future applications. *Materials Horizons, 7*, 2237–2257.
73. Jian, Y., et al., (2020). Gas sensors based on chemi-resistive hybrid functional nanomaterials. *Nano-Micro Letters, 12*, 71.
74. Ramgir, N. S., Sinju, K. R., Debnath, A. K., & Patil, S. J., (2021). Chapter 19 – Nanotechnology-based E-nose for smart manufacturing. In: Thomas, S., Nguyen, T. A., Ahmadi, M., Farmani, A., & Yasin, G. (eds.), *Micro- and Nano-Technologies* (pp. 417–444). Elsevier. doi: https://doi.org/10.1016/B978-0-12-823358-0.00021-6.
75. Conti, P. P., Andre, R. S., Mercante, L. A., Fugikawa-Santos, L., & Correa, D. S., (2021). Discriminative detection of volatile organic compounds using an electronic nose based on TiO_2 hybrid nanostructures. *Sensors and Actuators B: Chemical, 344*, 130124.

76. Sinju, K. R., et al., (2022). ZnO nanowires based e-nose for the detection of H_2S and NO_2 toxic gases. *Materials Science in Semiconductor Processing, 137*, 106235.
77. Germanese, D., et al., (2019). An E-nose for the monitoring of severe liver impairment: A preliminary study. *Sensors, 19* Preprint at https://doi.org/10.3390/s19173656.
78. Seesaard, T., Thippakorn, C., Kerdcharoen, T., & Kladsomboon, S., (2020). A hybrid electronic nose system for discrimination of pathogenic bacterial volatile compounds. *Analytical Methods, 12*, 5671–5683.
79. Thazin, Y., Eamsa-Ard, T., Pobkrut, T., & Kerdcharoen, T., (2019). Formalin adulteration detection in food using e-nose based on nanocomposite gas sensors. In: *2019 IEEE International Conference on Consumer Electronics-Asia (ICCE-Asia)* (pp. 64–67). IEEE.
80. Andre, R. S., et al., (2021). Nanocomposite-based chemiresistive electronic nose and application in coffee analysis. *ACS Food Science & Technology, 1*, 1464–1471.
81. Yin, F., et al., (2021). Carbon-based nanomaterials for the detection of volatile organic compounds: A review. *Carbon N Y, 180*, 274–297.
82. Baik, J. M., et al., (2010). Tin-oxide-nanowire-based electronic nose using heterogeneous catalysis as a functionalization strategy. *ACS Nano, 4*, 3117–3122.
83. Kong, H., et al., (2012). Protein discrimination using fluorescent gold nanoparticles on plasmonic substrates. *Analytical Chemistry, 84*, 4258–4261.
84. You, C. C., et al., (2007). Detection and identification of proteins using nanoparticle-fluorescent polymer "chemical nose" sensors. *Nature Nanotechnology, 2*, 318–323.
85. Park, S. Y., et al., (2019). Chemoresistive materials for electronic nose: Progress, perspectives, and challenges. *InfoMat, 1*, 289–316.
86. Hatfield, J. V., Neaves, P., Hicks, P. J., Persaud, K., & Travers, P., (1994). Towards an integrated electronic nose using conducting polymer sensors. *Sensors and Actuators B: Chemical, 18*, 221–228.
87. Hsieh, H., & Tang, K., (2012). *VLSI Implementation of a Bio-Inspired, 23*, 1065–1073.
88. Zampolli, S., et al., (2004). An electronic nose based on solid state sensor arrays for low-cost indoor air quality monitoring applications. *Sensors and Actuators B: Chemical, 101*, 39–46.
89. Song, K., Wang, Q., Liu, Q., Zhang, H., & Cheng, Y., (2011). A wireless electronic nose system using a Fe_2O_3 gas sensing array and least squares support vector regression. *Sensors (Basel), 11*, 485–505.
90. Qiu, X., & Hu, S., (2013). "Smart" materials based on cellulose: A review of the preparations, properties, and applications. *Materials, 6*, 738–781.

Index

D

E

F

N

O

Q

R

T